OFFICE FOR OUTER SPACE AFFAIRS
UNITED NATIONS OFFICE AT VIENNA

The Interoperable Global Navigation Satellite Systems Space Service Volume

UNITED NATIONS
Vienna, 2018

UNITED NATIONS PUBLICATION
Sales No. E.19.IV.1
ISBN 978-92-1-130355-1
eISBN 978-92-1-047440-5

ST/SPACE/75

The technical information contained in this booklet does not represent any formal commitment from the service providers contributing to this document. Formal commitments can only be obtained through the programme-level documents released by the different service providers. Neither organization contributing to this booklet makes any warranty or guarantee, or promise, expressed or implied, concerning the content or accuracy of the views expressed herein.

The designations employed and the presentation of material in this publication do not imply the expression of any opinion whatsoever on the part of the Secretariat of the United Nations concerning the legal status of any country, territory, city or area, or of its authorities, or concerning the delimitation of its frontiers or boundaries.

Information on uniform resource locators and links to Internet sites contained in the present publication are provided for the convenience of the reader and are correct at the time of issue. The United Nations takes no responsibility for the continued accuracy of that information or for the content of any external website.

Publishing production: English, Publishing and Library Section, United Nations Office at Vienna.

Contents

Executive summary

Global navigation satellite systems (GNSS), which were originally designed to provide positioning, velocity, and timing services for terrestrial users, are now increasingly utilized for autonomous navigation in space as well. Historically, most space users have been located at low altitudes, where GNSS signal reception is similar to that on the ground. More recently, however, users are relying on these signals at high altitudes, near to or above the GNSS constellations themselves.

High-altitude applications of GNSS are more challenging due to reduced signal power levels and visibility, potentially reduced pseudorange accuracy, less optimal geometric diversity, and in the case of elliptical orbits, highly dynamic motion. In these environments, an increased number of available GNSS signals of sufficient power and accuracy would substantially improve the potential signal visibility, and thus mission navigation performance. Via interoperability, multiple GNSS constellations can be used in combination to increase overall performance over any single constellation. The benefits of employing interoperable, multi-constellation GNSS at these higher altitudes are numerous, including more precise, real-time position, velocity, and timing knowledge on-orbit; increased resiliency due to multi-GNSS signal diversity; reduced reliance on ground support infrastructure; increased responsiveness to trajectory manoeuvres resulting in improved on-orbit agility; and the ability to utilize lower-cost components such as on-board clocks.

The availability and performance of GNSS signals at high altitude is documented as the GNSS Space Service Volume (SSV). While different definitions of the SSV exist and may continue to exist for the different service providers, within the context of this booklet it is defined as the region of space between 3,000 km and 36,000 km above the Earth's surface, which is the geostationary altitude. For space users located at low altitudes (below 3,000 km), the GNSS signal reception is similar to that for terrestrial users and can be conservatively derived from the results presented for the lower SSV in this booklet.

The SSV is itself divided in the context of this booklet into two regions, based on differing signal usage scenarios: the lower SSV, covering 3,000–8,000 km altitude, and the upper SSV, covering 8,000–36,000 km. Within these regions, the performance of a single GNSS constellation or combination of constellations for a particular mission is determined by three parameters:

- Pseudorange accuracy
- Received signal power
- Signal availability for one signal and four signals simultaneously

These three parameters are interrelated; if a signal is too weak, if Earth blocks the signal, or if the signal does not have sufficient accuracy, it is not considered as available. Signal availability in particular is critically important for all GNSS users; by using on-board navigation filters in combination with orbit knowledge, space users can achieve navigation and timing solutions with only one available signal at a time. The performance associated with each GNSS constellation is different, but within the lower SSV, single-signal availability from a single constellation is nearly 100% for the entire time, while within the upper SSV, which extends to geostationary altitude, it can be as low as 36% with long outages.

In addition to the global characterization of the GNSS availability and performances in the lower and upper SSV, the booklet also provides performance indications for specific mission profiles which cross the boundaries of the SSV defined in this booklet. The specific mission-specific performance assessments contained in this booklet are geostationary Earth orbit, highly elliptical Earth orbit, and lunar transfer cases.

Within the United Nations International Committee on GNSS (ICG), there is an initiative under way to ensure that GNSS signals within the SSV are available and interoperable across all international global constellations and regional augmentations. This initiative is being carried out within the ICG Working Group B (WG-B) on "Enhancement of GNSS Performance, New Services and Capabilities". The individual efforts led by the WG-B participants include documenting and publishing the SSV performance metrics for each individual constellation, developing standard assumptions and definitions to perform multi-GNSS SSV performance analyses, encouraging the design and manufacturing of GNSS receivers that can operate in the SSV, characterizing GNSS antenna performance to more accurately predict SSV mission performance, providing a reliable reference for space mission analysts, and working towards the formal specification of SSV performance by each GNSS provider.

The multi-constellation, multi-frequency analysis described in this booklet shows availability improvements over any individual constellation when all GNSS constellations are employed. Within the high-altitude SSV, single-signal availability reaches 99% for the L1 band when all GNSS constellations are employed, and four-signal availability jumps from a maximum of 5% for any individual constellation to 62% with all. For the L5 band, continuous signal availability with no outages is provided at geostationary altitude via use of the multi-GNSS SSV, leading to the potential for fully autonomous navigation on demand for these users. The simulations described in this document are based on the constellation-provided data shown in annex A and summarized in chapter 4, and are intended to be more conservative than actual on-orbit performance. In particular, the data provided derive from the main lobe of the transmit antenna patterns only, capture only minimum transmit power and worst-case pseudorange accuracy, and derive from a set of conservative assumptions as described in chapter 5. On-orbit users may see significantly higher performance.

These benefits are only possible through the continued cooperation of all GNSS providers. Through the ICG, all providers have agreed on the information presented in this booklet, and on a number of recommendations to continue development, support, and expansion of the multi-GNSS SSV concept. For the community of GNSS providers, there are three WG-B recommendations that have been formally endorsed by ICG, aimed at continuing development of the SSV, and providing the user community adequate data to utilize it.

- *GNSS providers are recommended to support the SSV outreach by making the booklet on "Interoperable GNSS Space Service Volume" available to the public through their relevant websites.*
- *Service Providers, supported by Space Agencies and Research Institutions, are encouraged to define the steps necessary and to implement them in order to support SSV in future generations of satellites. Service Providers and Space Agencies are invited to report back to WG-B on their progress on a regular basis.*
- *GNSS providers are invited to consider providing the following additional data if available:*
    - *GNSS transmit antenna gain patterns for each frequency, measured by antenna panel elevation angle at multiple azimuth cuts, at least to the extent provided in each constellation's SSV template.*
    - *In the long term, GNSS transmit antenna phase centre and group delay patterns for each frequency.*

For the user community, there is one recommendation to ensure that the full capabilities of the multi-GNSS SSV can be utilized:

- *The authors encourage the development of interoperable multi-frequency spaceborne GNSS receivers that exploit the use of GNSS signals in space.*

Humanity is now beginning to benefit from GNSS usage in the SSV, starting with applications that use only individual constellations, and ultimately expanding to multi-constellation GNSS. For example, weather satellites employing GNSS signals in the SSV will enhance weather prediction and public-safety situational awareness of fast-moving events, including hurricanes, flash floods, severe storms, tornadoes and wildfires. All participants in this study agree that there is the enormous potential of this capability in the future, including lives saved and critical infrastructure and property protected. When fully utilized, an interoperable multi-GNSS SSV will result in orders of magnitude return on investment to national Governments, as well as extraordinary societal benefits.

1. Introduction

The vast majority of Global Navigation Satellite System (GNSS) users are located on the ground, and the GNSS systems are designed to serve these users. However, the number of satellites utilizing on-board GNSS space receivers is steadily growing. Space receivers in the SSV operate in an environment significantly different than the environment of a classical terrestrial receiver or GNSS receiver in low Earth orbit. SSV users span very dynamic and changing environments when traversing above and below the GNSS constellation. Users located below the GNSS constellation can make use of direct line of sight (LoS) signals, while those above the orbit of the GNSS constellations must rely on GNSS signals transmitted from the other side of the Earth, passing over the Earth's limb. These space users experience higher user ranging error, lower user-received power levels, and significantly reduced satellite visibility.

An interoperable GNSS SSV can significantly enhance the GNSS performance. The International Committee on GNSS (ICG) defines interoperability as "the ability of global and regional navigation satellite systems, and augmentations and the services they provide, to be used together to provide better capabilities at the user level than would be achieved by relying solely on the open signals of one system".

This document has been produced by Working Group B (WG-B) of the ICG, with the objectives of defining, establishing, and promoting an interoperable GNSS SSV for the benefit of GNSS space users and GNSS space receiver manufacturers. The information in this document provides to GNSS space users and GNSS space receiver manufacturers a single resource with a concise overview on the characteristics provided by every GNSS as their contribution to an interoperable GNSS SSV.

Chapter 2 of this booklet illustrates the importance of interoperability of GNSS in the SSV by identifying some of the user benefits. Chapter 3 defines the SSV and provides an overview of relevant background information. GNSS constellation parameters relevant to the SSV are collected from each provider in chapter 4. WG-B has taken these parameters and simulated the service that users can expect in different regimes, both from individual constellations, and from the combination of constellations enabled by interoperability. Simulation results are presented in chapter 5, and the ICG WG-B conclusions and recommendations in chapter 6. Chapter 7 identifies potential topics that might be addressed in future releases of this booklet. Further details on the constellation parameters and the WG-B simulation results are contained in the annexes.

2. Benefits to users

The number and scope of GNSS-based space applications has grown significantly the since the first GNSS space receiver was flown. The vast majority of space users are operating in low Earth orbit (LEO), where use of GNSS receivers has become routine. For spacecraft in the SSV, however, the first demonstrated uses came in the late 1990s. Use of GNSS receivers aboard high-altitude spacecraft remains limited due to the challenges involved, including much weaker signals, reduced geometric diversity, and limited signal availability. By focusing on interoperability, the multi-GNSS SSV will provide numerous benefits, expanding the opportunity for full exploitation of the existing potential.

The potential benefits for space users in the SSV are numerous, and fall into several categories, such as navigation performance, mission-enabling technology advancement, and operational flexibility as well as resiliency.

In terms of spacecraft navigation performance, the interoperable multi-GNSS SSV will:

- Significantly increase the number of GNSS signals available to a given user, allowing nearly continuous generation of on-board navigation solutions and reducing "navigation jitter" for improved stability
- Improve the relative geometry between GNSS satellites and the user, improving overall navigation accuracy
- Foster the development of new concepts and algorithms to take advantage of the availability of multi-constellation, multi-frequency and multi-signal GNSS
- Allow higher accuracy for Position, Velocity and Time (PVT) determination, precise orbit determination (POD), and attitude determination
- Allow use of less expensive on-board clocks by reducing the need for time stability between GNSS signal measurements

Related to mission-enabling technology advancement, the interoperable multi-GNSS SSV will:

- Foster the development and availability of GNSS space receivers that can take advantage of the available high-altitude capabilities
- Enable new mission concepts, such as advanced weather observations, precise relative positioning, autonomous cislunar, agile proximity operations, and co-location of spacecraft in geostationary orbit (GEO) longitude boxes
- Promote use of combined antenna arrays for satellite orbit and attitude determination, allowing both states to be based on a single sensor

Enhancing operational flexibility and resiliency, the interoperable multi-GNSS SSV will:

- Enable development of new operations concepts with reduced ground interactions
- Increase feasibility of satellite on-board autonomy at high altitude
- Increase the operational robustness for spacecraft navigation due to the redundant use of multiple independent GNSS signals
- Reduce ground operational needs by reducing ranging requests, lowering mission costs, and allowing ground stations to focus on communications activities
- Simplify mission architectures, leading to the potential for standardization of satellite navigation design from LEO to GEO and beyond

These benefits are applicable to a wide range of mission classes and applications, including (but not limited to) the following examples:

- *Earth weather observation:* The United States' Geostationary Operational Environmental Satellite-R series of spacecraft (GOES-R) is designed to collect observations continually, with outages of less than 2 hours per year, even with daily station-keeping manoeuvres. To accomplish this, they rely on nearly continuous GNSS signals.
- *Precision formation flying:* The European Proba-3 solar occultation mission seeks to observe the Sun's corona by flying a solar-occulting spacecraft and an observing spacecraft in precise formation, in a highly elliptical Earth orbit. The highly precise relative positioning of the two spacecraft will rely on GNSS signals up to approximately 60,000 km altitude.
- *Cislunar trajectories:* Launch vehicle upper stages and cislunar exploration missions travel well beyond GEO altitude, with some travelling all the way to lunar distance. GNSS is planned to be used by these vehicles for its high accuracy and high cadence, which improve insertion accuracy when returning to Earth. Weak-signal receivers are enabling use of GNSS signals at extremely long distances as well, potentially allowing for use as a supplemental measurement source in lunar orbit.
- *Satellite servicing:* Satellite servicing missions are being developed for spacecraft at GEO, where they will need to autonomously rendezvous with their target

spacecraft. The precision and autonomy required for this type of mission will require continuous precise GNSS signals to be available.

- *New concepts for GEO co-location:* The most highly sought orbit for commercial users is in the GEO belt, where the current number of spacecraft is limited by the longitude spacing requirements put in place to avoid collisions. With GNSS, these spacecraft could reduce relative navigation errors, recover quickly from manoeuvres, and reduce burden on the ground control centre, even while utilizing the available space at GEO more efficiently.

3. Interoperable GNSS space service volume

Historically, most space users have been located at low altitudes, where GNSS signal reception is similar to that on the ground. More recently, however, users are relying on these signals at high altitudes, near to or above the GNSS constellations themselves. The availability and performance of GNSS signals at high altitude is documented as the GNSS SSV. While different definitions of the SSV exist and may continue to exist for the different service providers, within the context of this booklet it is defined as the region of space between 3,000 km and 36,000 km above the Earth's surface, which is the geostationary altitude. For space users located at low altitudes (below 3,000 km), the GNSS signal reception is similar to that for terrestrial users and can be conservatively derived from the results presented for the lower SSV in this booklet.

3.1 Definition

The GNSS SSV is defined in the context of this booklet as the region of space extending from 3,000 km to 36,000 km altitude, where terrestrial GNSS performance standards may not be applicable. GNSS system service in the SSV is defined by three key parameters:

- Pseudorange accuracy
- Minimum received power
- Signal availability

The SSV covers a large range of altitudes; the GNSS performance will degrade with increasing altitude. In order to allow for a more accurate reflection of the performance variations, the SSV itself is divided into two distinct areas that have different characteristics in terms of the geometry and quantity of signals available to users in those regions:

1. *Lower SSV for medium Earth orbits:* 3,000–8,000 km altitude. This area is characterized by reduced signal availability from a zenith-facing antenna alone, but increased availability if both a zenith and nadir-facing antenna are used.

2. *Upper SSV for geostationary and high Earth orbits:* 8,000–36,000 km altitude. This area is characterized by significantly reduced signal received power and availability, due to most signals travelling across the limb of the Earth.

Users with adequate antenna and signal processing capabilities will also be able to process GNSS signals above the identified altitude of 36,000 km.

The relevant regions of the GNSS SSV are depicted in figure 3.1, along with the altitude ranges of the contributing GNSS constellations that are located in medium Earth orbit (MEO). It is noted that some GNSS also offer satellites at geostationary orbits (GEO) and/or inclined geosynchronous orbits (IGSO).

Figure 3.1 The GNSS SSV and its regions

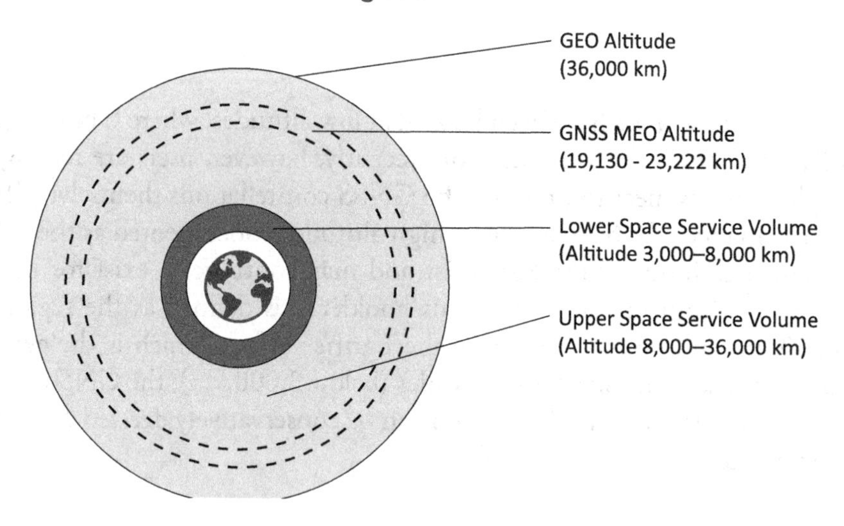

3.1.1 Lower space service volume

Figure 3.2 shows the signal reception geometry for a receiving spacecraft in the SSV for the lower SSV.

Figure 3.2 Signal reception geometry in the lower SSV

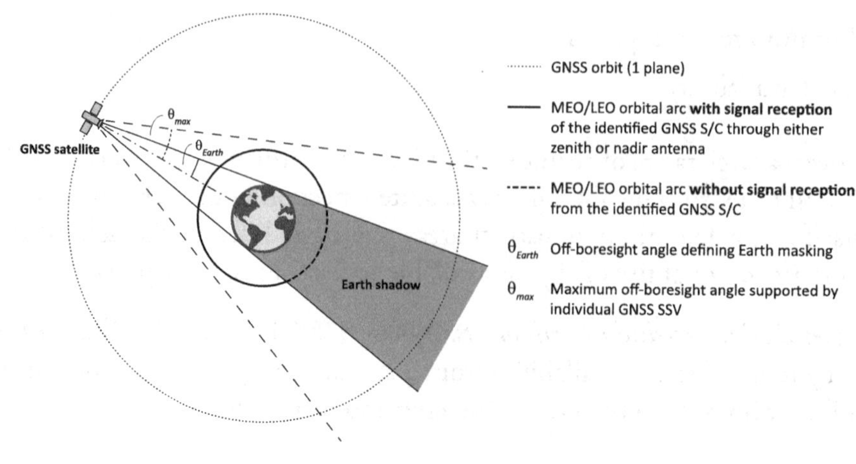

GNSS space receivers located between 3,000 km and 8,000 km altitude can receive GNSS signals from the spacecraft nadir direction and the spacecraft zenith direction with respect to the Earth. Zenith signals are received in-line with LEO spacecraft and Earth-based GNSS signal reception. The signals arriving from spacecraft nadir are emitted by GNSS satellites located at the opposite side of the Earth and pass the limb of the Earth before arriving at the receiver. This is highlighted in figure 3.2.

When employing an entire GNSS constellation, or multiple combined constellations, signal availability is expected to exceed four simultaneous signals when viewed from a spacecraft zenith-facing antenna, and even more with multiple spacecraft antennas.

3.1.2 Upper space service volume

Figure 3.3 shows the signal reception geometry for a receiving spacecraft in the upper SSV, defined as the region between 8,000 km and 36,000 km altitude.

Figure 3.3 Signal reception geometry in the upper SSV

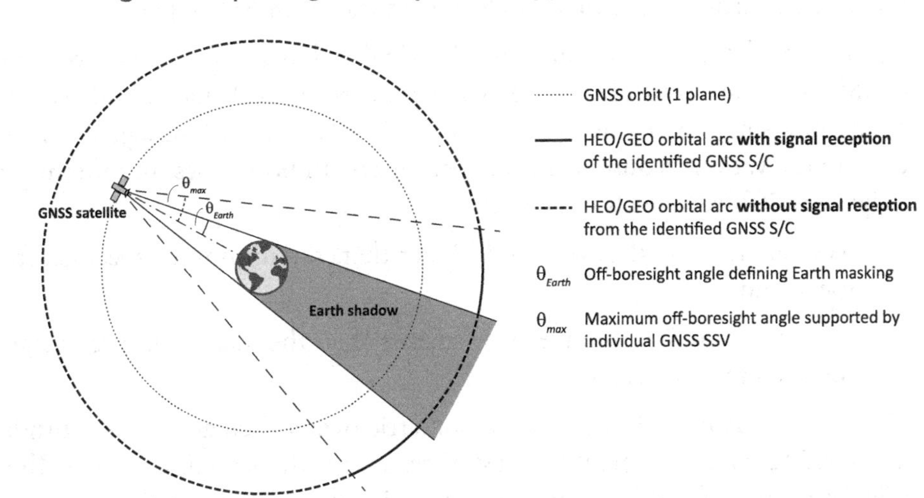

In the high-altitude SSV, especially at altitudes above the GNSS constellations, no signal reception from the spacecraft zenith direction is possible, necessitating all signals to be received from a nadir-facing antenna. Generally, all GNSS signals arrive from the opposite side of the Earth and pass over the limb of the Earth. As illustrated in figure 3.3, the Earth blocks a large portion of the signal for users within the upper SSV. The signal is further limited to the extent of usable signals from the GNSS transmitting antennas, which may be limited to approximately 16–34 degrees from the GNSS satellite nadir direction, depending on the constellation.

Although figure 3.3 only shows a single satellite out of a full constellation, it is evident that for GNSS space users located within the upper SSV that the availability of GNSS signals is significantly constrained. Thus, space users in the upper SSV will significantly benefit from an interoperable GNSS SSV, in which multiple GNSS signals from different constellations can be used simultaneously. The interoperable GNSS SSV will significantly improve the number of visible satellites and thus the availability of GNSS signals.

3.2 SSV performance characterization metrics

The characterization of the SSV performance of an individual GNSS constellation relates at a minimum to the characterization of the following three parameters for every ranging signal:

1. *Pseudorange accuracy:* Since users in the SSV do not typically generate PVT solutions using multiple simultaneous GNSS measurements, this instead measures the error in the ranging signal itself. This relates to the orbit determination and clock stability errors, and additional systematic errors.
2. *Received signal power:* This is the minimum user-received signal power obtained by a space user in the relevant orbit, assuming a 0 dBic user antenna. Generally, this power is calculated at the highest altitude in the given SSV region.
3. *Signal availability:* Signal availability is calculated as the percentage of time that GNSS signals are available for use by a space user. It is calculated both as the availability of a single signal in view, and as the availability of four signals in view, to capture the various requirements of space users. In both cases, in order to declare a signal available, it needs to be both:

    a. received at a signal power level higher than the minimum specified for SSV users, and

    b. observed with a user range error smaller than the maximum user range error specified for SSV users.

    The signal availability is measured as a metric over a shell at a given altitude (e.g. at 36,000 km) and is generated as a statistic over both location and time. The exact calculation used for this metric by an individual GNSS constellation is specified explicitly in annex A.

    A sub-metric to signal availability is maximum outage duration, defined as the maximum duration when a space user at a particular orbit will not obtain availability for at least one single signal or at least four signals simultaneously, depending on the exact metric being calculated. The definition of maximum outage duration is closely linked to the definition of signal availability.

These three parameters characterize at a minimum the contribution of an individual GNSS to an interoperable GNSS SSV. In addition to these parameters, constellation service providers may identify additional parameters useful to characterize their particular contribution to the interoperable GNSS SSV.

4. Individual constellation contributions to multi-GNSS space service volume

To convey a consistent set of capabilities across all GNSS constellations, an SSV capabilities template has been completed by each GNSS service provider to capture their contributions to each of the parameters identified in section 3.2. The full text of these completed templates, along with appropriate context, is available in annex A. This chapter presents an aggregated subset of the full data so that the individual SSV characteristics of each constellation can be readily compared and contrasted.

Note that the SSV service characteristics outlined here and in annex A represent the service documented by each individual GNSS service provider, either by formal specification or by characterization and analysis. On-orbit flight results will differ from these characteristics due to mission-specific geometry, receiver sensitivity, time-dependent service characteristics, and other factors. In all cases, only service provided by the main-lobe signal is captured here; the extent of this main-lobe service is documented in table 4.2 as the reference off-boresight angle. For full details, see annex A.

Table 4.1 presents an overview of the configuration of each constellation, including operational status, constellation configuration, and general orbit parameters. Further, table 4.2 aggregates SSV signal characteristics for each constellation, including signal, minimum received power and signal availability. Finally, table 4.3 aggregates the user range error for each constellation.

Table 4.1. Overview of global and regional navigation satellite systems

System name	Nation	Coverage	Status	No. frequencies / signals	No. spacecraft (nominal)/ orbital planes	Semi-major axis (km)	Inclination (°)	Comments
GPS	USA	Global	Operational	3/4	24/6	26560	55	
GLONASS	Russia	Global	Operational	2/6	24/3	25510	64.8	
Galileo	European Union	Global	Operational	5/10	24/3	29600	56	Initial service: 2016 FOC planned: 2020
BDS	China	Global	Operational (regional) In build-up (global)	3/5	MEO: 24/3 IGSO: 3/3 GEO: 5/1	27906 42164 42164	55 55 0	Service planned: Regional FOC: 2012 Global initial service: 2018 FOC: 2020
QZSS	Japan	Regional (Japan)	In build-up	4/7	HEO: 3/3 GEO: 1/1	42164	40 0	Service planned: 2018
NavIC	India	Regional (India)	In build-up	2/2	GSO: 4/2 GEO: 3/1	42164	29 0	Service planned: 2018

Table 4.2. SSV signal characteristics for each GNSS service provider

Band	Constellation	Frequency (MHz)	Minimum received civilian signal power		Signal availability (%)			
					Lower SSV		Upper SSV	
			0dBi RCP antenna at GEO (dBW)	Reference off-boresight angle (°)	At least 1 signal	4 or more signals	At least 1 signal	4 or more signals
L1/E1/B1	GPS	1575.42	-184 (C/A) -182.5 (C)	23.5	100	97	80	1
	GLONASS	1605.375[a]	-179	26	100	99.8	93.9	7.0
	Galileo	1575.42	-182.5	20.5	100	99	64	0
	BDS	1575.42	-184.2 (MEO) -185.9 (I/G)	25 19	99.90	96.20	97.40	24.10
	QZSS	1575.42	-185.5	22	100	N/A	54	N/A
L2/E6	GPS	1227.6	-183	26	100	100	92	6.5
	GLONASS	1248.6251	-178	34	100	66	100	29
	Galileo	1278.75	-182.5	21.5	100	100	72	0
	QZSS	1227.6	-188.7	24	100	N/A	54	N/A
	GPS	1176.45	-182	26	100	100	92	6.5
L5/L3/E5/B2	GLONASS	12011	-178	34	100	100	99.9	60.3
	Galileo	1206.45 (E5b)	-182.5	22.5	100	100	80	0
		1191.795 (E5ABOC)	-182.5	23.5	100	100	86	0
		1176.45 (E5a)	-182.5	23.5	100	100	86	0
	BDS	1191.795	-182.8 (MEO) -184.4 (I/G)	28 22	100	99.90	99.90	45.40
	QZSS	1176.45	-180.7	24	100	N/A	54	N/A
	NavIC	1176.45	-184.54	16	98	51.40	36.90	0.60

[a] Centre of FDMA band

Table 4.3. User range error as defined in annex A for each GNSS service provider

Constellation	GPS	GLONASS	Galileo	BDS	QZSS	NavIC
User range error	0.8 metres	1.4 metres	1.1 metres	2.5 metres	2.6 metres	2.11 metres

5. Simulated performance of interoperable space service volume

The Working Group B of the International Committee on GNSS (ICG WG-B), has simulated the GNSS single- and multiple-constellation performance expectations in the SSV, based on the individual constellation signal characteristics documented in chapter 4. As outlined in chapter 3, navigation performance in the SSV is primarily characterized by three properties: user range error (URE), received signal power, and signal availability. The focus of these simulations is on signal availability, which serves as a proxy for navigation capability.

An available signal from a GNSS satellite is one that a space user with adequate equipment is able to detect with sufficient strength to form a usable measurement, that is, above the carrier power to noise power spectral density (C/No) threshold value required to acquire and track the signal, and with unobstructed LoS. In addition to availability, the results include maximum outage duration (MOD), the longest duration that a user can expect to be without a signal. MOD is a critical parameter for space users employing GNSS for time or concerned with navigation stability and short-term navigation effects, such as during trajectory manoeuvres. Availability and MOD estimates are calculated for the case in which a single signal is detected by a user, as well as for the case in which four signals are available simultaneously. Four-signal-in-view coverage enables kinematic positioning and one-signal-in-view coverage is the minimum needed for GNSS to contribute to a navigation solution. For many users, signal availability and signal outages are the primary drivers for navigation performance.

Two types of performance estimates are provided: globally averaged, and mission-specific. Global performance is estimated by simulating signal availability at a fixed grid of points in space, at both the lower SSV altitude of 8,000 km, and the upper SSV at 36,000 km. This availability is then calculated by simulating navigation receiver operation over a two-week duration, and over all the points in each grid. This can be interpreted as a measure of the performance that space missions can expect while employing GNSS in the SSV.

Mission-specific performance estimates are obtained by estimating signal availability for a spacecraft on a particular trajectory within the SSV. Mission-specific scenarios considered in this study include: 1) geostationary orbit, 2) a highly elliptic orbit, and 3) a lunar trajectory. The purpose of this phase of analysis is to provide "real-world" estimates for a concrete mission using similar methods to those used for estimation of global performance. In total, this information will provide prospective SSV users simulation results that demonstrate the benefits and possibilities offered by an interoperable SSV.

The simulations described in this document are based on the constellation-provided data shown in annex A and summarized in chapter 4, and are intended to be more conservative than actual on-orbit performance. In particular, the provided data derive from the main lobe of the transmit antenna patterns only, capture only minimum transmit power and worst-case pseudorange accuracy, and derive from a set of conservative assumptions as described in chapter 4. The objective is to demonstrate the value of the multi-GNSS SSV in terms of combined performance, as compared to that provided by any specific constellation. It is not intended to validate or predict real-world flight results, or to validate the contents of chapter 4 or annex A, which may differ based on the assumptions used. See annex B for more details on the simulated simulation methodology and full results.

The characteristics of the constellations and signals being simulated are captured in chapter 4, and in the appropriate annexes. The transmit beamwidth specification (given in terms of 'reference off-boresight angle') and delivered power levels at GEO altitude are used to define the geometric reach and the minimum radiated transmit power (MRTP) in the simulation—see annex B for its definition and further details. Only the L1/E1/B1 and L5/L3/E5/B2 bands (see also table 4.3 for further details on the signals provided by each system in these bands) are used in the simulation. Additional simulation results and more in-depth descriptions and data on the specific simulation parameters are contained in annex B.

5.1 Global space service volume performance

Global performance estimates of availability and MOD are given in table 5.1. These results show available performance at GEO altitude (the upper limit of the upper SSV) considering a zero-gain user antenna. The user space is simulated in this case by a sphere at GEO altitude. Both availability and MOD are calculated at the worst-case grid point. An asterisk (*) marks cases in which an availability threshold is never reached over the full duration of the simulation for the worst-case grid location.

Simulations were performed using three different C/No thresholds. Performance results are provided for thresholds of C/No of 15, 20, and 25 dB-Hz. These thresholds roughly correspond to the performance levels of space GNSS receivers that exist or are in development.

In calculating availability, the MRTP value is assumed to be constant over the entire beamwidth of the transmit antenna. A zero-gain antenna is applied in the calculation of C/No, the effects of a Low Noise Amplifier (LNA) or any other aspects of the Radio Frequency/ Intermediate Frequency (RF/IF) are not considered. These simplifying assumptions lead to conservative Position, Navigation, Timing (PNT) performance estimates.

The performance values for signal availability and MOD in this chapter may not necessarily match the figures provided in chapter 4 by the different service providers for the following reasons:

1. Different receiver parameters may have been assumed.
2. The implementation of the availability figure of merit and the MOD figure of merit may have been realized differently.

5.1.1 Performance in the upper space service volume

Table 5.1 shows the signal availability and the MOD for a user in the upper SSV as a function of different C/No thresholds for each individual constellation and for all constellations combined. The C/No thresholds relate to the tracking threshold of the assumed space receiver and values of 15 dB-Hz, 20 dB-Hz and 25 dB-Hz are analysed.

Figure 5.1 shows an example of simulated signal availability for the 20 dB-Hz C/No threshold case. Note that the better availability estimated in the L5/L3/E5/B2 case over the L1/E1/B1 case is due to generally wider beamwidths for the lower frequency band for each constellation.

General observations concerning the results shown in table 5.1 indicate the following:

- One-signal availability significantly exceeds four-signal availability, underscoring the benefit of employing an on-board navigation filter, which can process individual measurements at a time, for missions in the SSV.
- At the highest threshold of 25 dB/Hz, availability is nearly 0%. This indicates the challenge of extremely low GNSS signal levels for missions in the upper SSV, and the importance of using specialized high-altitude receivers and high-gain antennas.
- When the constellations are used together, one-signal availability is nearly 100% for all but one case (25 dB-Hz threshold, L1). The abundance of signals available in an interoperable multi-GNSS SSV greatly reduces constraints imposed by navigation at high altitudes.

Figure 5.1. Estimated number of satellites visible, by individual constellation and combined, for sample L1/E1/B1 GEO user with 20 dB-Hz C/No threshold. Actual visibility changes with location and time

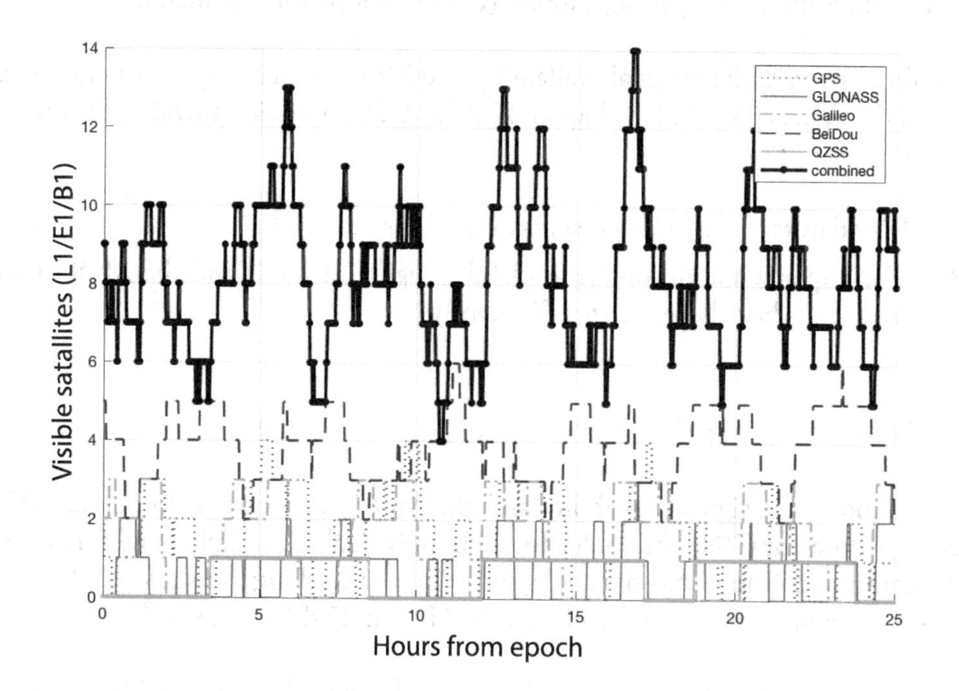

Table 5.1. Global performance estimates of availability and maximum outage duration for each constellation and all constellations together. Results for nadir-pointing antenna in the upper SSV

Band	Constellation	$C/N0_{min}$ = 15 dB-Hz				$C/N0_{min}$ = 20 dB-Hz				$C/N0_{min}$ = 25 dB-Hz			
		At least 1 signal		4 or more signals		At least 1 signal		4 or more signals		At least 1 signal		4 or more signals	
		Avail. (%)	MOD (min)	Avail. (%)	MOD (min)	Avail. (%)	MOD (min)	Avail. (%)	MOD (min)	Avail. (%)	MOD (min)	Avail. (%)	MOD (min)
L1/E1/B1	GPS	90.5	111	4.8	*	90.5	111	4.8	*	0.0	*	0	*
	GLONASS	93.9	48	7	*	93.9	48	7	*	93.9	48	7	*
	Galileo	78.5	98	1.2	*	78.5	98	1.2	*	0.0	*	0	*
	BDS	97.4	45	24.1	*		70	0.6	*	0.0	*	0	*
	QZSS	26.7	*	0.8	*	0.0	*	0	*	0.0	*	0	*
	Combined	**99.9**	**29**	**98.1**	**93**	**99.9**	**33**	**89.8**	**117**	**93.9**	**48**	**7**	*
L5/L3/E5a/B2	GPS	96.9	77	15.6	1180	96.9	77	15.6	1180	0.0	*	0	*
	GLONASS	99.9	8	60.3	218	99.9	8	60.3	218	99.9	8	60.3	218
	Galileo	93.4	55	4.2	*	93.4	55	4.2	*	0.0	*	0	*
	BDS	99.9	7	45.4	644	99.9	7	32.4	644	0.0	*	0	*
	QZSS	30.5	*	1.5	*	30.5	*	1.5	*	0.0	*	0	*
	NavIC	36.9	*	0.6	*	1.0	*	0	*	0.0	*	0	*
	Combined	**100**	**0**	**99.9**	**15**	**100**	**0**	**99.9**	**15**	**99.9**	**8**	**60.3**	**218**

*No signal observed for the worst-case grid location for maximum simulation

5.1.2 Performance in the lower space service volume

Global performance estimates of availability and MOD for the lower SSV (represented by a user sphere at 8,000 km altitude) are shown in table 5.2. These results were generated based on geometrical availability only. In this case, availability is constrained only by obstruction of the LoS visibility between the transmitter and the grid point.

Similar observations hold for these results as above. Performance in the lower SSV is estimated to be significantly better than that in the upper SSV, due to the improved geometric availability at the lower altitude. Single-satellite availability is nearly 100% for all individual systems and combined-constellation availability is 100% in all cases. For the lower SSV, the C/No is typically higher than the assumed 25dB-Hz minimum tracking threshold. Therefore, no sensitivity of the results against different receiver tracking thresholds is presented.

Table 5.2. Global performance estimates of availability and maximum outage duration for each constellation and all constellations together. Results for omni pointing antenna (nadir and zenith) in the lower SSV

Band	Constellation	Signal availability (%)		Max outage duration (min)	
		At least 1 signal	4 or more signals	At least 1 signal	4 or more signals
L1/E1/B1	GPS	100	99.6	0	45
	GLONASS	100	99.8	0	24
	Galileo	99.9	95.0	11	60
	BDS	100	100	0	0
	QZSS	99.6	79.4	197	*
	Combined	**100**	**100**	**0**	**0**
L5/L3/E5a/B2	GPS	100	99.9	0	16
	GLONASS	100	100	0	0
	Galileo	100	100	0	0
	BDS	100	100	0	0
	QZSS	99.6	79.4	197	*
	NavIC	98.0	51.4	348	*
	Combined	**100**	**100**	**0**	**0**

* No signal observed for the worst-case grid location for maximum simulation

5.2 Mission-specific performance

Mission-specific simulations use scenarios that are considered to be realistic use cases of GNSS space users. When defining the mission scenarios, particular care was taken to ensure that realistic assumptions were made, including selection of user antenna characteristics that are representative of existing space-qualified hardware. Three representative mission scenarios were selected for simulation, a geostationary orbit mission, a highly elliptical orbit mission, and a lunar mission.

For mission-specific analysis, an antenna beam pattern for the user spacecraft is included in the link power calculation. In particular, two different user antenna gain characteristics were used: a patch antenna with gain of approximately 2 dBi, and a "high-gain" antenna with gain of 8 to 9 dBi. The patch antenna would be used when a wider beam is desired, and the high-gain antenna would be chosen for longer-range missions.

5.2.1 Geostationary orbit mission

The GEO mission scenario analyses multi-GNSS signal reception for six geostationary satellites. The objective is to obtain more representative signal strength values than in the global analysis by using realistic user antenna patterns on-board the space users for receiving the B1/E1/L1 and B2/E5A/L5 signals.

Spacecraft trajectory

Six GEO satellites are simulated and share the same orbital plane apart from a 60-degree separation in longitude (see table 5.3). The right ascension of the ascending node (RAAN) angle is used to synchronize the orbit with the Earth rotation angle at the start of the simulation. The true anomaly is used to distribute the six GEO user receivers along the equator. This placement of the satellites was chosen to ensure that even signals from regional GNSS satellites in (inclined) geosynchronous orbits would be visible to at least one of the GEO user receivers (see figure 5.2).

Table 5.3. GEO osculating Keplerian orbital elements

Epoch	1 Jan 2016 12:00:00 UTC		
Semi-major axis	42164.0 km	**Right ascension of the ascending node**	100.379461 deg
Eccentricity	0.0	**Argument of perigee**	0.0 deg
Inclination	0.0 deg	**True anomaly**	0/60/120/180/240/300 deg

Spacecraft attitude and antenna configuration

The user antenna on-board the user spacecraft is a high-gain antenna that permanently points towards the nadir (centre of the Earth). The user antenna patterns used on the two signals are specified in table B10. The assumed acquisition threshold of the space user receiver is 20 dB-Hz.

Figure 5.2. Example for visibility of NavIC satellite from the GEO at 240 degree longitude

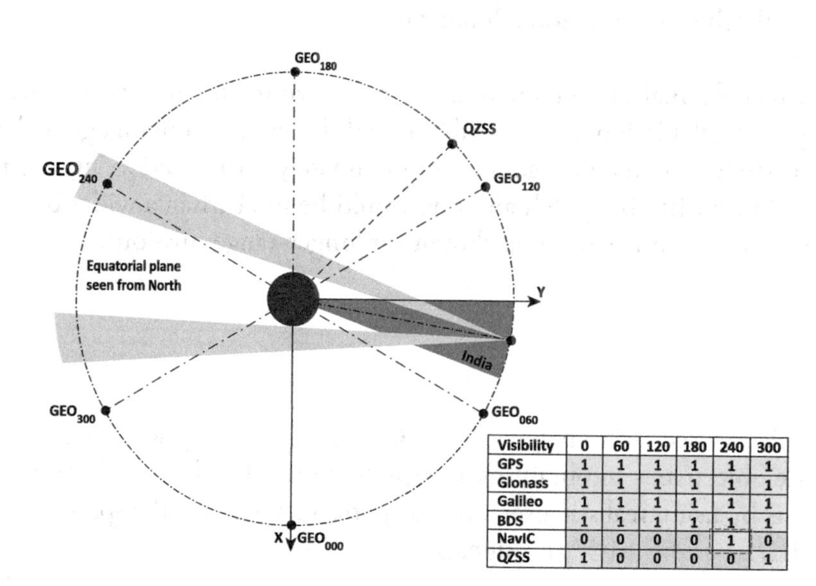

Visibility	0	60	120	180	240	300
GPS	1	1	1	1	1	1
Glonass	1	1	1	1	1	1
Galileo	1	1	1	1	1	1
BDS	1	1	1	1	1	1
NavIC	0	0	0	0	1	0
QZSS	1	0	0	0	0	1

Results

The six GEO satellites are all in the equatorial orbital plane but phased by 60 degrees in longitude, or four hours in time. The MEO GNSS satellites have orbital periods in the order of 12 - 14 hours, or about half that of the GEO. This means that the GEO and MEO orbits are almost in phase with each other, in such a way that the visibility patterns at the GEO receiver repeat almost exactly with periods of one day. The MEO satellites move 120 degrees during the four-hour interval between GEO satellites, but there are multiple GNSS MEO in each orbital plane. This means that the visibility patterns in terms of number of visible MEO signals are very similar to all six GEO receivers.

The situation is different for the inclined geosynchronous GNSS satellites of the Navigation with Indian Constellation (NavIC), Quasi-Zenith Satellite System (QZSS) and BDS constellations. The GEO and IGSO longitudes are frozen relative to each other. At most GEO longitudes, the GNSS satellites in IGSO orbits are never visible, either because the GEO is located outside the half-cone angle of the transmitting satellite, or because the signal is blocked by the Earth. This means that reception of the IGSO GNSS signals is an exception rather than the rule. However, those GEO receivers that do see signals from these transmitters will see them continuously, or at very regular patterns (see NavIC B2/E5A/L5 signal).

Examples are given in figure 5.3 and figure 5.4 for the simulated cases with the lowest number of visible satellites and the highest number of visible satellites. The difference is mainly caused by visible BDS and QZSS satellites in the second case. Even for the worst case, the combined constellations offer four visible satellites at L1 almost continuously. At L5, the combined constellations offer between 12 and 20 signals all the time. Complete visibility six GEO receivers and at both carrier frequencies are provided in annex B.

Table 5.4 to table 5.9 show the visibility of at least one or at least four satellites, as a percentage of time. For the combined GNSS constellations, four or more B2/E5A/L5 signals are available at every simulated GEO longitude for 100% of the time. The slightly weaker B1/E1/L1 signal drops to around 93% visibility for four satellites, but there is always at least one signal available. This is a considerably better result than for any of the individual MEO constellations (GPS, Galileo, GLONASS individual solutions), which reach at most 53% visibility at GEO height for four signals, individually.

The conclusion is that when using the combined GNSS constellations, it is possible to continuously form an on-board PVT solution. In addition to this, it is also possible to perform a real-time kinematic orbit determination process on-board the GEO satellite. This may allow real-time positioning of GEO at a few metres accuracy level. This enables new concepts for GEO co-location due to more accurate positioning information from GNSS than from terrestrial ranging.

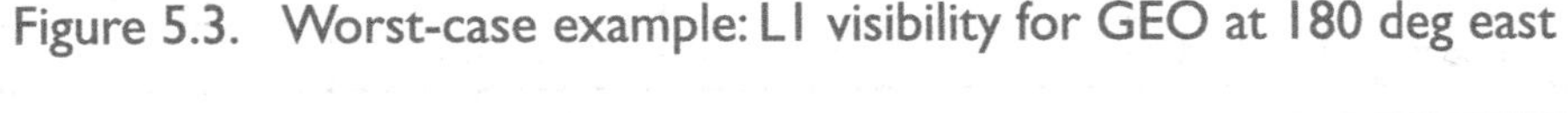
Figure 5.3. Worst-case example: L1 visibility for GEO at 180 deg east

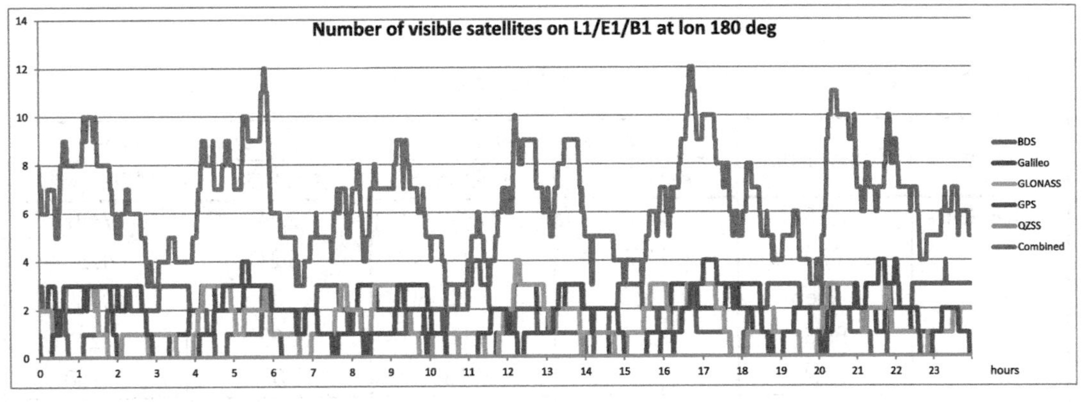

Figure 5.4. Best-case example: L5 visibility for GEO at 60 deg west

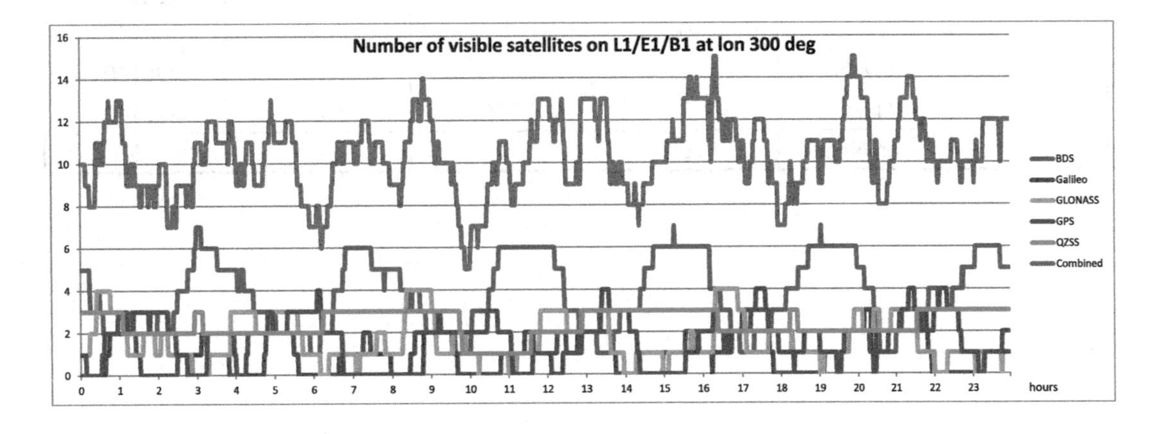

Table 5.4. Performance for GEO receiver at longitude 0 deg

Band	Constellation	At least 1 signal		4 or more signals	
		Avail. (%)	MOD (min)	Avail. (%)	MOD (min)
L1/E1/B1	GPS	82.24	72	3.29	697
	GLONASS	84.84	38	0.71	3808
	Galileo	63.4	82	0	20160
	BDS	100	0	28.47	200
	QZSS	64.31	210	0	20160
	Combined	**100**	**0**	**99.7**	**11**
L5/L3/E5a/B2	GPS	94.29	50	14.55	425
	GLONASS	100	0	43.99	189
	Galileo	86.43	41	0	20160
	BDS	100	0	96.44	21
	QZSS	90.26	117	0	20160
	NavIC	0	0	0	20160
	Combined	**100**	**0**	**100**	**0**

Table 5.5 Performance for GEO receiver at longitude 60 deg

Band	Constellation	At least 1 signal		4 or more signals	
		Avail. (%)	MOD (min)	Avail. (%)	MOD (min)
L1/E1/B1	GPS	80.01	89	2.44	748
	GLONASS	90.7	34	5.21	475
	Galileo	62.83	82	0	20160
	BDS	92.27	28	0.92	724
	QZSS	0	20160	0	20160
	Combined	**100**	**0**	**94.93**	**36**
L5/L3/E5a/B2	GPS	94.77	55	10.04	402
	GLONASS	100	0	53.14	90
	Galileo	86.87	42	0	20160
	BDS	100	0	10.65	262
	QZSS	0	20160	0	20160
	NavIC	0	20160	0	20160
	Combined	**100**	**0**	**100**	**0**

Table 5.6. Performance for GEO receiver at longitude 120 deg

Band	Constellation	At least 1 signal		4 or more signals	
		Avail. (%)	MOD (min)	Avail. (%)	MOD (min)
L1/E1/B1	GPS	79.56	103	3.68	462
	GLONASS	90.48	37	5.07	474
	Galileo	63.12	82	0	20160
	BDS	92.38	26	0.93	732
	QZSS	0	20160	0	20160
	Combined	**100**	**0**	**94.62**	**51**
L5/L3/E5a/B2	GPS	91.85	66	9.17	419
	GLONASS	100	0	52.81	94
	Galileo	86.55	41	0	20160
	BDS	100	0	11.72	255
	QZSS	0	20160	0	20160
	NavIC	0	20160	0	20160
	Combined	**100**	**0**	**100**	**0**

Table 5.7. Performance for GEO receiver at longitude 180 deg

Band	Constellation	At least 1 signal		4 or more signals	
		Avail. (%)	MOD (min)	Avail. (%)	MOD (min)
L1/E1/B1	GPS	82.15	73	3.31	697
	GLONASS	84.93	41	0.72	3805
	Galileo	63.4	82	0	20160
	BDS	91.85	27	0.94	722
	QZSS	0	20160	0	20160
	Combined	**100**	**0**	**93.24**	**41**
L5/L3/E5a/B2	GPS	94.28	50	14.55	425
	GLONASS	100	0	44.08	195
	Galileo	86.43	41	0	20160
	BDS	100	0	10.37	263
	QZSS	0	20160	0	20160
	NavIC	100	0	0	20160
	Combined	**100**	**0**	**100**	**0**

Table 5.8. Performance for GEO receiver at longitude 240 deg

Band	Constellation	At least 1 signal		4 or more signals	
		Avail. (%)	MOD (min)	Avail. (%)	MOD (min)
L1/E1/B1	GPS	79.92	89	2.46	748
	GLONASS	90.85	34	5.19	476
	Galileo	62.83	82	0	20160
	BDS	100	0	28.63	200
	QZSS	0	20160	0	20160
	Combined	**100**	**0**	**99.34**	**19**
L5/L3/E5a/B2	GPS	94.75	55	10.06	402
	GLONASS	100	0	53.12	91
	Galileo	86.87	42	0	20160
	BDS	100	0	54.16	118
	QZSS	0	20160	0	20160
	NavIC	100	0	5.18	371
	Combined	**100**	**0**	**100**	**0**

Table 5.9. Performance for GEO receiver at longitude 300 deg

Band	Constellation	At least 1 signal		4 or more signals	
		Avail. (%)	MOD (min)	Avail. (%)	MOD (min)
L1/E1/B1	GPS	79.43	103	3.68	462
	GLONASS	90.47	33	5.02	939
	Galileo	63.12	82	0	20160
	BDS	100	0	50.5	142
	QZSS	100	0	0	20160
	Combined	**100**	**0**	**100**	**0**
L5/L3/E5a/B2	GPS	91.81	66	9.19	419
	GLONASS	100	0	52.96	94
	Galileo	86.55	41	0	20160
	BDS	100	0	100	0
	QZSS	100	0	0	20160
	NavIC	67.78	216	0	20160
	Combined	**100**	**0**	**100**	**0**

5.2.2 *Scientific highly elliptical orbit mission*

Spacecraft trajectory

A highly elliptical orbit (HEO) mission scenario with apogee altitude of about 58,600 km and perigee altitude of 500 km is used to demonstrate the GNSS visibility performance through all the GNSS SSV altitudes, both below and above the GNSS constellations.

GNSS visibility conditions near the perigee are similar to those of space user receivers in LEO, with the important difference that the spacecraft is moving very fast – around 8 km/s to 11 km/s – so extreme Doppler shifts occur on the GNSS signals, and visibility times between any particular GNSS satellite and the HEO space user receiver are much shorter than for terrestrial receivers.

Table 5.10. Osculating Keplerian HEO orbital elements

Epoch	1 Jan 2016 12:00:00 UTC		
Semi-major axis	35937.5 km	RAAN	0 deg
Eccentricity	0.80870	Argument of perigee	270 deg
Inclination	63.4 deg	True anomaly	0 deg

Spacecraft attitude and antenna configuration

The on-board GNSS antennas are configured in both nadir and zenith-facing sides of the spacecraft. As shown in figure 5.5 the nadir-pointing antenna with high gain and narrow beam-width can ensure the GNSS signal link from the opposite side of the Earth, including when flying above the GNSS altitude and during the apogee period. The zenith-pointing patch antenna can provide visibility during the perigee period. The antenna patterns for both type of antennas are given in table B10. The acquisition and tracking thresholds of the user receiver were both set to 20 dB-Hz when evaluating the signal availability in the HEO simulation.

Figure 5.5. Schematic of the HEO mission with nadir and zenith-pointing antennas

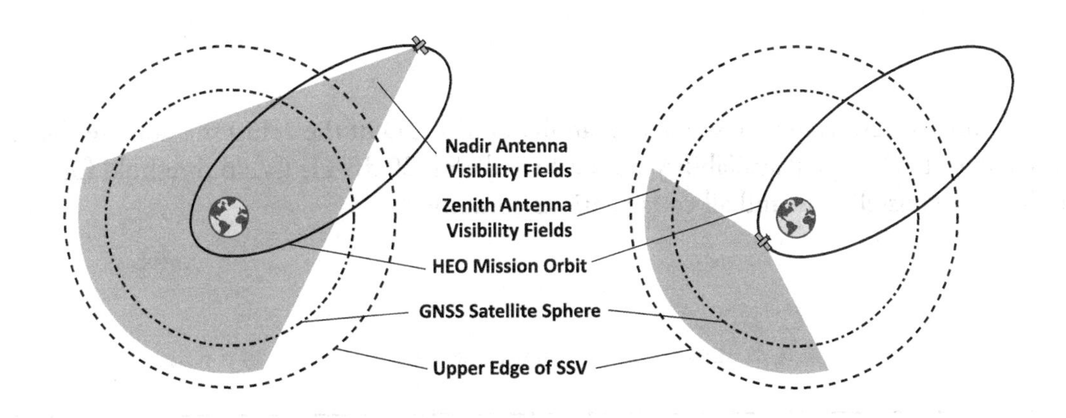

Results

Figure 5.6 shows the GNSS signal availability of all GNSS constellations and L5/L3/E5a/B2 signal for the HEO nadir and zenith-pointing antennas over the time of 1.5 HEO orbital periods. Note that when the spacecraft is below the GNSS constellation altitude, the visibility can be significantly improved by combining the signals from both nadir and zenith antennas at the same time. However, within this simulation only the strongest signal from either is employed at a given time. Around apogee, only the nadir-pointing antenna provides signal availability.

Figure 5.6. Visible GNSS satellites over 1.5 orbital periods of HEO (L5/L3/E5a/B2)

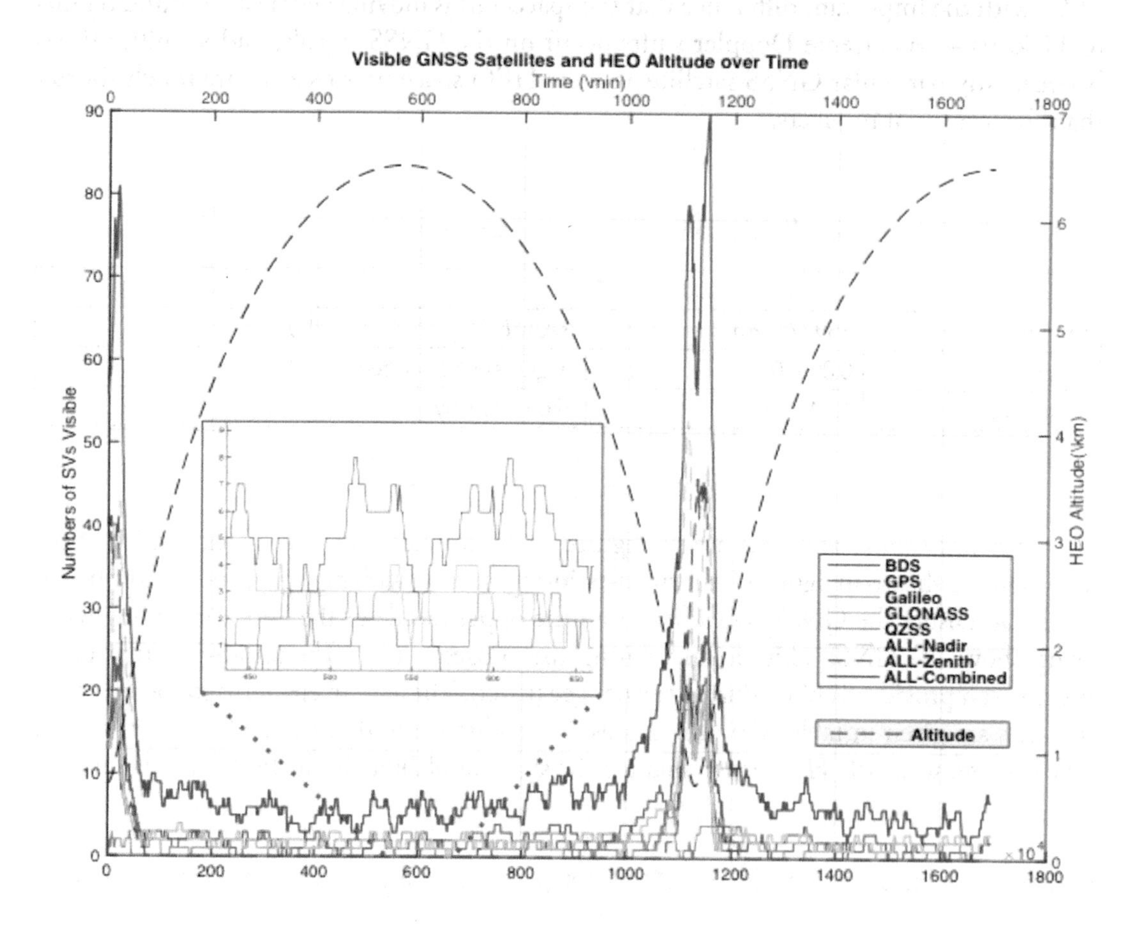

The simulated results for the signal availability and MOD of the HEO mission are shown in table 5.11. The signal availability was evaluated with 20 dB-Hz C/No threshold for each individual constellation and all constellations combined.

Table 5.11. HEO mission simulated performance result

Band	Constellation	At least 1 signal		4 or more signals	
		Avail. (%)	MOD (min)	Avail. (%)	MOD (min)
L1/E1/B1	GPS	87.3	70	12.7	1036
	GLONASS	98.8	12	14.1	986
	Galileo	74.3	85	9.9	1026
	BDS	88.1	51	16.1	1008
	QZSS	27.5	1031	2.5	2175
	Combined	**100**	**0**	**94.5**	**47**

L5/L3/E5a/B2	GPS	94.7	53	17.4	911
	GLONASS	100	0	55.5	133
	Galileo	87.1	63	11.6	980
	BDS	96.9	30	26.0	925
	QZSS	32.1	1021	5.8	1091
	NavIC	35.1	989	5.8	1091
	Combined	**100**	**0**	**100**	**0**

For both L1/E1/B1 and L5/L3/E5/B2 the one-signal availability can reach 100% with all constellations combined. In case of L1, four-signal availability is below 20% and the MOD is around 1,000 minutes, which is close to the HEO orbital period of 1,130 minutes, for an individual constellation. The performance is significantly improved by receiving signals from all constellations combined to nearly 100%. The result of L5 case is similar and the four-signal availability is 100% with all constellations combined. The table also shows that signal availability for the L5 case is better than the L1 case.

5.2.3 Lunar mission

The lunar mission case models a simple ballistic cislunar trajectory from LEO to lunar orbit insertion, similar to the trajectories flown by the 1968 United States Apollo 8 mission and many others. This case seeks to explore the boundaries of the GNSS SSV beyond Earth orbit.

Figure 5.7. Lunar trajectory phases; only the outbound trajectory segment is analysed.

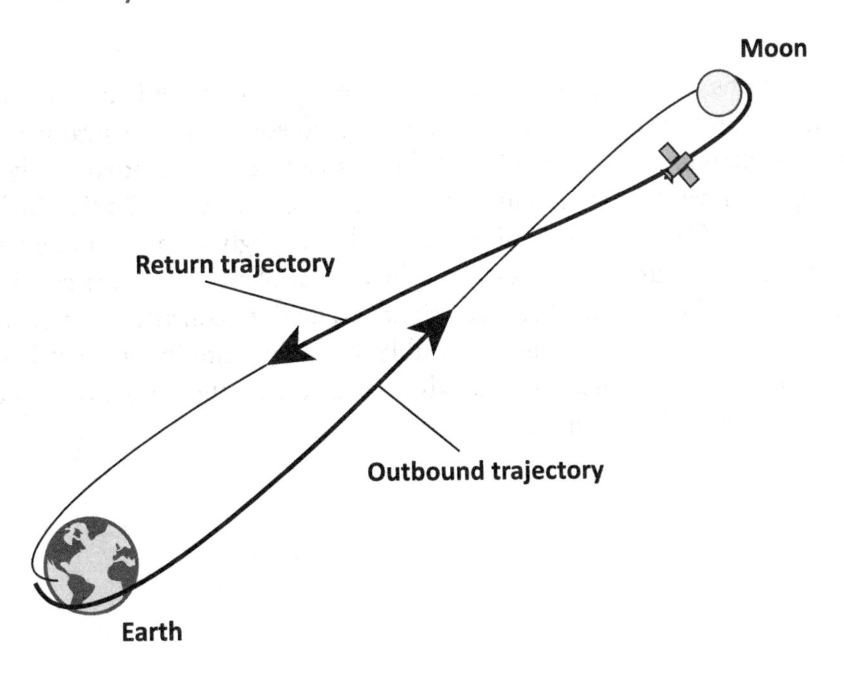

Spacecraft trajectory

Figure 5.7 shows a diagram of the trajectory being modelled; only the outbound portion is being modelled for this analysis. Earth orbit, lunar orbit, and return trajectories are projected to have known or similar performance. Table 5.12 shows the fundamental characteristics of the simulation.

Table 5.12. Lunar simulation parameters

Parameter	Earth departure	Lunar arrival
Epoch (UTC)	1 Jan 2016 12:00:00.000	5 Jan 2016 22:07:59.988
Altitude	185 km	100 km
Eccentricity	0	0
Inclination (body-centred J2000)	32.5°	75°
RAAN	30°	165°
Argument of perigee (AOP)	32°	319°
True anomaly	0°	0°

Spacecraft attitude and antenna configuration

A generic spacecraft is modelled, with two GNSS antennas: one zenith-pointing with peak gain less than 5 dB for reception at low altitudes, and one nadir-pointing with peak gain of approximately 10 dB for reception above the GNSS constellations. The results presented assume that the antenna with the greatest number of tracked satellites is used. As in the other HEO and GEO cases, the acquisition and tracking thresholds were both set to 20 dB-Hz.

Results

Table 5.13 contains the simulated performance results for this mission. In the case of both L1 and L5 bands, the availability of four simultaneous signals is nearly zero for any individual constellation, though combined there is coverage to approximately 30 Earth radii (RE) (approximately half the distance to the Moon) near 10–15%. Single-satellite availability reaches 36% for the combined case at L5, though as shown in figure 5.8, this availability primarily occurs at low altitudes. The benefit of the combined case is best seen above 10 RE, where the combined case has signal availability consistently higher than any individual constellation, and often nearly double. Notably, combining constellations does not increase the altitude at which such signals are available; rather, it increases the number of signals available at a given altitude.

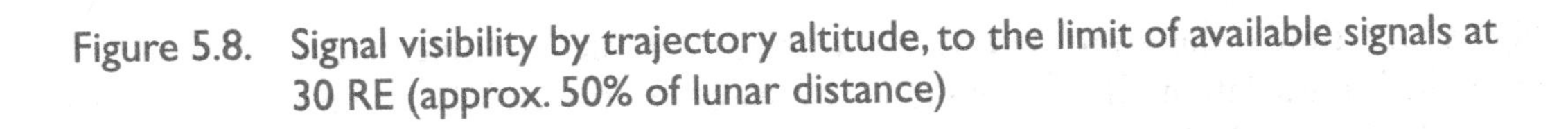

Figure 5.8. Signal visibility by trajectory altitude, to the limit of available signals at 30 RE (approx. 50% of lunar distance)

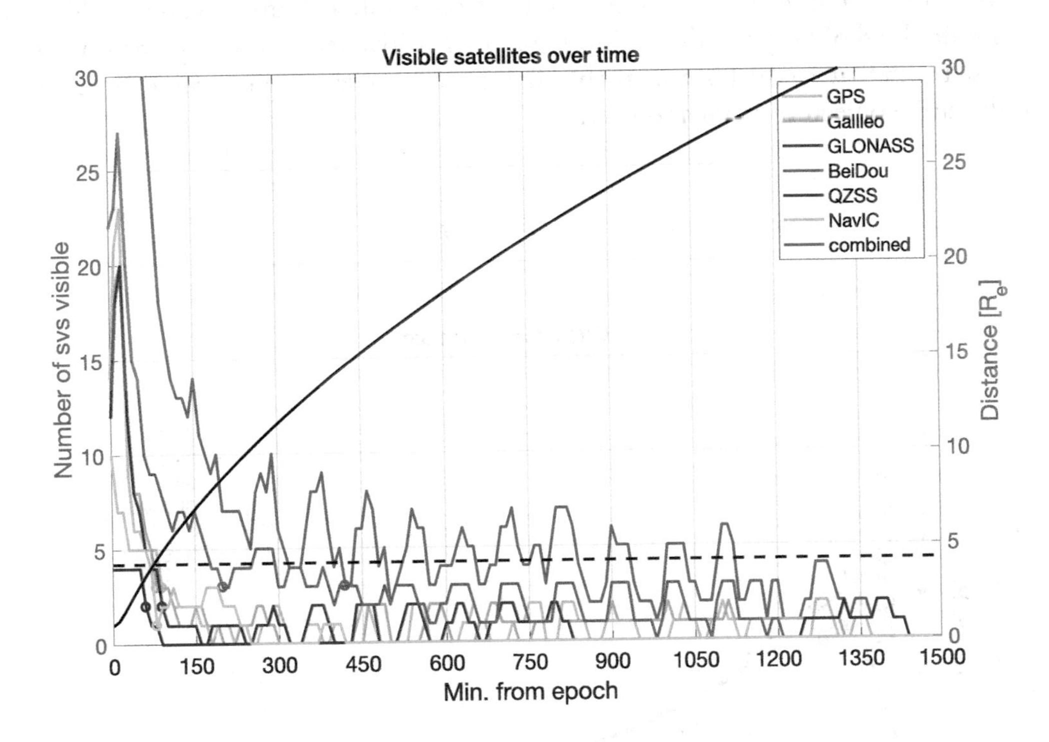

Table 5.13. Lunar mission simulated performance results

Band	Constellation	Signal availability (%)	
		At least 1 signal	4 or more signals
L1/E1/B1	GPS	9%	1%
	GLONASS	8%	0%
	Galileo	14%	1%
	BDS	14%	3%
	QZSS	1%	0%
	Combined	**21%**	**9%**
L5/L3/E5a/B2	GPS	12%	1%
	GLONASS	33%	1%
	Galileo	16%	1%
	BDS	18%	5%
	QZSS	4%	0%
	NavIC	4%	1%
	Combined	**36%**	**16%**

Figure 5.9 shows the simulated C/No received by the example spacecraft for each individual constellation. The figure shows the reason for the visibility drop-off near 30 RE shown in figure 5.8: the C/No of signals at the receive antenna drops below the 20 dB-Hz minimum threshold beyond that point. If a more sensitive receiver or higher-gain antenna was used such that 15 dB-Hz were usable, however, signal availability would be achievable for the entire trajectory to lunar distance.

Figure 5.9. Simulated C/No for lunar trajectory

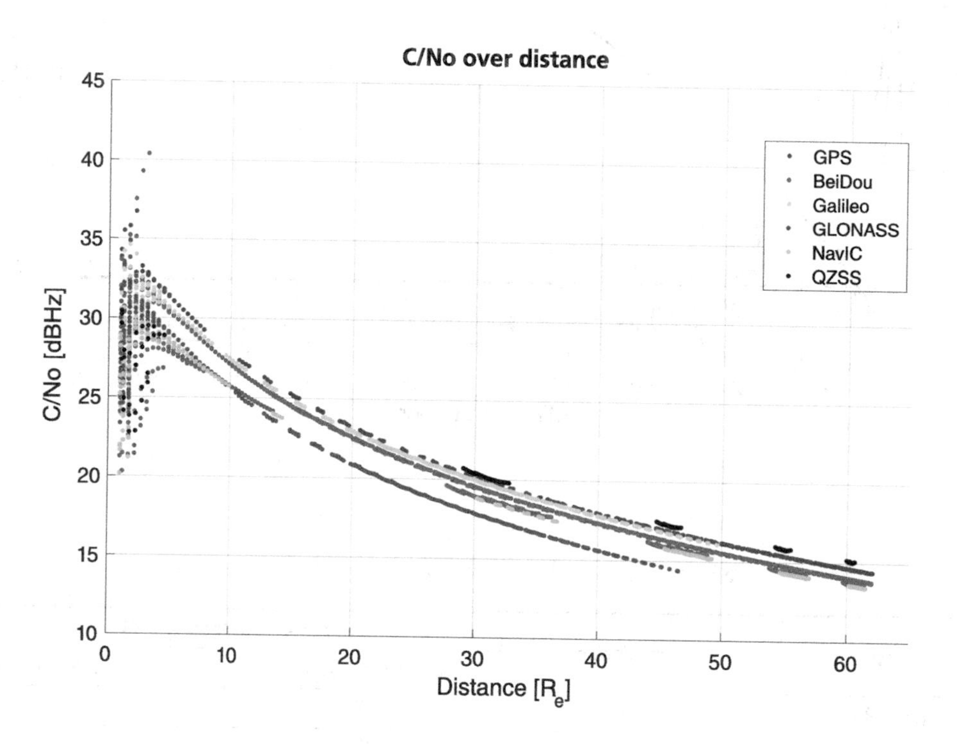

These results show that GNSS-based navigation with the combined interoperable GNSS SSV is feasible for nearly half the duration of a lunar outbound trajectory, well beyond the formal definition the upper bound of the SSV, and possibly a solution for navigation beyond the outbound trans-lunar injection (TLI) burn and return trajectory correction manoeuvres (TCMs). With further user modifications, it could provide on-board navigation at even higher altitudes.

6. Conclusions and recommendations

GNSS, which were originally designed to provide positioning and timing services to users on the ground, are increasingly being utilized for on-board autonomous navigation in space. While use of GNSS in LEO has become routine, its use in higher orbits has historically posed unique and difficult challenges, including limited geometric visibility and reduced signal strength. Only recently have these been overcome by high-altitude users through weak-signal processing techniques and on-board estimation filters.

The SSV was defined to provide a framework for documenting and specifying GNSS constellation performance for these users, up to an altitude of 36,000 km. The United Nations International Committee on GNSS (ICG) has worked on a collaborative basis to publicize the performance of each GNSS constellation in the SSV, and to promote the establishment of an interoperable multi-GNSS SSV in which all existing GNSS constellations can be utilized together to improve mission performance.

There are many benefits to an interoperable SSV, including increased signal availability for high-altitude users over that provided by any individual constellation alone, increased geometric diversity and thus accuracy in the final navigation solution, increased responsiveness and potential autonomy due to reduced signal outages, and increased resiliency due to the diversity of signals and constellations used. These benefits are truly enabling for classes of emerging advanced users, including ultra-stable remote sensing from geostationary orbit (GEO), agile and responsive formation flying, and more efficient utilization of valuable slots in the GEO belt.

This booklet captures SSV characteristics of each individual GNSS constellation, in terms of pseudorange accuracy, minimum received signal power, and signal availability (including MOD). In addition, the multi-constellation analysis documented here shows the benefits of the interoperable multi-GNSS SSV. In particular, there are significant availability improvements over any individual constellation when all GNSS constellations are employed. Within the high-altitude SSV, single-signal availability reaches 99% for the L1 band, and four-signal availability jumps from a maximum of 7% for any individual constellation to 89% with all, with a maximum signal outage duration of only 33 minutes. Further, similar benefits are shown explicitly for geostationary, highly elliptical, and lunar use cases.

The analyses presented clearly show the benefit and importance of interoperability of GNSS for high-altitude space users. To fully realize this benefit, the ICG makes the following recommendations:

1. *The authors encourage the development of interoperable multi-frequency space-borne GNSS receivers that exploit the use of GNSS signals in space.*
2. *GNSS providers are recommended to support the SSV outreach by making the booklet on "Interoperable GNSS Space Service Volume" available to the public through their relevant websites.*
3. *Service providers, supported by space agencies and research institutions, are encouraged to define the necessary steps and to implement them in order to support SSV in the future generations of satellites. Service providers and space agencies are invited to report back to WG-B on their progress on a regular basis.*
4. *Looking ahead, GNSS providers are invited to consider supplying the following additional data if available:*
   - *GNSS transmit antenna gain patterns for each frequency, measured by antenna panel elevation angle at multiple azimuth cuts, at least to the extent provided in each constellation's SSV template*
   - *In the long term, GNSS transmit antenna phase centre and group delay patterns for each frequency*

7. Potential future evolutions of this SSV booklet

To promote the multi-GNSS SSV for the purpose of safe robotic or manned missions in SSV as defined in this booklet and beyond including cislunar space it will be necessary to update this booklet, extend efforts on simulation and modelling as well as elaborating further on recommendations for GNSS providers. Some potential evolutions of the booklet could include:

- More accurate models of transmit antenna patterns and transmit power, based on provider published data and on-orbit derived observations
- Improved simulation models of end-to-end antenna systems to more accurately compute link analyses
- Recommended antenna system types for specific missions and orbits
- Improved simulations models, based on flight observed data, to more accurately represent the expected performance of missions for various orbits
- Expanding the user benefits and mission types, based on a more in-depth understanding of international use of GNSS in the SSV
- Expanding the SSV and improving user performance and SSV resiliency through trade studies on additional beacons or augmentations and service provider upgrades

The following process will be used to update the booklet contents:

- Updates of data can be provided by the service providers and other ICG members and will be processed by the ICG via WG-B.
- New releases of this booklet will be issued periodically, as necessary, after endorsement by ICG and all service providers.

Annex A. Description of individual GNSS support to SSV

A1. Global Positioning System SSV characteristics

Introduction

The Global Positioning System (GPS) is a United States-owned utility that provides users with positioning, navigation, and timing (PNT) services. GPS represents a "system of systems" consisting of three segments: a space segment, employing a nominal constellation of 24 space vehicles (SV) transmitting one-way signals with the GPS satellite's position and time; a control segment consisting of a global network of ground facilities that track the GPS satellites, monitor their transmissions, perform analyses, and send commands and data to the constellation; and a user segment that consists of GPS receiver equipment, which receives the signals from at least four GPS satellites and uses the transmitted information to calculate in real-time the three dimensional position and the time. The United States Air Force develops, maintains, and operates the space and control segments. Official United States Government information about GPS and related data topics is available at the National Coordination Office (https://www.gps.gov/).

Space segment

The United States is committed to maintaining the availability of at least 24 operational GPS satellites, 95% of the time to support PNT operations between the surface and 3,000 km altitude. In June 2011, the Air Force successfully completed a GPS constellation expansion known as the "Expandable 24" configuration. Three of the 24 slots were expanded, and six satellites were repositioned, so that three of the extra satellites became part of the constellation baseline. As a result, GPS now effectively operates as a 27-slot constellation with improved coverage in most parts of the world. To ensure this commitment, the Air Force is flying 31 operational GPS satellites.

The first satellite of what is now the GPS constellation was launched in 1978. Since then, the GPS space segment has evolved through three block architectures and multiple upgrades (Block I, Block IIA, Block IIR, Block IIR-M, Block IIF, Block III SV 1-10). At the time of writing, GPS is currently on-bid for its newest block upgrade—Block III SV11+.

GPS is configured in six orbital planes, inclined at 55 degrees and at an altitude of 20,182 km above the Earth. These orbital parameters result in an orbit period of a half-sidereal day (11 hours, 58 minutes) and a ground track that repeats every sidereal day.

Control segment

The current Operational Control Segment (OCS) includes a master control station, an alternate master control station, 11 command and control antennas, and 16 monitoring sites. OCS acquires the GPS signals, checks signal integrity and uplinks PNT correction and satellite ephemeris data. The control segment is currently undergoing a systems modernization to support Next Generation Operational Control System (OCX) operations. OCX will be delivered in increments, with increasingly more capable and sophisticated operations support. Block 0 will support launch and checkout of the GPS III satellites. Block 1 will operate and manage the GPS constellation. It will replace the Architecture Evolution Plan (AEP) system that is currently operational, and it will add modernized operational capabilities. Block 2 will enable the modernized civilian and military signals to become fully operational. This includes the civilian L1C, L2C & L5 signals and the military M-code signal.

Signal structure

GPS signal capabilities and structure have evolved with the evolution of the constellation block architecture. At full operational capability (mid-1990s), the GPS signal structure included an L1 C/A signal downlink at 1575.42 MHz for civilian applications and an L1/L2 P(Y) signal downlink at 1575.42 MHz/1227.6 MHz for military applications.

Subsequent improvements to the GPS signal structure have evolved to support GNSS interoperability and safety-of-life needs. These employ the long-used L1 (1575.42 MHz) and L2 (1227.6 MHz) frequencies with augmented modulations to support interoperability, enhanced civilian use and more robust military application. L5 was added using the 1176.45 MHz frequency to support safety-of-life operations. Block IIR-M (2005) inaugurated the second GPS civilian signal (L2C) designed specifically to meet commercial needs (for example, surveying), and also jam-resistant M (military) codes. Block IIF (2010) inaugurated the third civilian signal (L5) designed to meet demanding requirements for safety-of-life transportation and other high-performance applications. Block III satellites, the first of which at the time of writing is ready for launch, include a fourth civilian signal (L1C) designed to enable interoperability between GPS and international satellite navigation systems. L2C, L5, and M-code are currently pre-operational. These will become fully operational after control segment upgrades (e.g. OCX) and constellation replenishment results in sufficient signals to support full operations.

Space service volume

The signal information shown in the template conforms to the SSV requirements that are embedded in the GPS III vehicle specification. To date, GPS is the only GNSS constellation with a formal SSV specification. The current SSV specification addresses performance supplied by the spacecraft main-lobe signals.

Table A1. GPS III SSV characteristics

Definitions	Notes
Lower SSV: 3,000 to 8,000 km altitude	Four GPS signals available simultaneously a majority of the time, but GPS signals over the limb of the Earth become increasingly important. One-metre orbit accuracies are feasible (post-processed).
Upper SSV: 8,000 to 36,000 km altitude	Nearly all GPS signals received over the limb of the Earth. Users will experience periods when no GPS satellites are available. Accuracies ranging from 10 to 100 metres are feasible (post-processed) depending on receiver sensitivity and local oscillator stability.

Parameters	Value	
User range error [1]	0.8 metres	
Signal centre frequency		
L1 C/A	1575.42 MHz	
L1C	1575.42 MHz	
L2 (L2C or C/A)	1227.60 MHz	
L5 (I5 or Q5)	1176.45 MHz	
Minimum received civilian signal power	0 dBi RCP antenna at GEO	Reference off-boresight angle
L1 C/A	-184.0 dBW	23.5 deg
L1C	-182.5 dBW	23.5 deg
L2 (L2C or C/A)	-183.0 dBW	26 deg
L5 (I5 or Q5)	-182.0 dBW	26 deg
Signal availability [2]		
Lower SSV	**At least 1 signal**	**4 or more signals**
L1	100%	> 97%
L2, L5	100%	100%
Upper SSV	**At least 1 signal**	**4 or more signals**
L1	≥ 80% [3]	≥ 1%
L2, L5	≥ 92% [4]	≥ 6.5 %
Note 1: This value represents pseudorange accuracy, not the final user position error, which is dependent on many mission-specific factors such as orbit geometry and receiver design.		
Note 2: Assumes a nominal, optimized 27-satellite constellation and no GPS spacecraft failures. Signal availability at 95% of the areas at a specific altitude within the specified SSV.		
Note 3: Assumes less than 108 minutes of continuous outage time.		
Note 4: Assumes less than 84 minutes of continuous outage time.		

Figure A1. GPS geometry for SSV characteristics

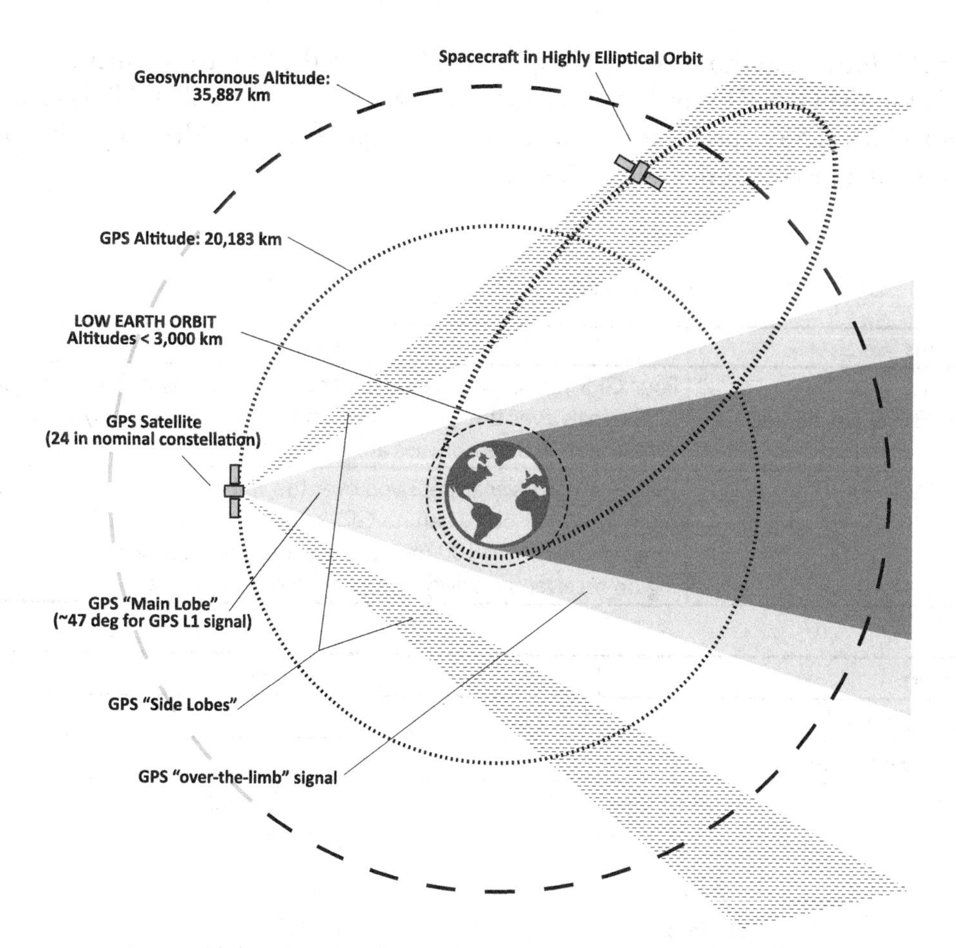

A2. GLONASS SSV characteristics

Introduction

GLONASS has three main segments: a space segment, generating and broadcasting navigation signals; a ground control segment, performing the functions of satellites operation control, continuous orbits and clocks parameters correction, delivering temporal programs, control commands and navigation data to satellites; and a user segment.

Space segment

The first GLONASS satellite was launched in 1982. Since then, there have been three generations of GLONASS satellites: GLONASS, GLONASS-M and GLONASS-K. The next generation of satellites being currently developed is GLONASS-K2. The additional L-band code division multiple access (CDMA) mission payload including the dedicated antenna is planned to be installed on-board satellites of the second phase of GLONASS modernization.

The current orbital constellation consists of GLONASS-M and GLONASS-K satellites. The GLONASS satellites are placed in roughly circular orbits with an altitude of 18,840...19,440 km (the nominal orbit altitude is 19,100 km) and the orbital period of 11h 15 min 44 sec ±5 sec. The orbital planes are separated by the 120° right ascension of the ascending node. Eight navigation satellites are equally spaced in each plane with the 45° argument of latitude. The orbital planes have an argument of latitude displacement of 15° relative to each other. With full orbital constellation, the repetition interval of satellites ground tracks and radio coverage zones for ground users is 17 orbit passes (7 days 23 hours 27 minutes 28 seconds).

The GLONASS orbital constellation is highly stable and does not demand additional corrections during satellites' life cycle. So maximum satellite drift of the ideal satellite orbital position does not exceed ±5° at a 5-year interval, while the average orbital planes precession rate is $0.59251 \cdot 10^{-3}$ rad/s.

A nominal orbital constellation consists of 24 satellites. The current orbital constellation has 24 operational satellites.

Control segment modernization

Ground Control Segment (GCS) Development Plans before 2020 involve all basic GCS elements for the purpose of their performance improvement (including upgrading one-way measuring and computing stations, master clock, measuring and laser ranging stations network extension).

The modernized ground control segment will additionally include:

Annex A: On-board intersatellite measurement equipment ground control loop providing orbit and clock data insertion to navigation satellite

Annex B: One-way measuring stations network for generating orbit and clock data to improve accuracy and integrity

Signal structure

The existing GLONASS constellation is comprised of GLONASS-M and GLONASS-K satellites broadcasting five navigation signals: L1OF (open Frequency Division Multiple Access (FDMA) in L1); L2OF (open FDMA in L2); L1SF (secured FDMA in L1); L2SF (secured FDMA in L2); L3OC (open CDMA in L3).

Space service volume

The GLONASS contribution to the interoperable GNSS SSV is provided in the following table.

Table A2. GLONASS SSV characteristics

Definitions	Notes
Lower SSV: 3,000 to 8,000 km altitude	Four GLONASS signals available simultaneously a majority of the time, but GLONASS signals over the limb of the Earth become increasingly important. One-metre orbit accuracies are feasible (post-processed).
Upper SSV: 8,000 to 36,000 km altitude	Nearly all GLONASS signals received over the limb of the Earth. Accuracies ranging from 20 to 200 metres are feasible (post-processed) depending on receiver sensitivity and local oscillator stability.

Parameters	Value	
User range error[1]	1.4 m	
Signal centre frequency		
L1	1605.375 MHz	
L2	1248.625 MHz	
L3	1201 MHz	
Minimum received civilian signal power (GEO)	0 dBi RCP antenna at GEO	Reference off-boresight angle
L1[2,3]	-179 dBW	26 deg
L2	-178 dBW	34 deg
L3[4]	-178 dBW	34 deg
Signal availability [5]		
MEO at 8,000 km	**At least 1 signal**	**4 or more signals**
L1	59.1%	64%
L2, L3	100%	66%
Upper SSV	**At least 1 signal**	**4 or more signals**
L1	70%	2.7%
L2, L3	100%	29%
Note 1: This value represents pseudorange accuracy, not the final user position error, which is dependent on many mission-specific factors such as orbit geometry and receiver design.		
Note 2: FDMA signals in L1 and L2 and CDMA signals in L3		
Note 3: L1, L2 signals are transmitted by GLONASS-M and GLONASS-K satellites. At present, the L3 signal is transmitted by the GLONASS-K satellite. Furthermore, the final seven GLONASS-M satellites will also transmit L3 signal (starting with the GLONASS-M No. 55 satellite).		
Note 4: L3 signals for GLONASS-K satellites.		
Note 5: Assumes at least one GLONASS satellite in view in the high-orbit service volume.		

Figure A2. GLONASS geometry for SSV characteristics

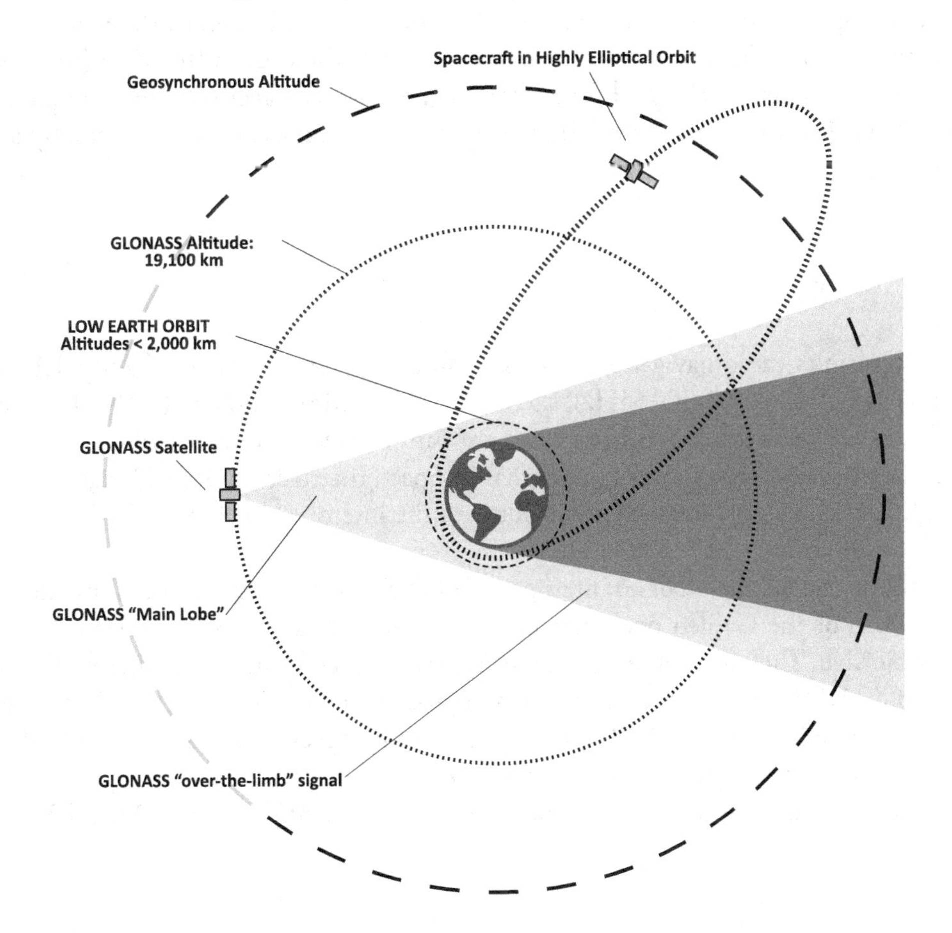

A3. Galileo full operational capability SSV characteristics

Galileo space segment

The nominal Galileo space segment consists of a constellation of 24 satellites, plus six active in-orbit spares, spaced evenly in three circular MEO planes inclined at 56 degrees relative to the equator. Their orbits have a nominal altitude of about 29,600 km and an orbital period of approximately 14 hours. Today the Galileo space segment consists of four in-orbit-validation (IOV) satellites and a series of full-operational-capability (FOC) satellites for which the number is continuously increasing thanks to the ongoing deployment process with the objective to reach full operational capability by 2020. Both IOV and FOC type of satellites belong to the operational Galileo constellation.

Ground segment

The Galileo ground segment controls the Galileo satellite constellation, monitors the health of the satellites, provides core functions of the navigation mission (satellite orbit determination, clock synchronization), performs the statistical analysis of the signal-in-space ranging error, determines the navigation messages, and uploads the navigation data for subsequent broadcast to users. The key elements of the transmitted data (such as satellite orbit ephemeris, clock synchronization, signal-in-space accuracy and the parameters for the NeQuick ionospheric model) are calculated from measurements made by a global network of reference sensor stations.

Galileo signals

Galileo transmits radio-navigation signals in four different frequency bands: E1 (1,559-1,594 MHz), E6 (1,260-1,300 MHz), E5a (1,164-1,188 MHz) and E5b (1,195-1,219 MHz). The details of the Galileo signal structure are summarized in the following tables and are specified in the Galileo Open Service Signal-in-Space Interface Control Document. Signals highlighted with (*) in these tables contribute to the interoperable GNSS SSV.

In relation to the definition of an interoperable GNSS SSV, it is to be noted that during the design phase of the Galileo open service signals interoperability with other GNSS was a major objective. The open service signal in E1, the so-called composite binary offset carrier or CBOC(6,1,1/11) signal, was originally designed in cooperation with the United States to aid interoperability with the GPS L1C signal. Similar spectral shapes have later also been adopted by BDS and QZSS, paving the way for multi-constellation interoperability. Also, the Galileo E5a signal is fully interoperable with GPS L5, BDS B2 and QZSS L5.

Table A3. Galileo E1 signal characteristics overview

Service name	E1 OS*		PRS
Centre frequency	1575.42 MHz		
Spreading modulation	CBOC(6,1,1/11)		BOCcos(15,2.5)
Sub-carrier frequency	1.023 MHz and 6.138 (Two sub-carriers)		15.345 MHz
Code frequency	1.023 MHz		2.5575 MHz
Signal component	Data	Pilot	Data
Primary PRN code length	4092		N/A
Secondary PRN code length	-	25	N/A
Data rate	250 sps	-	N/A

Table A4. Galileo E6 signal characteristics overview

Service name	E6 CS data*	E6 CS pilot*	E6 PRS
Centre frequency	1278.75 MHz		
Spreading modulation	BPSK(5)	BPSK(5)	BOCcos(10,5)
Sub-carrier frequency	-	-	10.23 MHz
Code frequency	5.115 MHz		

Signal component	Data	Pilot	Data
Primary PRN code length	5115	5115	N/A
Secondary PRN code length	-	100	N/A
Data rate	1,000 sps	-	N/A

Table A5. Galileo E5 signal characteristics overview

Service name	E5a data*	E5a pilot*	E5b data*	E5b pilot*
Centre frequency	1191.795 MHz			
Spreading modulation	AltBOC(15,10)			
Sub-carrier frequency	15.345 MHz			
Code frequency	10.23 MHz			
Signal component	Data	Pilot	Data	Pilot
Primary PRN code length	10230			
Secondary PRN code length	20	100	4	100
Data rate	50 sps	-	250 sps	-

Typical characteristics of Galileo FOC satellites for SSV

The typical characteristics of Galileo FOC satellites to support the interoperable GNSS SSV are provided in this section. Detailed and exhaustive measurement campaigns during the satellite ground testing were conducted for FOC-class satellites in order to characterize the typical emissions at SSV-relevant off-boresight angles. The results as obtained from different FOC-class satellites are summarized in the following tables.

The typical characteristics provided next shall not be interpreted as commitment from the Galileo Programme for existing or future Galileo FOC-class satellites. Official information related to SSV characteristics of Galileo will be published in the future through the Galileo Open Service - Service Definition Document.

In order to ensure the support of Galileo to SSV users, actions are put in place in to maintain and enforce these capabilities in the future.

The support of Galileo FOC satellites to the interoperable GNSS SSV is provided in the following table.

Table A6. Galileo SSV characteristics

Definition	Notes
Lower SSV: 3,000 to 8,000 km altitude	Four Galileo signals available simultaneously a majority of the time, but Galileo signals over the limb of the Earth become increasingly important. Capability of the user to receive both from nadir and from zenith is considered.
Upper SSV: 8,000 to 36,000 km altitude	Nearly all Galileo signals received over the limb of the Earth. Users will experience periods when no Galileo satellites are available.

Parameters	Typical characteristics of nominal GSAT02xx satellites	
User range error [1]	1.1 metres	
Signal centre frequency		
E1B/C	1575.42 MHz	
E6B/C	1278.75 MHz	
E5b	1206.45 MHz	
E5ABOC	1191.795 MHz	
E5a	1176.45 MHz	
Minimum received civilian signal power	0 dBi RCP antenna at GEO	Reference off-boresight angle
E1B/C	-182.5 dBW	20.5 deg
E6B/C	-182.5 dBW	21.5 deg
E5b	-182.5 dBW	22.5 deg
E5ABOC	-182.5 dBW	23.5 deg
E5a	-182.5 dBW	23.5 deg
Signal availability [2]		
Lower SSV	**At least 1 signal**	**4 or more signals**
E1B/C	100%	> 99% [7]
E6B/C	100%	100%
E5b	100%	100%
E5a or E5ABOC	100%	100%
Upper SSV	**At least 1 signal**	**4 or more signals**
E1B/C	>= 64% [3]	0%
E6B/C	>= 72% [4]	0%
E5b	>= 80% [5]	0%
E5a or E5ABOC	>= 86% [6]	0%
Note 1: This value represents pseudorange accuracy, not the final user position error, which is dependent on many mission-specific factors such as orbit geometry and receiver design.		
Note 2: Assumes a nominal, Galileo Walker 24/3/1 constellation, full navigation message availability and no Galileo spacecraft failures. Signal availability is provided at 95% of the areas within the specific altitude.		
Note 3: Assumes less than 93 minutes of continuous outage time.		
Note 4: Assumes less than 75 minutes of continuous outage time.		
Note 5: Assumes less than 64 minutes of continuous outage time.		
Note 6: Assumes less than 54 minutes of continuous outage time.		
Note 7: >99% at 21.5 deg (-182.5 dBW).		

A4. BDS SSV characteristics

BDS constellation

The current regional BeiDou Navigation Satellite System (BDS) space segment consists of five geostationary orbit satellites (GEO), five IGSO and four medium Earth orbit satellites (MEO). The GEO satellites are operating in orbit with an altitude of 35,786 kilometres and positioned at 58.75°E, 80°E, 110.5°E, 140°E and 160°E respectively. The IGSO satellites are operating in orbit with an altitude of 35,786 kilometres and an inclination of 55° to the equatorial plane. The phase difference of right ascensions of ascending nodes of those orbital planes is 120°. The sub-satellite tracks for three of those IGSO satellites are coincided while the longitude of the intersection point is at 118°E. The sub-satellite tracks for the other two IGSO satellites are coincided while the longitude of the intersection point is at 95°E. The MEO satellites are operating in orbit with an altitude of 21,528 kilometres and an inclination of 55° to the equatorial plane. The satellite recursion period is 13 rotations within seven days. The phase is selected from the Walker24/3/1 constellation, and the right ascension of ascending node of the satellites in the first orbital plane is 0°. The current four MEO satellites are in the seventh and eighth phases of the first orbital plane, and in the third and fourth phases of the second orbital plane respectively. The 5 GEO + 5 IGSO constellation provides regional coverage, and the MEO satellites were deployed for performance improvement, system redundancy and flight test for global service.

By 2020, the space constellation of BDS will consist of 5 GEO satellites, 3 IGSO satellites and 27 MEO satellites. Stationary positions of the 5 GEO satellites are consistent with the regional system. The crossing longitude of 3 IGSO satellites is 118°E. A total of 24 out of the 27 MEO satellites shape up into Walker 24/3/1 constellation, and the remaining 3 are separately taken as spare satellites in each orbit plane. The GEO and IGSO satellites are deployed to offer better anti-shielding capabilities, regional augmentation, short message communication and other active services.

BDS OS signals

The current regional BDS transmit two operational open service (OS) signals: B1I and B2I. The nominal frequency of the B1I signal is 1561.098 MHz, and the nominal frequency of the B2I signal is 1207.140 MHz. The detailed signal characteristics are specified in the BDS SIS-ICD 2.0.

The performance of modernized OS signals B1-C (1575.42MHz), B2-a and B2-b (1191.795MHz) broadcast by new-generation navigation satellites is enhanced significantly compared to the operational OS signals. The modernized signals of BDS can provide better compatibility and interoperability with other navigation satellite systems. Related documents will be updated and published in step with BDS construction and development.

Typical characteristics of BDS satellites for SSV

In this section the typical characteristics of BDS satellite are provided.

The parameters were measured from pre-flight ground test of the new-generation navigation satellites deployed in 2015. The parameters are provided to support the assessment of the interoperable GNSS SSV and do not represent a specification for existing or future

BDS satellites. BDS is taking actions in SSV performance characterization and specification. In future, official information related to SSV characteristics of BDS will be published through the BDS Open Service Performance Standard Document.

The signal availability below is evaluated by assuming a BDS constellation consists of 5 GEO satellites, 3 IGSO satellites and 24 MEO satellites. (The 3 spare MEO satellites are not incorporated.)

Table A7. BDS MEO/GEO/IGSO SSV characteristics

Parameters	Value	
User range error [1]	2.5 metres [9]	
Signal centre frequency		
B1	1575.42 MHz	
B2	1191.795MHz	
Minimum received civilian signal power	0 dBi RCP antenna at GEO	Reference off-boresight angle
B1 (MEO)	-184.2 dBW	25 deg
B1 (GEO/IGSO)	-185.9 dBW	19 deg
B2 (MEO)	-182.8 dBW	28 deg
B2 (GEO/IGSO)	-184.4 dBW	22 deg
Signal availability [2]		
Lower SSV [7]	At least 1 signal	4 or more signals
B1	99.9%	96.2%
B2	100%	99.9%
Upper SSV [8]	At least 1 signal	4 or more signals
B1	97.4% [3]	24.1% [4]
B2	99.9% [5]	45.4% [6]
Note 1: This value represents pseudorange accuracy, not the final user position error, which is dependent on many mission-specific factors such as orbit geometry and receiver design.		
Note 2: Signal Availability is evaluated by averaging performance over the 8,000km sphere for lower SSV and 36,000km for upper SSV.		
Note 3: Assumes less than 45 minutes of continuous outage time.		
Note 4: Partial region will be not visible for four signals.		
Note 5: Assumes less than 7 minutes of continuous outage time.		
Note 6: Assumes less than 644 minutes of continuous outage time.		
Note 7: The antenna for a user in the Lower SSV is considered to be omnidirectional.		
Note 8: The antenna for a user in the upper SSV is considered to be nadir-pointing.		
Note 9: The URE value is from specification of current regional BDS and will be enhanced significantly with the construction of global system.		

Figure A3. BDS geometry for SSV characteristics (left: MEO, right: IGSO/GEO)

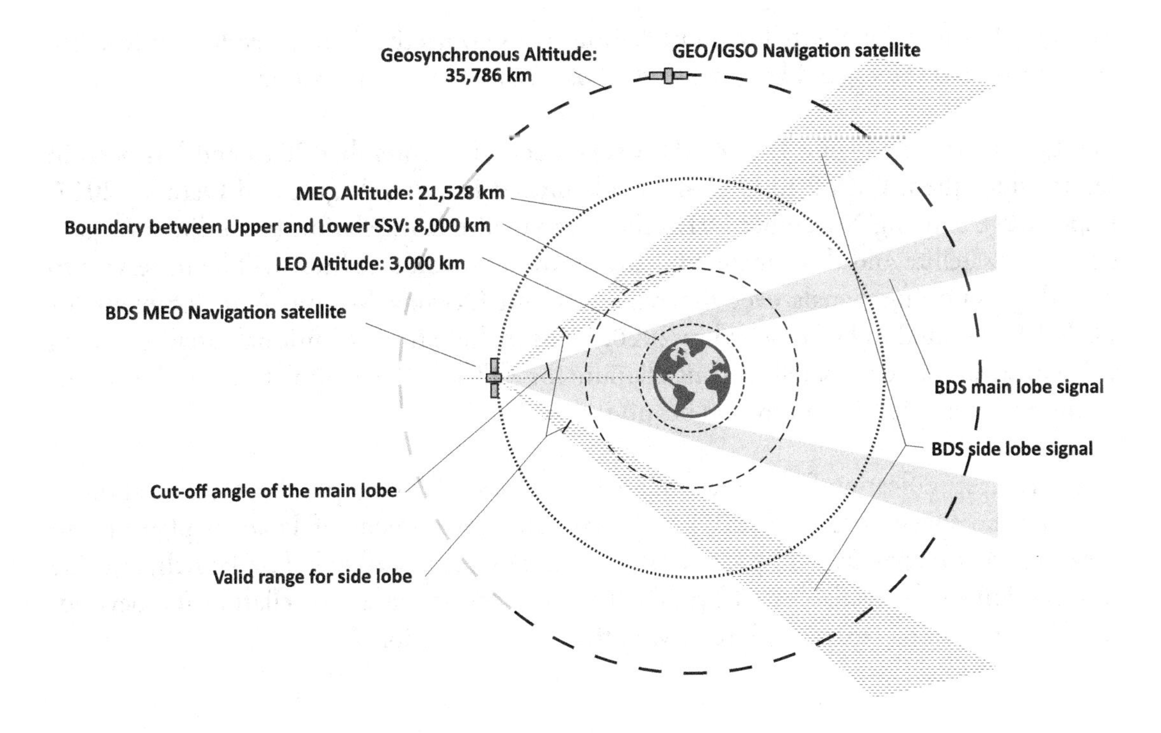

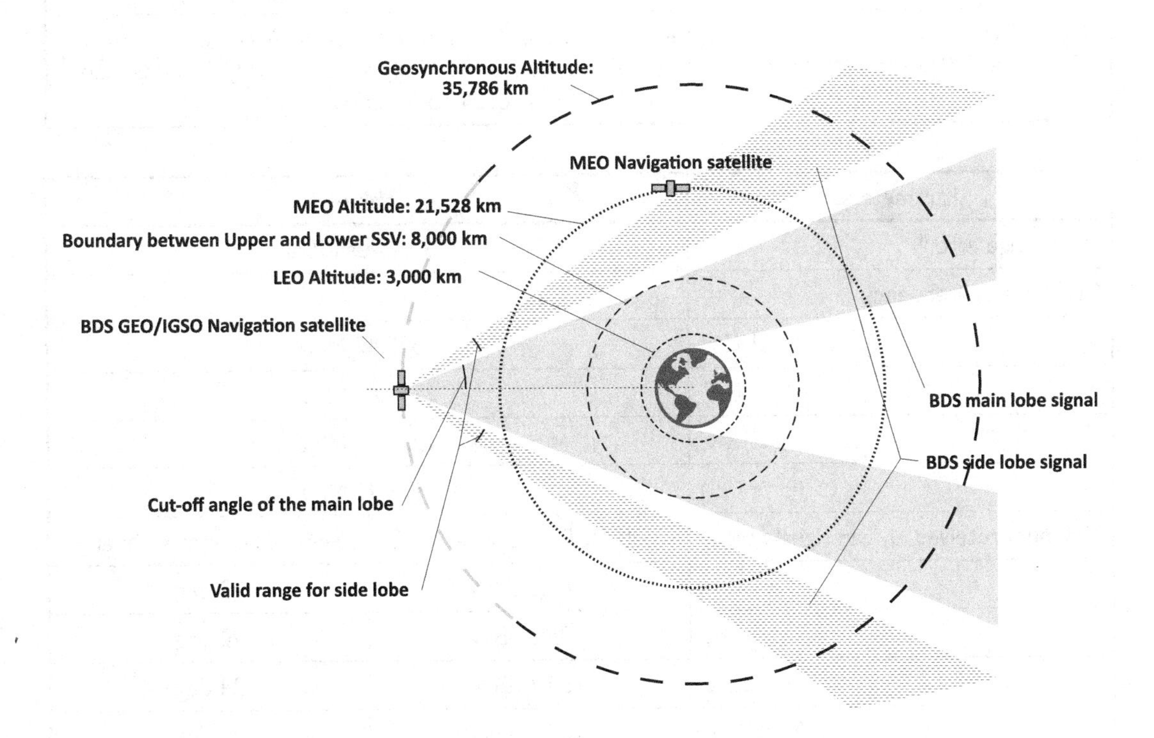

A5. Quasi-zenith satellite system SSV characteristics

The QZSS is a regional satellite constellation whose objective is to provide a fully GPS-compatible and interoperable signal to the East Asia and Oceania region.

The Quasi-Zenith Satellite 1 (QZS-1) was launched in September 2010 and has been in service since then. QZS-2 to QZS-4 were launched in June, August and October 2017, respectively. Starting November 2018, the four-satellite constellation (including one geostationary satellite and three inclined geosynchronous orbit satellites) will be in service to provide positioning signals over the East Asia and Oceania Region. A replacement for QZS-1 is expected to be launched in 2020. Plans include three additional satellites which will constitute a seven-satellite constellation for QZSS. The completion of the seven-satellite constellation is expected to be around 2023.

The current specification for a four-satellite constellation is not applicable for SSV application (i.e. no specification for SSV.) However, the Government of Japan is planning to measure antenna pattern and phase characteristics of each satellite before launch, and the information will be available to the public. For the seven-satellite constellation and beyond, the Government of Japan is still reviewing the SSV application.

Table A8. QZSS SSV characteristics

Definition	Notes
Lower SSV: 3,000 to 8,000 km altitude	QZS-1 signals are available above the East Asia and Oceania region. Signal-in-space user range error accuracy is 2.6 metres (95%).
Upper SSV: 8,000 to 36,000 km altitude	QZS-1 signals received over the limb of the Earth. Accuracies ranging from 10 to 100 metres are feasible (post-processed) depending on receiver sensitivity and local oscillator stability.

Parameters	Value	
User range error[1]	2.6 metres (95%)	
Signal centre frequency		
L1 C/A	1575.42 MHz	
L1C	1575.42 MHz	
L2 C	1227.60 MHz	
L5 (I5 or Q5)	1176.42 MHz	
Minimum received civilian signal power	0 dBi RCP antenna at GEO	Reference off-boresight angle
L1 C/A	-185.3 dBW	22 deg
L1C	-185.3 dBW	22 deg
L2 C	-188.7 dBW	24 deg
L5 (I5 or Q5)	-180.7 dBW	24 deg

Signal availability[2]		
Lower SSV	At least 1 signal	4 or more signals
L1	100%[3]	N/A
L2, L5	100%[3]	N/A
Upper SSV	At least 1 signal	4 or more signals
L1	≥ 54%[4]	N/A
L2, L5	≥ 54%[4]	N/A
Note 1: This value represents pseudorange accuracy, not the final user position error, which is dependent on many mission-specific factors such as orbit geometry and receiver design.		
Note 2: Assumes a nominal, no QZS-1 spacecraft failures and no orbit manoeuvre. Signal availability at 95% of the areas within the specific altitude.		
Note 3: Assumes user satellites between 20 degrees (east) and 120 degrees (west).		
Note 4: Assumes user satellites between 9 degrees (east) and 99 degrees (west).		

Figure A4. QZSS geometry for SSV characteristics

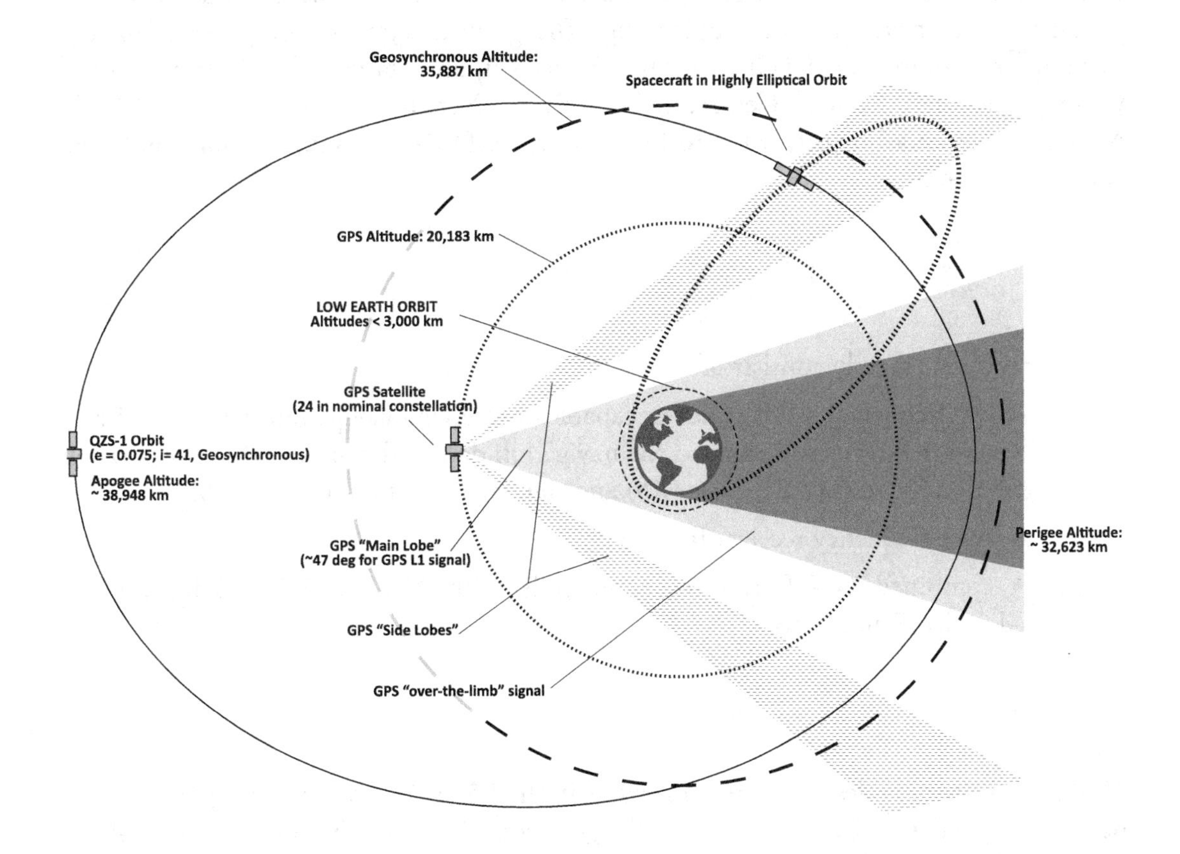

A6. Navigation with Indian Constellation SSV characteristics

The NavIC is an ISRO initiative to build an independent satellite navigation system to provide precise PVT to users over the Indian region. The system is designed to provide position accuracy better than 20 metres (2 σ) and time accuracy better than ± 40 ns (2 σ) over the Indian subcontinent and a region extending to about 1,500 km around India for a dual frequency user. The NavIC system mainly consists of space segment, ground segment and user segment.

NavIC space segment

The space segment consists of seven satellites, three satellites in GEO and four satellites in IGSO with inclination of 29° to the equatorial plane. Along with these seven satellites, an additional four IGSO satellites are planned. These additional four satellites are yet to be coordinated. All the satellites will be visible in the service region for 24 hours and will transmit navigation signals in both L5 and S bands.

Ground segment

The ground segment is responsible for the maintenance and operation of the NavIC constellation. It provides the monitoring of the constellation status, correction to the orbital parameters and navigation data uploading. The ground segment comprises telemetry, tracking and command (TTC) & navigation data uplink stations, Navigation Control Centre, Spacecraft Control Centre, IRNSS/NavIC Network Timing Centre, IRNSS/NavIC Range and Integrity Monitoring Stations, CDMA Ranging Stations and data communication links.

User segment

The user segment mainly consists of:

1. A dual frequency NavIC receiver capable of receiving navigation signals in L5 and S band frequencies, download the navigation data and compute the user position solution for restricted service (RS) and standard positioning service (SPS)
2. A single frequency receiver for SPS
3. A combined GNSS receiver compatible with NavIC, BDS, Galileo, GPS, GLONASS and QZSS

NavIC signals

NavIC basically provides two types of services in the L5 (1176.45 MHz) frequency band, namely SPS and RS. The NavIC L5 SPS signal contributes to the interoperable GNSS SSV. The NavIC signal parameters in the L5 band are provided below.

Table A9. NavIC L5 signal parameters

Parameters	NavIC L5 signal parameters
Carrier frequency	1176.45MHz
Signal bandwidth	±12MHz
Modulation type	BPSK-R(1)
Chip rate	1.023 Mcps
Data rate	25 bps/50 sps
Spreading code type	Gold
Spreading code period	1 ms

Typical characteristics of NavIC SSV

The NavIC L5 SPS signal contributes to the interoperable GNSS SSV and the SSV parameters are provided in the table below.

The typical characteristics provided next shall not be interpreted as commitment from the NavIC system. Official information related to SSV will be published in the future through the NavIC SIS ICD.

Table A10. NavIC SSV characteristics

Definitions
Lower SSV: 3,000 to 8,000 km altitude
Upper SSV: 8,000 to 36,000 km altitude. The signals of all GNSS services together play a major role in ensuring accuracy in this service volume.

Parameters	Value	
User range error (without Iono)[1]	2.1 metres	
Minimum received civilian signal power, in dBW	**0 dBi RCP antenna at GEO**	**Reference off-boresight angle**
L5	-184.54	16 deg
Signal availability[2]	**At least 1 signal**	**4 or more signals**
Lower SSV[3]		
L5	98.00%[4]	51.40%[5]
Upper SSV[6]		
L5	36.9%[7]	0.6%[8]
Note 1: This value represents pseudorange accuracy, not the final user position error, which is dependent on many mission-specific factors such as orbit geometry and receiver design.		
Note 2: Assumes a nominal, optimized NavIC constellation of 11 satellites and no NavIC spacecraft failures.		

Note 3: The antenna for a user in the Lower SSV is considered to be omnidirectional.
Note 4: Maximum continuous outage time of the constellation is 348 mins (scenario duration of 14 days), signal availability at 96.5% of the areas at a specific altitude within the specified SSV.
Note 5: Maximum continuous outage time of the constellation is 20,160 mins (scenario duration of 14 days), signal availability at 47.4% of the areas at a specific altitude within the specified SSV.
Note 6: The antenna for a user in the upper SSV is considered to be nadir-pointing, signal availability at 35% of the areas at a specific altitude within the specified SSV.
Note 7: Maximum continuous outage time of the constellation is 20,160 mins in upper SSV (scenario duration of 14 days).
Note 8: Maximum continuous outage time of the constellation is 20,160 mins in upper SSV (scenario duration of 14 days), signal availability at 0% of the areas at a specific altitude within the specified space service volume.

Annex B. Detailed simulation configuration and results

This chapter provides the full set of SSV simulation results, as well as the configuration and methodology used to execute the simulations themselves. This information should allow the simulations to be independently implemented and the results to be independently reproduced.

B1. Global SSV simulations

This section will cover the globally averaged SSV simulations. These simulations analyse the SSV using both geometrical access constraints alone as well as combined geometrical and radio frequency access constraints. In both cases, a fixed grid of points is used to represent the set of receiver locations.

Geometrical analysis configuration

The geometrical access-only simulations are based on the orbit propagation set-up and access considerations specified in table B1, utilizing the orbital parameters specified for each constellation in annex C. Note that the effective Earth radius used when determining access is taken as the sum of the spherical Earth radius and the atmospheric radius.

Table B1. Keplerian orbital simulation assumptions

Parameter	Value
Initial simulation date and time (UTC)	1 January 2016 12:00:00
Simulation duration (days)	14
Simulation time step (minutes)	1
Earth universal gravitational parameter (m^3/s^2)	3.986004415e14

Parameter	Value
π (standard Matlab π)	3.141592653589793
Spherical Earth radius (km)	6378
Atmospheric radius (km)	50
Geostationary grid altitude (km)	36,000
Earth rotation rate (rad/day)	2π(1.002737811911135448)
Earth rotation angle at reference epoch (rad)	2π(0.7790572732640)

The global analysis represents the SSV receiver locations using an equal-area grid of points, as illustrated in figure B1. Each point represents a receiver's fixed ground track location on the Earth's surface from its target MEO or GEO altitude. The grid is specifically equal-area so that results computed using the points are not biased to regions containing many more points. It has roughly 4° spacing near the equator and comprises 2562 points.

Figure B1. User grid locations over Earth's surface

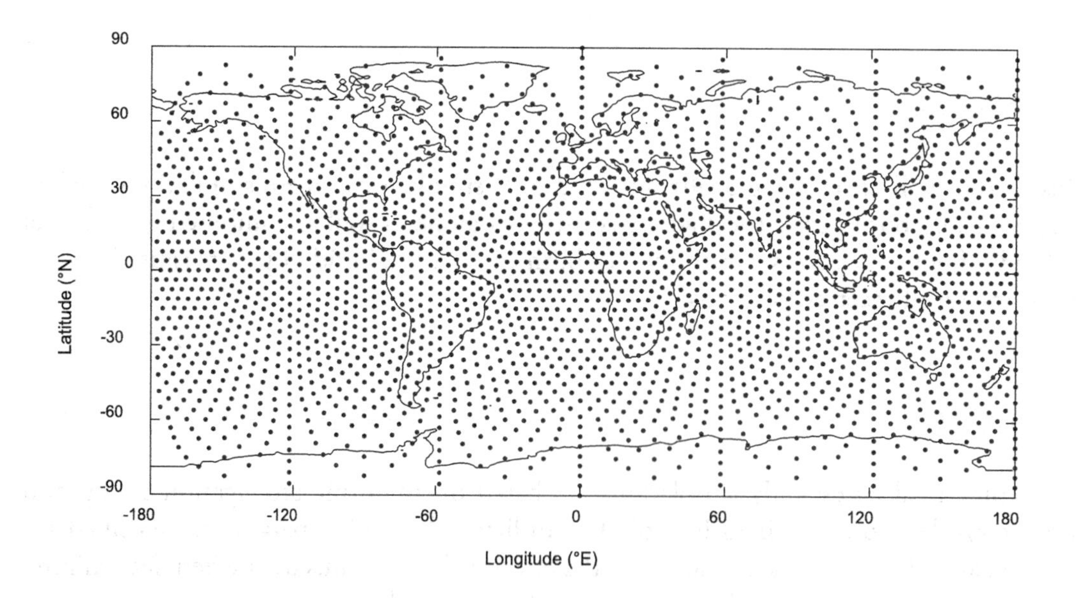

Table B2 summarizes the GNSS transmit beamwidths for both the L1 and L5 frequency bands that are studied in the simulations. Note that for the BDS constellation, the beamwidth is defined separately for the satellites in MEO and the satellites in GEO/IGSO. Also note that the NavIC L1 beamwidth is not applicable, as NavIC does not transmit in the L1 frequency band. It is also important to note that all L5 beamwidths are larger than their constellation's L1 beamwidth as a result of antenna physics, as that will directly impact performance results.

Table B2. GNSS transmitter beamwidths

GNSS constellation	L1 beamwidth (°)	L5 beamwidth (°)
BDS	25 (MEO) 19 (GEO/IGSO)	28 (MEO) 22 (GEO/IGSO)
Galileo	20.5	23.5
GLONASS	20	28
GPS	23.5	26
NavIC	N/A	16
QZSS	22	24

The attitude of each GNSS transmitting antenna is determined according to table B3, depending on which constellation the spacecraft belongs to. Additionally, depending on the simulation, the receiving antenna's boresight is pointed either nadir or zenith relative to the centre of the Earth, and its field of view is defined as either hemispherical or omnidirectional.

Table B3. Boresight pointing direction for GNSS transmit antenna

GNSS constellation	Transmitter boresight
NavIC	5°N, 83°E
All others	Nadir (Earth's centre)

Geometrical analysis methodology

The overall simulation methodology is performed in multiple steps, which are listed below:

1. Propagate orbit position vectors into Earth-centred Earth-fixed frame coordinates over scenario time instances.
2. Calculate angle off-GNSS-boresight vector to all SSV grid points over scenario time instances.
3. Calculate angle off-SSV-boresight vector to all GNSS orbit positions over scenario time instances.
4. Determine geometrical access using maximum GNSS beamwidth consideration, Earth blockage consideration, and SSV hemispherical/omnidirectional beamwidth consideration over scenario time instances for all SSV grid points.
5. Calculate figures of merit from access determination over scenario time instances over all SSV grid points.

Geometrical analysis results

Results in table B4 and table B5 provide the globally averaged SSV expected system performance when considering only geometrical access constraints. Please note that the reported availability figures are evaluated as the average availability over all grid points and all time epochs. Since the grid points are defined as having equal area pertaining to each grid point, averaging of performance over the grid points can be done using a pure mean calculation, without additional scale factors needing to be applied.

Note that all system availability metrics are rounded down to the next lowest tenths decimal place, and outage time is limited to integer numbers of minutes due to the nature that the simulations were performed on one-minute intervals.

Table B4. Geometrical access performance with GEO and MEO/omnidirectional scenarios

		Upper SSV (nadir antenna)				Lower SSV (omni antenna)			
Band	Constellation	At least 1 signal		4 or more signals		At least 1 signal		4 or more signals	
		Avail. (%)	MOD (min)	Avail. (%)	MOD (min)	Avail. (%)	MOD (min)	Avail. (%)	MOD (min)
L1/E1/B1	GPS	90.5	111	4.8	*	100	0	99.6	45
	GLONASS	93.9	48	7	*	100	0	99.8	24
	Galileo	78.5	98	1.2	*	99.9	11	95	60
	BDS	97.4	45	24.1	*	100	0	100	0
	QZSS	26.7	*	0.8	*	99.6	197	79.4	*
	Combined	**99.9**	**29**	**98.1**	**93**	**100**	**0**	**100**	**0**
L5/L3/E5a/B2	GPS	96.9	77	15.6	1180	100	0	99.9	16
	GLONASS	99.9	8	60.3	218	100	0	100	0
	Galileo	93.4	55	4.2	*	100	0	100	0
	BDS	99.9	7	45.4	644	100	0	100	0
	QZSS	30.5	*	1.5	*	99.6	197	79.4	*
	NavIC	36.9	*	0.6	*	98	348	51.4	*
	Combined	**100**	**0**	**99.9**	**15**	**100**	**100**	**0**	**0**

*No signal observed for the worst-case grid location for maximum simulation

Table B5. Geometrical access performance with MEO/zenith and MEO/nadir scenarios

Band	Constellation	Lower SSV with zenith antenna				Lower SSV with nadir antenna			
		At least 1 signal		4 or more signals		At least 1 signal		4 or more signals	
		Avail. (%)	MOD (min)	Avail. (%)	MOD (min)	Avail. (%)	MOD (min)	Avail. (%)	MOD (min)
L1/E1/B1	GPS	84.0	*	0	*	100	0	95.9	93
	GLONASS	80.8	195	0	*	100	0	95.5	97
	Galileo	84.0	*	0	*	99.8	13	71.5	262
	BDS	97.5	181	34.2	*	100	0	99.6	31
	QZSS	51.0	*	15.4	*	84.1	*	28.3	*
	Combined	**99.9**	**37**	**91**	*	**100**	**0**	**100**	**0**
L5/L3/E5a/B2	GPS	94.3	*	0.1	*	100	0	99.9	25
	GLONASS	100	0	78.4	245	100	0	100	0
	Galileo	96.0	*	2.4	*	100	0	97.4	40
	BDS	99.9	10	62.4	*	100	0	100	0
	QZSS	51.0	*	15.4	*	84.1	*	28.3	*
	NavIC	25.5	*	15.3	*	92.8	*	33.5	*
	Combined	**100**	**0**	**99.9**	**9**	**100**	**0**	**100**	**0**

*No signal observed for the worst-case grid location for maximum simulation

RF access analysis configuration

Please note that for the calculation of the user-received power along the arc where the GNSS satellite is visible, the following assumption has been applied: The minimum radiated transmit power (MRTP) resulting from the inverse link budget calculation is based on the user minimum received civilian signal power as established via the SSV template (annex A). The MRTP is constant for all off-boresight angles smaller than the reference off-boresight angle.

Table B6 provides the minimum received power level per GNSS constellation, along with maximum beamwidth and specific centre frequency, used to derive the MRTP to be considered over the beamwidth following an inverse link budget calculation. Note that for the BDS constellation, the beamwidth is defined separately for satellites in MEO than for those in GEO or IGSO. Table B7 provides additional parameters pertaining to general radio frequency (RF) assumptions used for these calculations and the simulations performed in this analysis.

Table B6. GNSS RF parameters

GNSS constellation	Signal name	Frequency (MHz)	Max beamwidth (°)	Minimum received power (dBW)	MRTP (dBW)
GPS	L1 C/A	1575.42	23.5	-184	9.1
Galileo	E1 B/C	1575.42	20.5	-182.5	10.9
GLONASS	L1	1605.375	20	-179	14.1
BDS MEO	B1	1575.42	25	-184.2	9
BDS GEO/IGSO	B1	1575.42	19	-185.9	9
QZSS	L1 C/A	1575.42	22	-186.1	9.0
GPS	L5	1176.45	26	-182	8.5
Galileo	E5a	1176.45	23.5	-182.5	8.4
GLONASS	L3	1201	28	-178	12.6
BDS MEO	B2	1191.795	28	-182.8	8
BDS GEO/IGSO	B2	1191.795	22	-184.4	8.1
QZSS	L5	1176.45	24	-183.4	9.2
NavIC	L5	1176.45	16	-184.54	7.8

Table B7. General RF simulation assumptions

Parameter	Value
Speed of light (m/s)	299792458
Boltzmann's constant (m^2 kg s^{-2} K^{-1})	$1.38064852 \times 10^{-23}$
Receiver antenna gain (dBi)	0
System noise temperature (K)	290

MRTP inverse link budget calculation

Because MRTP is not included in the SSV template completed by the GNSS service providers, this value must be derived for each constellation using an inverse link budget calculation with the constellation's specified minimum received power. The overall situation for the link budget calculation and the terms taken into account is outlined in figure B2.

Figure B2. Link Budget calculation scenario, where Tx is transmitter on-board the GNSS satellite, LNA is the low noise amplifier and Rx is the user receiver

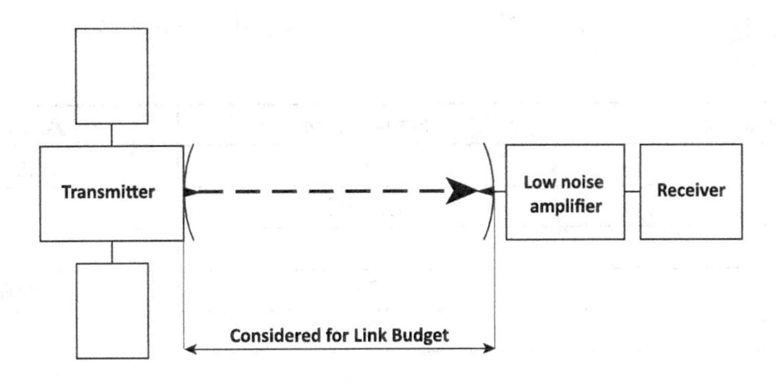

For the transmitting antenna pattern, on-board the GNSS spacecraft, figure B3 visualizes the basic assumption.

Figure B3. Simplified GNSS satellite antenna pattern, as used in the simulations for GNSS SSV Phase 3

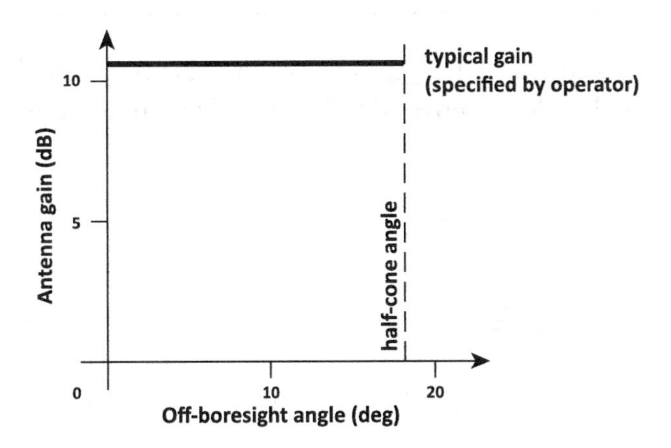

The inverse link budget is defined as

$$MRTP = P_{min} + L_S$$

where P_{min} is the specified minimum received power at GEO and L_S is the free space path loss at the worst-case Earth-limb distance:

$$L_S = 20 \log_{10} \frac{4\pi(\theta_{limb})f}{c}$$

In this equation, f is the centre frequency of the signal from table B6, c is the speed of light from table B7 and $R(\theta_{limb})$ is the distance from the worst-case apogee altitude of the GNSS constellation (see table B8) to a GEO user at 36,000 km altitude, along the line that intersects the Earth's limb.

Table B8. Worst-case apogee altitude used for each constellation in MRTP calculation

GNSS constellation	Signal name	Altitude (km)
GPS	L1 C/A	20181.80
Galileo	E1 B/C	23221.80
GLONASS	L1	19140.33
BDS MEO	B1	21611.86
BDS GEO/IGSO	B1	35912.69
QZSS	L1 C/A	38948.48
GPS	L5	20181.80
Galileo	E5a	23221.80
GLONASS	L3	19140.33
BDS MEO	B2	21611.86
BDS GEO/IGSO	B2	35912.69
QZSS	L5	38948.48
NavIC	L5	35815.71

The geometry used in calculating R(θ_{limb}) is shown in figure B4. Note that the Earth's radius from table B1 should be added to the GNSS and GEO altitudes to obtain R_{GNSS} and $R_{GEO,}$ respectively.

Figure B4. Geometry used in MRTP calculation

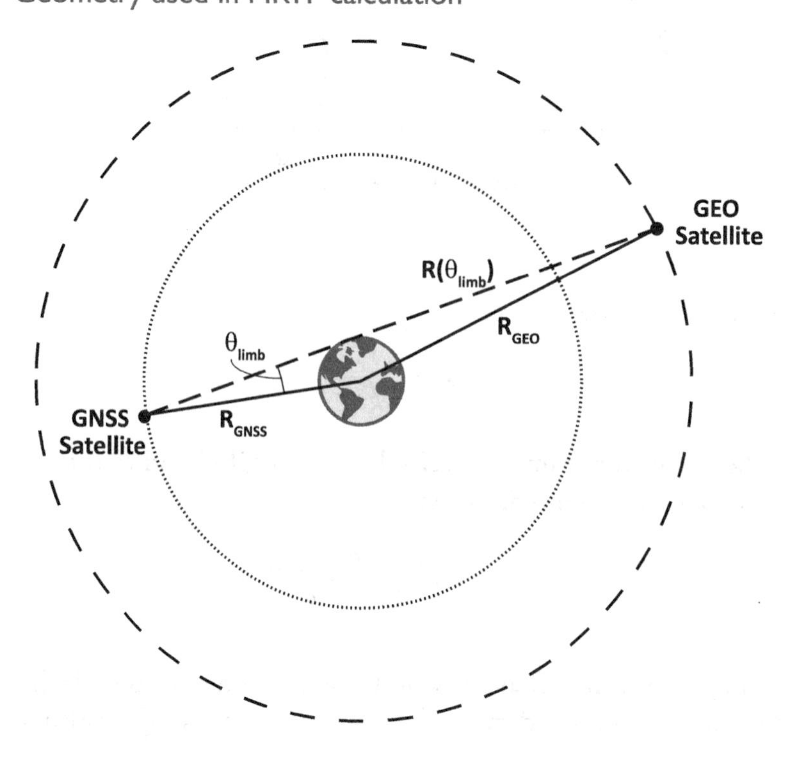

Using this geometry, the Earth-limb angle can first be calculated with

$$\theta_{limb} = \arcsin \frac{R_{EARTH}}{R_{GNSS}}$$

This angle can then be used to calculate the Earth-limb distance using the following formula:

$$R(\theta_{limb}) = R_{GNSS}\cos(\theta_{limb}) + \sqrt{R^2_{GEO} - R^2_{GNSS}\sin(\theta_{limb})^2}$$

The resulting MRTPs calculated with this method are shown in table B6 for each GNSS constellation.

RF access analysis methodology

The overall simulation methodology adds additional steps compared to the geometrical-only analysis to take into account the RF constraints. The full set of analysis steps are listed below:

- Propagate orbit position vectors into Earth-centred Earth-fixed frame coordinates over scenario time instances.
- Calculate angle off-GNSS-boresight vector to all SSV grid points over scenario time instances.
- Calculate angle off-SSV-nadir-boresight vector to all GNSS orbit positions over scenario time instances.
- Determine geometric access using maximum GNSS beamwidth consideration, Earth blockage consideration, and SSV hemispherical beamwidth consideration over scenario time instances for all SSV grid points.
- Calculate received signal to noise ratio to all SSV grid points from all GNSS transmitters, where geometrical access is available, over scenario time instances.
- Determine RF access comparing received signal-to-noise ratio with minimum threshold signal-to-noise ratio.
- Calculate figures of merit from RF-augmented access determination over scenario time instances over all SSV grid points.

RF access analysis results

Results in table B9 provide the average globalized upper SSV expected system performance when RF-based signal strength constraints are applied to geometrical-0nly access calculations. As stated previously, all system availability metrics provided are rounded down to the next lowest tenths decimal place, and maximum outage time is limited to integer numbers of minutes, due to the nature that the simulations are performed on one minute intervals. The lower SSV was only simulated under geometric conditions; see table 5.2 for details.

Table B9. Upper SSV performance with RF constraints, for various C/No thresholds

Band	Constellation	C/NO_{min} = 15 dB Hz				C/NO_{min} = 20 dB Hz				C/NO_{min} = 25 dB Hz			
		At least 1 signal		4 or more signals		At least 1 signal		4 or more signals		At least 1 signal		4 or more signals	
		Avail. (%)	MOD (min)	Avail. (%)	MOD (min)	Avail. (%)	MOD (min)	Avail. (%)	MOD (min)	Avail. (%)	MOD (min)	Avail. (%)	MOD (min)
L1/E1/B1	GPS	90.5	111	4.8	*	90.5	111	4.8	*	0.0	*	0	*
	GLONASS	93.9	48	7	*	93.9	48	7	*	93.9	48	7	*
	Galileo	78.5	98	1.2	*	78.5	98	1.2	*	0.0	*	0	*
	BDS	97.4	45	24.1	*		70	0.6	*	0.0	*	0	*
	QZSS	26.7	*	0.8	*	0.0	*	0	*	0.0	*	0	*
	Combined	**99.9**	**29**	**98.1**	**93**	**99.9**	**33**	**89.8**	**117**	**93.9**	**48**	**7**	*
L5/L3/E5a/B2	GPS	96.9	77	15.6	1180	96.9	77	15.6	1180	0.0	*	0	*
	GLONASS	99.9	8	60.3	218	99.9	8	60.3	218	99.9	8	60.3	218
	Galileo	93.4	55	4.2	*	93.4	55	4.2	*	0.0	*	0	*
	BDS	99.9	7	45.4	644	99.9	7	32.4	644	0.0	*	0	*
	QZSS	30.5	*	1.5	*	30.5	*	1.5	*	0.0	*	0	*
	NavIC	36.9	*	0.6	*	1.0	*	0	*	0.0	*	0	*
	Combined	**100**	**0**	**99.9**	**15**	**100**	**0**	**99.9**	**15**	**99.9**	**8**	**60.3**	**218**

*No signal observed for the worst-case grid location for maximum simulation

General observations concerning the availability estimates given in table B9 indicate the following:

- Comparing availability estimates between L5 and L1 bands, for the same system, indicates that L5 availability estimates are consistently better than those associated with L1 transmission when comparing codes from the same constellation. For one-signal coverage, L5 availability is 6% to 18% higher (relatively) than for L1 and for four-signal coverage, L5 availability is about 10% to 20% higher (absolutely) than L1. For MOD comparisons that are valid, L5 shows shorter MOD numbers by about 40 minutes. These improvements are averaged over all systems and vary by receiver C/No_{min}.
- Comparing availability estimates between one-signal and four-signal coverage, for the same system, indicates that one-signal availability estimates significantly exceed those associated with fourfold coverage when comparing codes from the same constellation. For C/No_{min} = 15 dB-Hz or 20 dB-Hz one-signal availability exceeds fourfold availability by 60% to 70% and in the L1 band and by about 50% in the L5 band. Insufficient data exist for comparisons of MOD between one-signal and four-signal coverage.
- However, an informal comparison of MOD and availability estimates (where valid) indicates a coarse inverse relationship between MOD and availability. For one-signal coverage when availability falls below about 50%, and for four-signal coverage when availability falls below about 10%, the MOD is likely to be equal to the simulation duration.
- At the threshold of 25 dB/Hz, performance drops to 0% availability for all but the GLONASS system. This set of results show that the required receiver capabilities are quite demanding in order to be able to utilize these extremely low GNSS signal levels.
- The most salient feature in all scenarios is the improvements in availability and MOD brought by the use of multiple constellations. For nearly all cases, L1 and L5 bands, one-signal and four-signal coverage, C/No_{min} = 15 dB-Hz, 20 dB-Hz and 25 dB-Hz, availability for the multi-constellation case is nearly 90% or better and MOD is limited to less than 120 minutes. Not until we get to the L1 band with four-signal coverage with C/No_{min} = 25 dB-Hz do availability and MOD drop precipitously (7% and "*"). These improvements for the multiple-system receiver are realized even in cases where individual systems are providing availability of less than 10% and MOD is at "*" (e.g. L1 band with four-signal coverage with C/No_{min}= 20 dB-Hz). For the global constellations (GPS, GLONASS, Galileo and BDS) a one-signal availability is indicated at a very high level: higher than 90% for the L5 band and C/No_{min} = 15 dB-Hz. However, only multi-constellation allows very high availability (> 99.5%) for four-signal coverage.

B2. Mission-specific SSV simulations

This section describes the detailed assumptions, methodology, and results associated with the three mission-specific SSV performance simulations performed: a geostationary mission, a highly elliptical Earth orbiting mission, and a lunar mission. These simulations are intended to illustrate the benefits of the multi-GNSS SSV to specific mission classes, beyond the globally characterized performance of the GNSS constellations themselves.

Common assumptions and methods

In all three mission simulations, certain common assumptions and methods were used for consistency.

The mission spacecraft was modelled in its mission-specific trajectory via either propagation from an initial state using the same assumptions as shown in table B1. The spacecraft attitude is modelled as nadir-pointing in all cases, though in the case of the HEO and lunar cases a zenith antenna is also simulated.

Two receiver antennas were modelled: a patch antenna (used for both L1 and L5 bands), and two different high-gain antennas, one each for L1 and L5. The antenna characteristics are captured in table B10 and correspond to characteristics of readily available antennas available on the open market.

Table B10. Antenna gain patterns for mission-specific simulations

Elevation angle [deg]	Patch antenna gain, L1 and L5 [dBi]	High-gain antenna	
		L1 [dBi]	L5 [dBi]
0	2.8	9.00	8.25
5	2.9	8.97	8.24
10	3.3	8.90	8.05
15	3.6	8.56	7.75
20	4.0	8.02	7.33
25	4.4	7.32	6.79
30	4.5	6.46	6.15
35	4.4	5.45	5.42
40	4.1	4.33	4.61
45	3.7	3.11	3.74
50	2.8	1.82	2.81
55	1.8	0.46	1.83
60	0.8	-0.93	0.83
65	-0.7	-2.34	-0.21
70	-1.7	-3.74	-1.25
75	-3.2	-5.12	-2.31
80	-5.2	-6.47	-3.38
85	-6.2	-7.78	-4.45
90	-8.7	-9.03	-5.52

The GNSS constellations and transmitter models were held identical to those used in the global simulations described above. The link budget characteristics and metrics for visibility were also held constant, with the notable exception of realistic receiver antenna models.

GEO mission

The GEO mission scenario analyses multi-GNSS signal reception for six geostationary satellites. The objective is to obtain more representative signal strength values than in the global analysis by using realistic user antenna patterns on-board the space users for receiving the B1/E1/L1 and B2/E5A/L5 signals.

Spacecraft trajectory

Six GEO satellites are simulated and share the same orbital plane apart from a 60-degree separation in longitude (see table B11). The right ascension of the ascending node (RAAN) angle is used to synchronize the orbit with the Earth rotation angle at the start of the simulation. The true anomaly is used to distribute the six GEO user receivers along the equator. This placement of the satellites was chosen to ensure that even signals from regional GNSS satellites in (inclined) geosynchronous orbits would be visible to at least one of the GEO user receivers (see figure B5).

Table B11. GEO osculating Keplerian orbital elements

Epoch	1 Jan 2016 12:00:00 UTC		
Semi-major axis	42164.0 km	Right ascension of the ascending node	100.379461 deg
Eccentricity	0.0	Argument of perigee	0.0 deg
Inclination	0.0 deg	True anomaly	0/60/120/180/240/300 deg

Spacecraft attitude and antenna configuration

The user antenna on-board the user spacecraft is a high-gain antenna that permanently points towards the nadir (centre of the Earth). The user antenna patterns used on the two signals are specified in table B10. The assumed acquisition threshold of the space user receiver is 20 dB-Hz.

Figure B5. Example for visibility of NavIC satellite from the GEO at 240-degree longitude

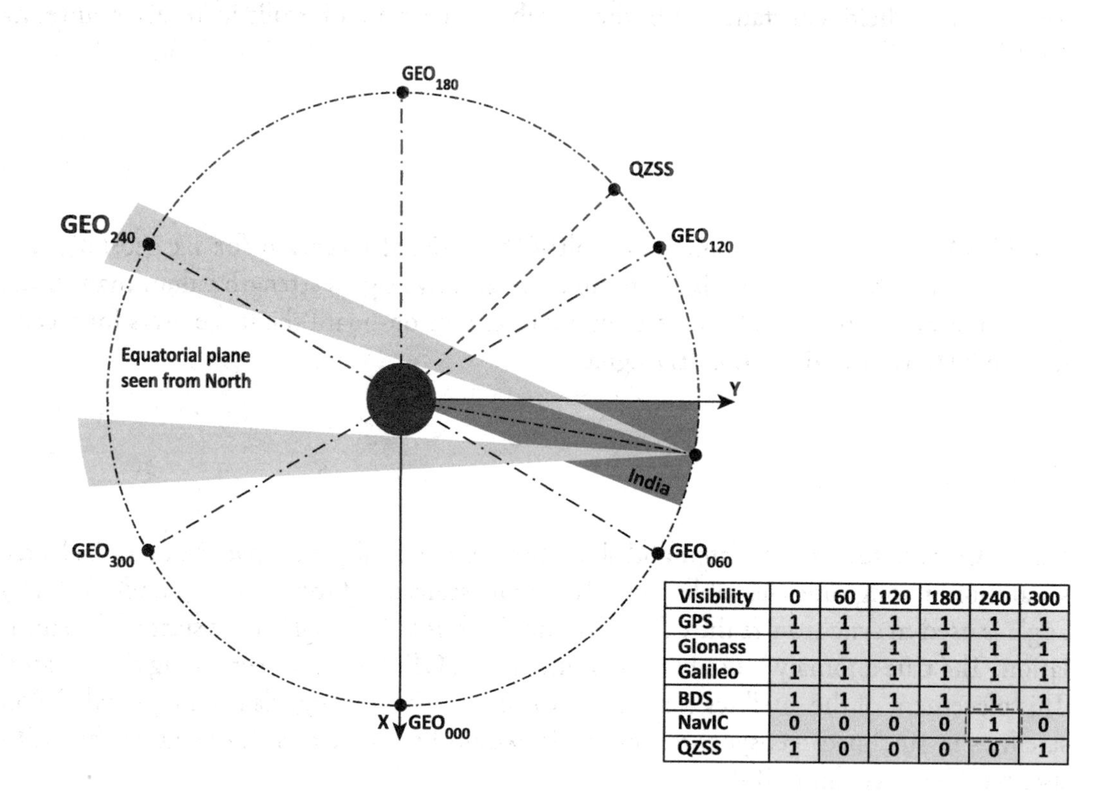

Visibility	0	60	120	180	240	300
GPS	1	1	1	1	1	1
Glonass	1	1	1	1	1	1
Galileo	1	1	1	1	1	1
BDS	1	1	1	1	1	1
NavIC	0	0	0	0	1	0
QZSS	1	0	0	0	0	1

Results

The six GEO satellites are all in the equatorial orbital plane but phased by 60 degrees in longitude, or four hours in time. The MEO GNSS satellites have orbital periods in the order of 12-14 hours, or about half that of the GEO. This means that the GEO and MEO orbits are almost in phase with each other, in such a way that the visibility patterns at the GEO receiver repeat almost exactly with periods of one day. The MEO satellites move 120 degrees during the four hours interval between GEO satellites, but there are multiple GNSS MEO in each orbital plane. This means that the visibility patterns in terms of number of visible MEO signals are very similar to all six GEO receivers.

The situation is different for the inclined geosynchronous GNSS satellites of the NavIC, QZSS and BDS constellations. The GEO and IGSO longitudes are frozen relative to each other. At most GEO longitudes, the GNSS satellites in IGSO orbits are never visible, either because the GEO is located outside the half-cone angle of the transmitting satellite, or because the signal is blocked by the Earth. This means that reception of the IGSO GNSS signals is an exception rather than the rule. However, those GEO receivers that do see signals from these transmitters will see them continuously, or at very regular patterns (see figures below for B2/E5A/L5 signal).

For all six GEO receivers, and at L1 and L5 frequencies, the satellite visibility is shown in the figure B6, figure B7, figure B8 and figure B9 below. The differences are mainly caused by the visibility of BDS, QZSS and NavIC regional geosynchronous satellites at certain GEO longitudes. Notably the GEO at 300-degree and 0-degree longitude appear to benefit from the Asian regional GNSS systems; these are GEO longitudes that are of specific interest to Europe and the North American East coast.

For GEO longitudes where no BDS, QZSS or NavIC geosynchronous satellites are visible, there are typically not more than three L1 signals available from any individual GNSS constellation. Combined, there are almost always four or more signals, and often up to ten signals.

For L5, the individual constellations are slightly better than for L1, and often provide four signals. The combined constellations almost always provide six or more signals. The red lines in figures B-8 and B-9 provide the signal visibility numbers for GEO. The number of signals is constantly varying, sometimes significantly, along an orbit and also at GEO longitude locations. The highest number of signals observed during the simulations was 21 signals at 300 degrees longitude. Note in particular the presence of BDS signals at GEO 300, which brings the combined visibility above 15 satellites through most of the simulation period.

Figure B6. L1/E1/B1 visibility for GEO at 0 deg, 60 deg and 120 deg

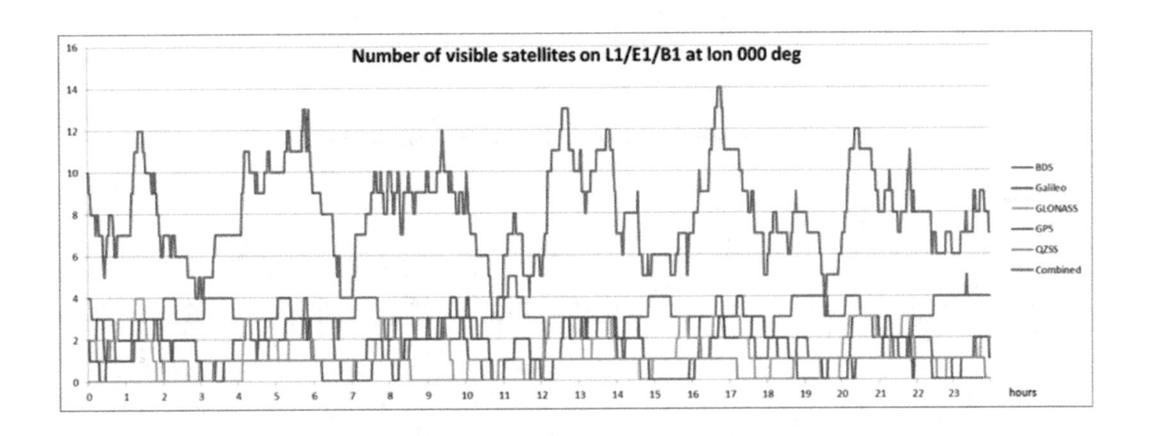

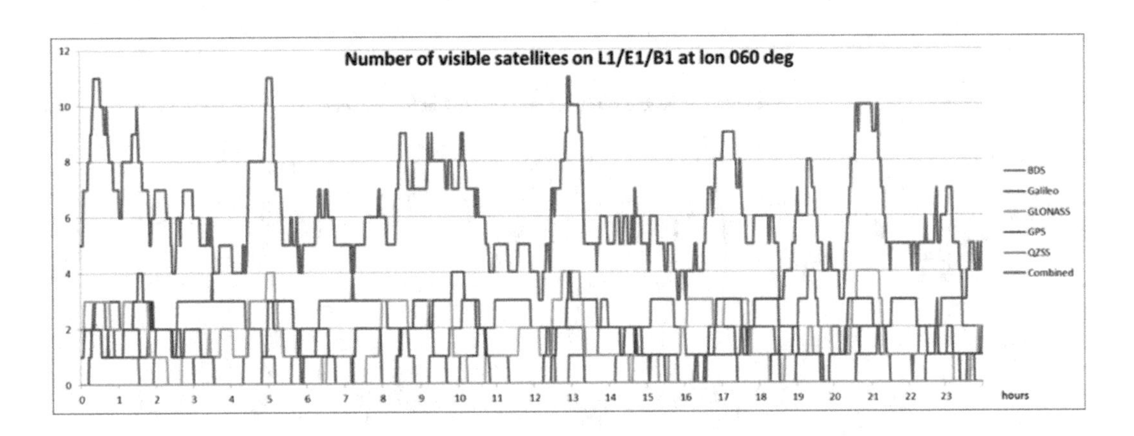

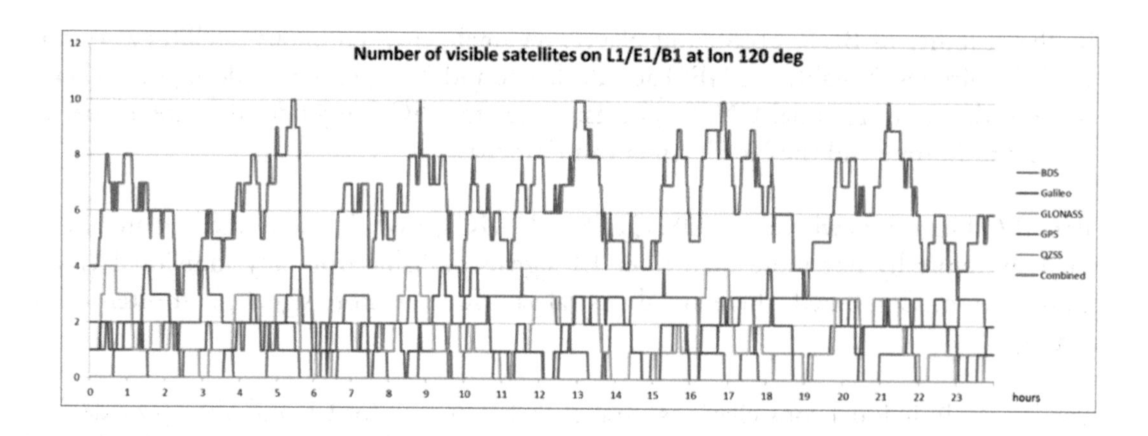

Figure B7. L1/E1/B1 visibility for GEO at 180 deg, 240 deg and 300 deg

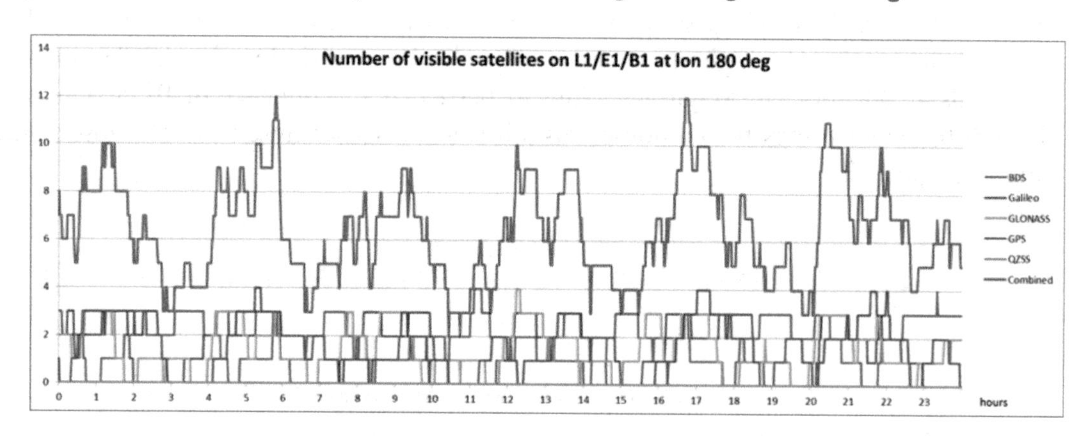

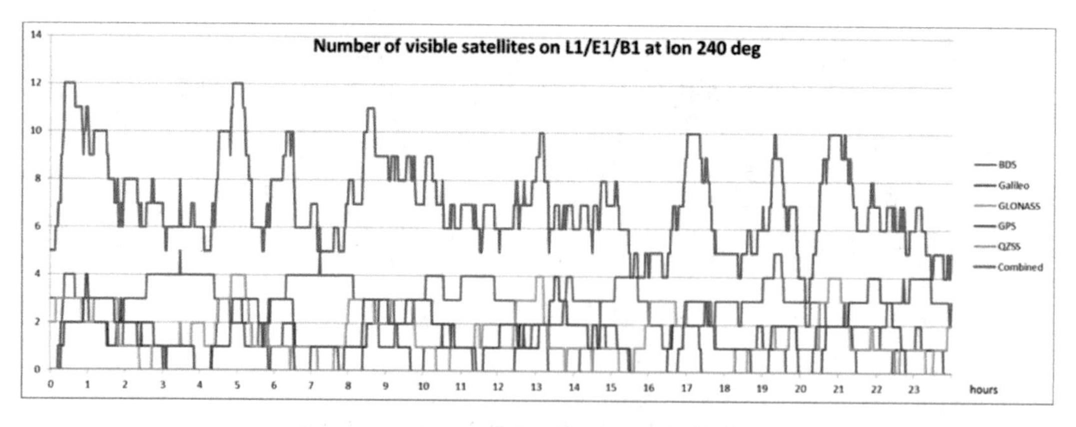

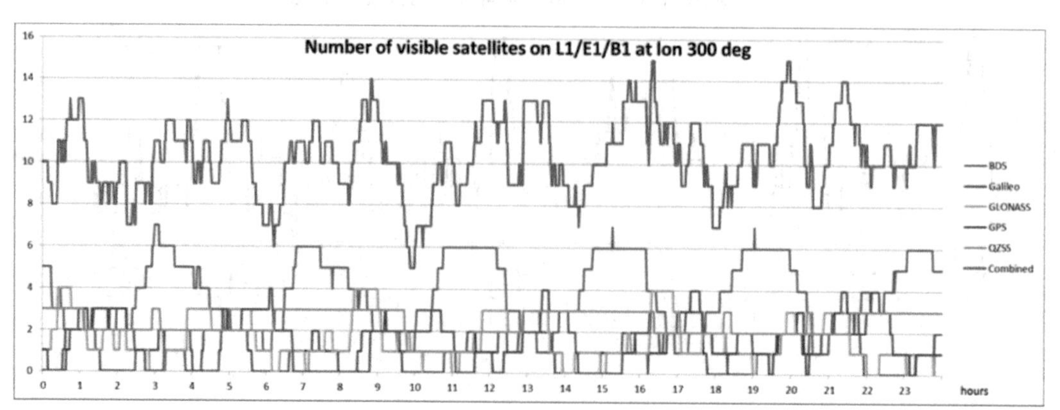

Figure B8. L5/E5a/B2 visibility for GEO at 0 deg, 60 deg and 120 deg

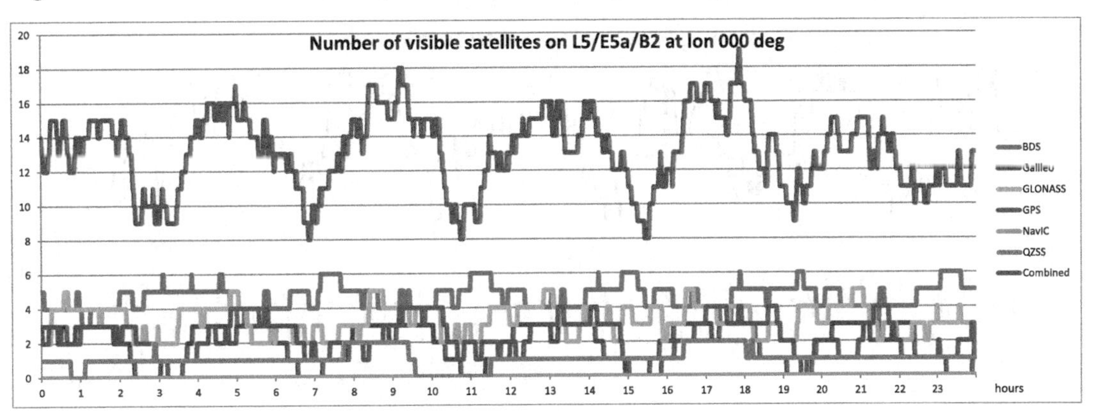

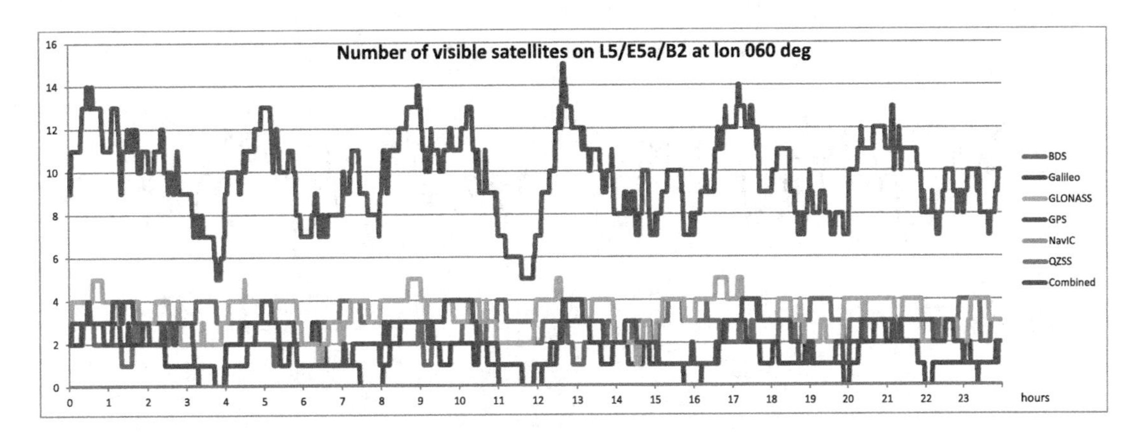

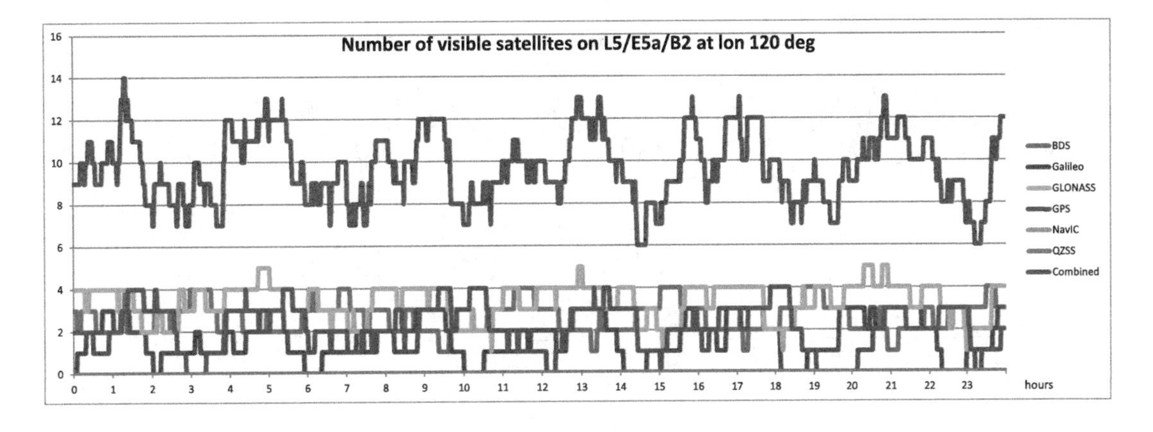

Figure B9. L5/E5a/B2 visibility for GEO at 180 deg, 240 deg and 300 deg

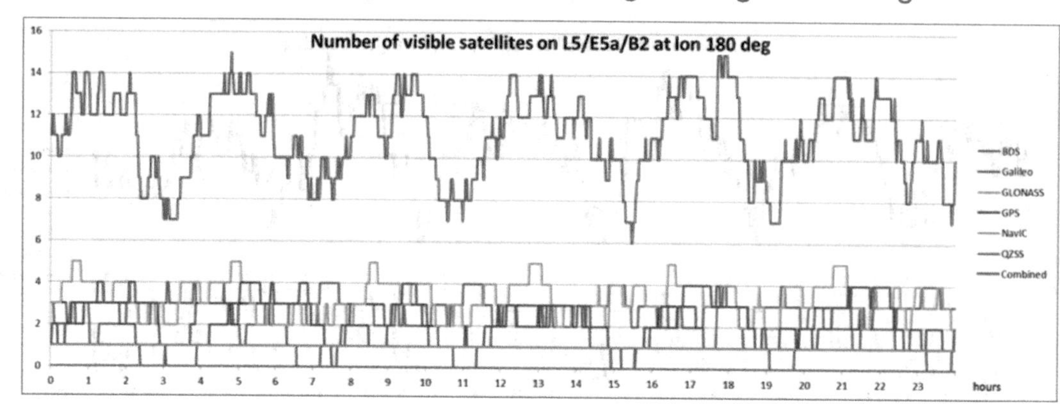

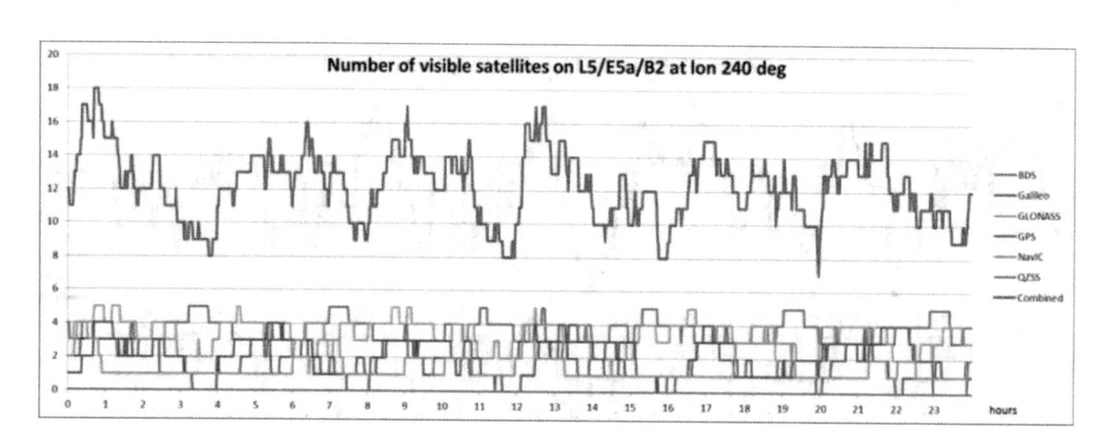

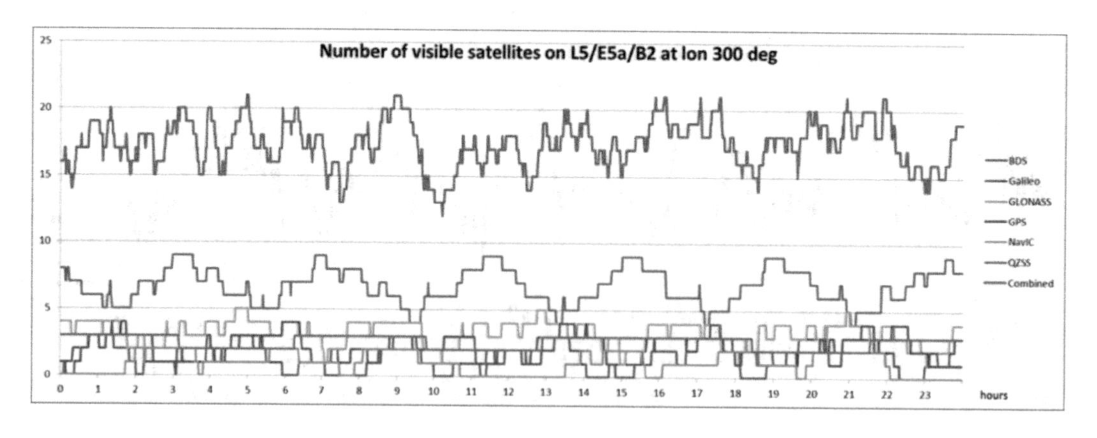

Scientific highly elliptical orbit mission

Spacecraft trajectory

An HEO mission scenario with apogee altitude of about 58,600 km and perigee altitude of 500 km is used to demonstrate the GNSS visibility performance through all the GNSS SSV altitudes, both below and above the GNSS constellations. GNSS visibility conditions near the perigee are similar to those of space user receivers in LEO, with the important difference that the spacecraft is moving very fast – around 8 km/s to 11km/s – so that extreme Doppler shifts occur on the GNSS signals, and visibility times between any particular GNSS satellite and the HEO space user receiver are much shorter than for terrestrial receivers.

Table B12. Osculating Keplerian HEO orbital elements

Epoch	1 Jan 2016 12:00:00 UTC		
Semi-major axis	35937.5 km	RAAN	0 deg
Eccentricity	0.80870	Argument of perigee	270 deg
Inclination	63.4 deg	True anomaly	0 deg

Spacecraft attitude and antenna configuration

The on-board GNSS antennas are configured in both nadir and zenith-facing sides of the spacecraft. As shown in figure B10 the nadir-pointing antenna with high-gain and narrow-beamwidth can ensure the GNSS signal link from the opposite side of the Earth, including when flying above the GNSS altitude and during the apogee period. The zenith-pointing patch antenna can provide visibility during the perigee period. The antenna patterns for both type of antennas are given in table B10. The acquisition and tracking thresholds of the user receiver were both set to 20 dB-Hz when evaluating the signal availability in the HEO simulation.

Figure B10. Schematic of the HEO mission with nadir and zenith-pointing antennas

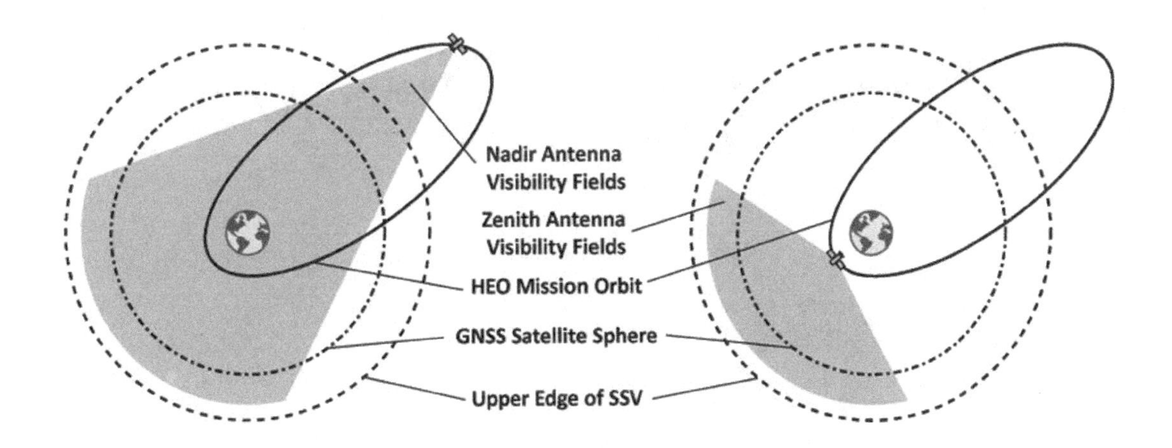

Results

Figure B11 and figure B12 shows the GNSS signal availability of all GNSS constellations for the HEO nadir and zenith-pointing antennas over the time of 1.5 HEO orbital periods. Note that when the spacecraft is below the GNSS constellation altitude, the visibility can be significantly improved by combining the signals from both nadir and zenith antennas at the same time. However, within this simulation only the strongest signal from either is employed at a given time. Around apogee, only the nadir-pointing antenna provides signal availability.

Figure B11. Visible GNSS satellites over 1.5 orbital periods of HEO (L1/E1/B1)

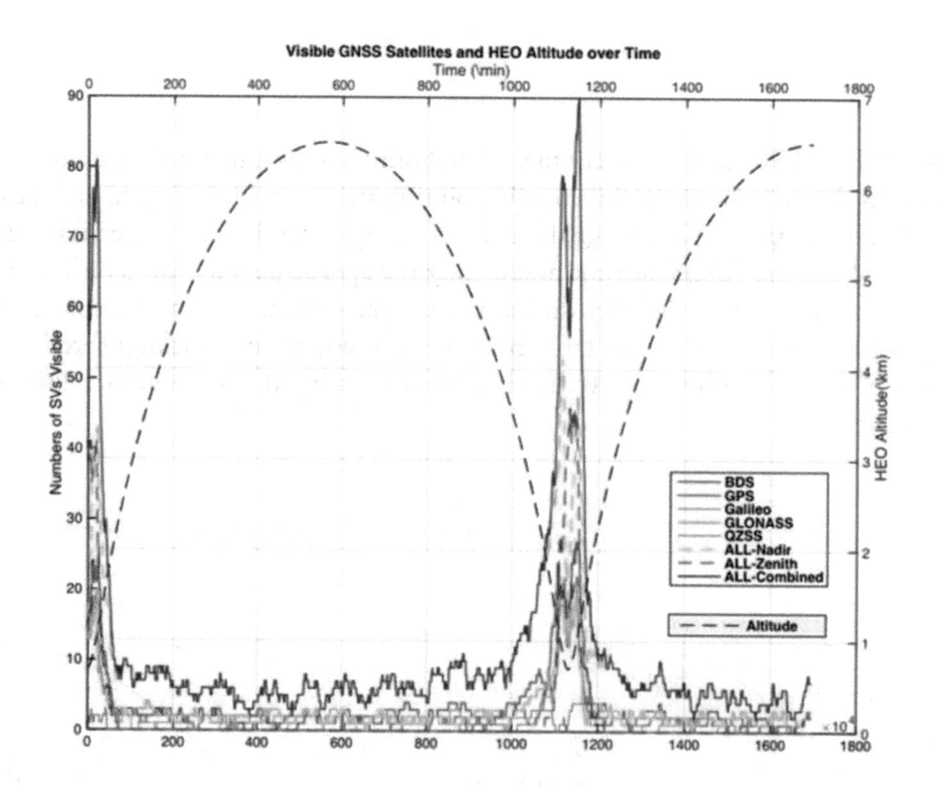

Figure B12. Visible GNSS satellites over 1.5 orbital periods of HEO (L5/L3/E5a/B2)

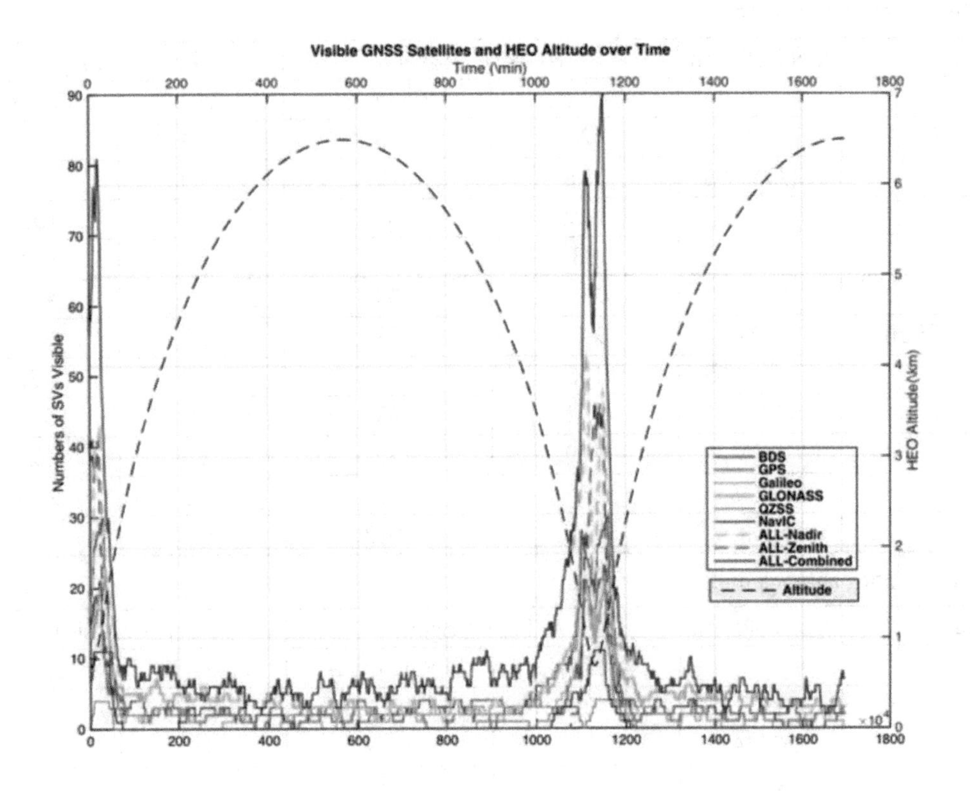

Figure B13. Visible satellites over HEO mission altitude (14 days)

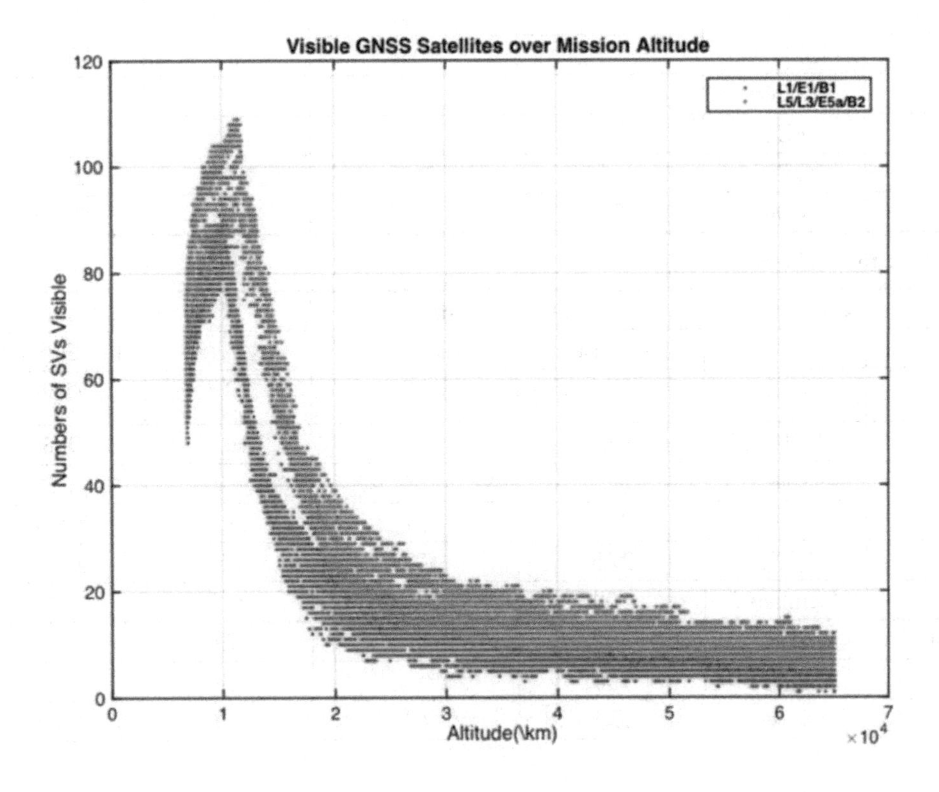

Table B13. HEO mission simulated performance result

Nadir-pointing antenna only					
Band	**Constellation**	**At least 1 signal**		**4 or more signals**	
		Avail. (%)	MOD (min)	Avail. (%)	MOD (min)
L1/E1/B1	GPS	87.3	70	14.4	1036
	GLONASS	98.8	12	13.4	992
	Galileo	74.2	85	9.2	1027
	BDS	88	51	15.3	1031
	QZSS	22.1	1042	1.1	4537
L5/L3/E5a/B2	GPS	94.7	53	17.3	911
	GLONASS	100	0	55.2	133
	Galileo	87.1	63	11	990
	BDS	96.9	30	25.6	925
	QZSS	27.5	1083	2.2	2238
	NavIC	32.7	1023	3.1	1098
Zenith-pointing antenna only					
Band	**Constellation**	**At least 1 signal**		**4 or more signals**	
		Avail. (%)	MOD (min)	Avail. (%)	MOD (min)
L1/E1/B1	GPS	7.4	1066	4.3	1086
	GLONASS	7.1	1059	4.4	1080
	Galileo	7.4	1059	4.1	1085
	BDS	10.1	1031	6	1076
	QZSS	5.7	2264	1.4	2296
L5/L3/E5a/B2	GPS	7.9	1059	4.9	1079
	GLONASS	8.3	1046	6.1	1061
	Galileo	8	1051	4.9	1075
	BDS	10.7	1026	7	1065
	QZSS	6.5	1130	2	2284
	NavIC	5.7	1130	3.3	2262
Nadir and zenith combined					
Band	**Constellation**	**At least 1 signal**		**4 or more signals**	
		Avail. (%)	MOD (min)	Avail. (%)	MOD (min)
L1/E1/B1	GPS	87.3	70	12.7	1036
	GLONASS	98.8	12	14.1	986
	Galileo	74.3	85	9.9	1026
	BDS	88.1	51	16.1	1008
	QZSS	27.5	1031	2.5	2175
	Combined	**100**	**0**	**94.5**	**47**
L5/L3/E5a/B2	GPS	94.7	53	17.4	911
	GLONASS	100	0	55.5	133
	Galileo	87.1	63	11.6	980

	BDS	96.9	30	26	925
	QZSS	32.1	1021	5.8	1091
	NavIC	35.1	989	5.8	1091
	Combined	**100**	**0**	**100**	**0**

Figure B13 shows the visible satellites over the HEO mission altitude in the 14-day simulation timespan with all constellations combined for L1/E1/B1 and L5/L3/E5/B2. As shown in the figure B13, visibility for the L5 case is better than the L1 case.

The simulated results for the signal availability and MOD of the HEO mission are shown in table B13. The signal availability was evaluated with 20 dB-Hz C/No threshold for each individual constellation and all constellations combined.

For both L1/E1/B1 and L5/L3/E5/B2 the one-signal availability can reach 100% with all constellations combined. In case of L1, four-signal availability is below 20% and the MOD is around 1,000 minutes, which is close to the HEO orbital period of 1130 minutes, for an individual constellation. The performance is significantly improved by receiving signals from all constellations combined to nearly 100%. The result of L5 case is similar and the four-signal availability is 100% with all constellations combined. It also shows in the table that signal availability for the L5 case is better than the L1 case.

Lunar mission

The lunar mission case models a simple ballistic cislunar trajectory from LEO to lunar orbit insertion, similar to the trajectories flown by the 1968 United States Apollo 8 mission and many others. This case seeks to explore the boundaries of the GNSS SSV beyond Earth orbit.

Spacecraft trajectory

A full lunar mission trajectory contains four phases:

1. Earth parking orbit
2. Outbound trajectory
3. Lunar orbit
4. Return trajectory

For the purposes of this analysis, only the outbound trajectory is modelled to illustrate the GNSS signal availability with increasing altitude. This is illustrated in figure B14.

Figure B14. Lunar trajectory phases

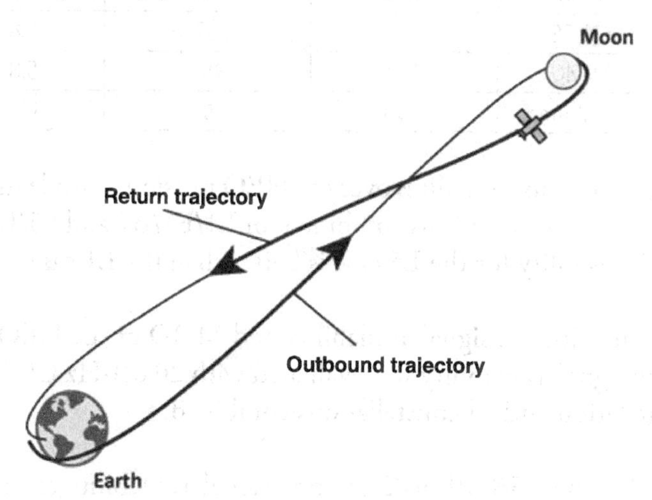

Unlike the GEO and HEO cases, an ephemeris was used to model the outbound trajectory. The trajectory was generated using the following parameters, starting at an Earth altitude of 185 km, and arriving in lunar vicinity at an altitude of 100 km.

Table B14. Lunar trajectory parameters

Parameter	Earth departure	Lunar arrival
Epoch (UTC)	1 Jan 2016 12:00:00.000	5 Jan 2016 22:07:59.988
Altitude	185 km	100 km
Eccentricity	0	0
Inclination (body-centred J2000)	32.5°	75°
RAAN	30°	165°
Argument of perigee (AOP)	32°	319°
True anomaly	0°	0°

The choice of Earth departure epoch fixes the required RAAN and argument of perigee (AOP) to reach lunar orbit. Therefore, there is a choice of epoch that will result in different inertial orientations of the trajectory, which may influence the predicted GNSS visibility. The simulated trajectory is one of these possibilities and is intended to be representative. The parameters listed in table B14 result in a trajectory aligned nearly along the inertial -Y axis.

Spacecraft attitude and antenna configuration

For this simplified lunar mission, the spacecraft attitude is fixed as nadir-pointing. Two GNSS antennas are used: one patch antenna that is permanently zenith-pointing (spacecraft -Z direction) and therefore relevant during the low-altitude portion of the mission, and one high-gain antenna that is permanently nadir-pointing (spacecraft +Z direction) and therefore relevant during the high-altitude portion of the mission. The patch and high-gain antenna characteristics are common to all mission-specific simulations and are shown in table B10. The assumed acquisition threshold of the receiver is 20 dB-Hz. Otherwise, all link budget calculations and parameters are as described in the global analysis.

Results

Table B15 contains the full simulated performance results for this mission. In the case of both L1 and L5 bands, the availability of four simultaneous signals is nearly zero for any individual constellation, though in the combined case there is coverage to approximately 30 RE (approximately half the distance to the Moon) near 10–15%. Single-satellite availability reaches 36% for the combined case at L5, though this availability occurs primarily at low altitudes. The benefit of the combined case is best seen above 10 RE, where the combined case has signal availability consistently higher than any individual constellation, and often nearly double. Notably, combining constellations does not increase the altitude at which such signals are available; rather, it increases the number of signals available at a given altitude. As noted in chapter 5, if a more sensitive receiver or higher-gain antenna were used such that signals at a C/No of 15 dB-Hz were usable, signal availability would be achievable for the entire trajectory to lunar distance.

Table B15. Lunar mission simulated performance results

Band	Constellation	Signal availability (%)	Max outage duration (min)	
		At least 1 signal	4 or more signals	At least 1 signal
L1/E1/B1	GPS	9%	1%	5330
	GLONASS	8%	0%	5200
	Galileo	14%	1%	4870
	BDS	14%	3%	5350
	QZSS	1%	0%	6300
	Combined	**21%**	**9%**	**4870**
L5/L3/E5a/B2	GPS	12%	1%	5010
	GLONASS	33%	1%	3420
	Galileo	16%	1%	5060
	BDS	18%	5%	5170
	QZSS	4%	0%	4940
	NavIC	4%	1%	5960
	Combined	**36%**	**16%**	**3420**

Note that the MOD is of limited utility in these results, as it is measured for the duration of the mission and there is no visibility achieved above 30 RE. Other outage duration metrics could be explored here instead in a more detailed simulation, such as outage duration within the visible range, or under a particular altitude.

Annex C. Constellation specification for simulations

This annex provides the orbital parameters used for every constellation for the SSV simulations reported in this booklet. These parameters are defined at the simulation start epoch, 2016/01/01 12:00:00 UTC.

GPS orbital parameters

Table C1. GPS orbital state definition

Satellite	Semi-major axis (m)	Eccentricity	Inclination (°)	Right ascension (°)	Argument of perigee (°)	Mean anomaly (°)
1	26559800	0	55	273.056	0	11.676
2	26559800	0	55	273.056	0	41.806
3	26559800	0	55	273.056	0	161.786
4	26559800	0	55	273.056	0	268.126
5	26559800	0	55	333.056	0	66.356
6	26559800	0	55	333.056	0	94.916
7	26559800	0	55	333.056	0	173.336
8	26559800	0	55	333.056	0	204.376
9	26559800	0	55	333.056	0	309.976
10	26559800	0	55	33.056	0	111.876
11	26559800	0	55	33.056	0	241.556
12	26559800	0	55	33.056	0	339.666
13	26559800	0	55	33.056	0	11.796
14	26559800	0	55	93.056	0	135.226
15	26559800	0	55	93.056	0	167.356
16	26559800	0	55	93.056	0	257.976
17	26559800	0	55	93.056	0	282.676

18	26559800	0	55	93.056	0	35.156
19	26559800	0	55	153.056	0	197.046
20	26559800	0	55	153.056	0	302.596
21	26559800	0	55	153.056	0	333.686
22	26559800	0	55	153.056	0	66.066
23	26559800	0	55	213.056	0	238.886
24	26559800	0	55	213.056	0	334.016
25	26559800	0	55	213.056	0	0.456
26	26559800	0	55	213.056	0	105.206
27	26559800	0	55	213.056	0	135.346

GLONASS orbital parameters

Table C2. GLONASS orbital state definition

Satellite	Semi-major axis (m)	Eccentricity	Inclination (°)	Right ascension (°)	Argument of perigee (°)	Mean anomaly (°)
1	25508200	0.000397	64.16	201.81	28.75	295.76
2	25505500	0.001181	64.64	202.16	229.92	47.69
3	25507000	0.001152	64.47	202.24	242.46	349.96
4	25509600	0.000341	64.49	202.16	229.04	317.62
5	25508200	0.000593	64.15	201.75	71.12	71.67
6	25505600	0.000838	64.14	201.75	134.53	321.03
7	25507100	0.001027	64.48	202.28	239.38	172.44
8	25509600	0.00154	64.48	202.27	282.44	85.43
9	25509000	0.002309	64.93	322.43	13.68	322.87
10	25506000	0.001662	65.73	322.85	160.86	131.88
11	25506000	0.001846	65.34	322.22	357.58	250.24
12	25509100	0.003395	64.93	322.44	167.5	34.48
13	25509000	0.000449	65.33	322.18	95.45	60.38
14	25505900	0.001493	65.71	322.79	163.14	306.41
15	25505700	0.002211	65.71	322.78	345.48	85.17
16	25509300	0.001967	64.91	322.37	149.58	229.88
17	25509600	0.000831	64.79	82.98	220.69	132.36
18	25507100	0.001346	65.06	82.75	338.33	331.94
19	25505300	0.000102	65.28	83.61	167.08	95.28
20	25508100	0.00106	65.29	83.67	344.69	231.89
21	25509600	0.000685	65	82.79	185.84	348.48
22	25507200	0.002793	65.19	82.78	356.75	132.46
23	25505600	0.000142	65.17	82.74	135.86	306.03
24	25508000	0.000779	65.18	82.77	84.59	315.17

Galileo orbital parameters

Table C3. Galileo orbital state definition

Satellite	Semi-major axis (m)	Eccentricity	Inclination (°)	Right ascension (°)	Argu-ment of perigee (°)	Mean anomaly (°)
1	29599801.224	0.0000001	56	326.60209225	0	107.1899147499
2	29599801.224	0.0000001	56	326.60209225	0	152.1899147499
3	29599801.224	0.0000001	56	326.60209225	0	197.1899147499
4	29599801.224	0.0000001	56	326.60209225	0	242.1899147499
5	29599801.224	0.0000001	56	326.60209225	0	287.1899147499
6	29599801.224	0.0000001	56	326.60209225	0	332.1899147499
7	29599801.224	0.0000001	56	326.60209225	0	17.1899147499
8	29599801.224	0.0000001	56	326.60209225	0	62.1899147499
9	29599801.224	0.0000001	56	86.60209225	0	122.1899147499
10	29599801.224	0.0000001	56	86.60209225	0	167.1899147499
11	29599801.224	0.0000001	56	86.60209225	0	212.1899147499
12	29599801.224	0.0000001	56	86.60209225	0	257.1899147499
13	29599801.224	0.0000001	56	86.60209225	0	302.1899147499
14	29599801.224	0.0000001	56	86.60209225	0	347.1899147499
15	29599801.224	0.0000001	56	86.60209225	0	32.1899147499
16	29599801.224	0.0000001	56	86.60209225	0	77.1899147499
17	29599801.224	0.0000001	56	206.60209225	0	137.1899147499
18	29599801.224	0.0000001	56	206.60209225	0	182.1899147499
19	29599801.224	0.0000001	56	206.60209225	0	227.1899147499
20	29599801.224	0.0000001	56	206.60209225	0	272.1899147499
21	29599801.224	0.0000001	56	206.60209225	0	317.1899147499
22	29599801.224	0.0000001	56	206.60209225	0	2.1899147499
23	29599801.224	0.0000001	56	206.60209225	0	47.1899147499
24	29599801.224	0.0000001	56	206.60209225	0	92.1890000000

BDS orbital parameters

Table C4. BDS orbital state definition

Satellite	Semi-major axis (m)	Eccentricity	Inclination (°)	Right ascension (°)	Argument of perigee (°)	Mean anomaly (°)
1	27906137	0.003	55	0	0	225.631
2	27906137	0.003	55	0	0	270.631
3	27906137	0.003	55	0	0	315.631
4	27906137	0.003	55	0	0	0.631

Table C4. BDS orbital state definition (*cont'd*)

Satellite	Semi-major axis (m)	Eccentricity	Inclination (°)	Right ascension (°)	Argument of perigee (°)	Mean anomaly (°)
5	27906137	0.003	55	0	0	45.631
6	27906137	0.003	55	0	0	90.631
7	27906137	0.003	55	0	0	135.631
8	27906137	0.003	55	0	0	180.631
9	27906137	0.003	55	120	0	240.631
10	27906137	0.003	55	120	0	285.631
11	27906137	0.003	55	120	0	330.631
12	27906137	0.003	55	120	0	15.631
13	27906137	0.003	55	120	0	60.631
14	27906137	0.003	55	120	0	105.631
15	27906137	0.003	55	120	0	150.631
16	27906137	0.003	55	120	0	195.631
17	27906137	0.003	55	240	0	255.631
18	27906137	0.003	55	240	0	300.631
19	27906137	0.003	55	240	0	345.631
20	27906137	0.003	55	240	0	30.631
21	27906137	0.003	55	240	0	75.631
22	27906137	0.003	55	240	0	120.631
23	27906137	0.003	55	240	0	165.631
24	27906137	0.003	55	240	0	210.631
25	42164200	0.003	0	0	2.204	336.229
26	42164200	0.003	0	0	23.459	336.229
27	42164200	0.003	0	0	54.082	336.229
28	42164200	0.003	0	0	83.582	336.229
29	42164200	0.003	0	0	103.582	336.229
30	42164200	0.003	55	61.445	0	336.229
31	42164200	0.003	55	301.445	0	96.229
32	42164200	0.003	55	181.445	0	216.229

QZSS orbital parameters

Table C5. QZSS orbital state definition

Satellite	Semi-major axis (m)	Eccentric-ity	Inclination (°)	Right ascension (°)	Argument of perigee (°)	Mean anomaly (°)
1	42164169.45	0.075	40	165	270	341.58
2	42164169.45	0.075	40	295	270	211.58
3	42164169.45	0.075	40	35	270	111.58
4	42164169.45	0	0	0	0	47.58

NavIC orbital parameters

Table C6. NavIC orbital state definition

Satellite	Semi-major axis (m)	Eccentricity	Inclination (°)	Right ascension (°)	Argument of perigee (°)	Mean anomaly (°)
1	42164200	0.0007	28.1	124.08	0	211.3
2	42164200	0.0007	29.97	303.04	0	32.32
3	42164200	0.0007	4.01	264.62	0	98.963
4	42164200	0.0007	29.98	303.19	0	88.964
5	42164200	0.0007	28.1	124.08	0	267.8
6	42164200	0.0007	5	270	0	42.663
7	42164200	0.0007	5	270	0	139.568
8*	42164200	0.0007	42	318.5	0	8.7629
9*	42164200	0.0007	42	110	0	235.5129
10*	42164200	0.0007	42	290	0	84.5129
11*	42164200	0.0007	42	279	0	121.2629

* Note: These additional four IGSO satellites are yet to be coordinated, and some parameters may change.

Annex D. References

Interface control documents/interface specifications

GPS interface specifications. http://www.gps.gov/technical/icwg/

- IS-GPS-200: Defines the requirements related to the interface between the GPS space and user segments for radio frequency L1 (L1 C/A) and L2 (L2C).
- IS-GPS-705: Defines the requirements related to the interface between the GPS space and user segments for radio frequency L5.
- IS-GPS-800: Defines the characteristics of GPS signal denoted L1 Civil (L1C).

GLONASS Interface Control Document Navigational Radio Signal in Bands L1, L2 (Edition 5.1) http://russianspacesystems.ru/bussines/navigation/glonass/interfeysnyy-kontrolnyy-dokument/

GLONASS Interface Control Document General Description of Code Division Multiple Access Signal System (Edition 1.0) http://russianspacesystems.ru/bussines/navigation/glonass/interfeysnyy-kontrolnyy-dokument/

GLONASS Interface Control Document Code Division Multiple Access Open Service Navigation Signal in L1 Frequency Band (Edition 1.0) http://russianspacesystems.ru/wp-content/uploads/2016/08/ICD-GLONASS-CDMA-L1.-Edition-1.0-2016.pdf

GLONASS Interface Control Document Code Division Multiple Access Open Service Navigation Signal in L2 Frequency Band (Edition 1.0) http://russianspacesystems.ru/wp-content/uploads/2016/08/ICD-GLONASS-CDMA-L2.-Edition-1.0-2016.pdf

GLONASS Interface Control Document Code Division Multiple Access Open Service Navigation Signal in L3 Frequency Band (Edition 1.0) http://russianspacesystems.ru/bussines/navigation/glonass/interfeysnyy-kontrolnyy-dokument/

European GNSS (Galileo) Open Service Signal in Space Interface Control Document https://www.gsc-europa.eu/electronic-library/programme-reference-documents

BeiDou Navigation Satellite System Signal In Space Interface Control Document http://www.beidou.gov.cn

NavIC(IRNSS) Signal-in-Space ICD for SPS (Standard Position Service). https://www.isro.gov.in/sites/default/files/irnss_sps_icd_version1.1-2017.pdf

QZSS Interface Specification (IS-QZSS) http://qzss.go.jp/en/technical/index.html

Conferences/papers

W. Enderle, M. Schmidhuber, E. Gill, O. Montenbruck, A. Braun, N. Lemke, O. Balbach, B. Eisfeller, "GPS performance for GEOs and HEOs: the EQUATOR-S spacecraft mission", Thirteenth International Symposium on Space Flight Dynamics, Goddard Space Flight Center, Greenbelt Maryland, United States, 1998.

O. Balbach, B. Eisfeller, G.-W. Hein, T. Zink, W. Enderle, M. Schmidhuber, N. Lemke, "Tracking GPS above GPS satellite altitude: results of the GPS experiment on the HEO mission EQUATOR-S", ION, United States, 1998.

W. Enderle, "Attitude determination of an user satellite in a Geo Transfer Orbit (GTO) using GPS measurements", the Fourth ESA International Conference on Spacecraft Guidance, Navigation and Control Systems, ESTEC, Noordwijk, the Netherlands, 18–21 October 1999.

M. Moreau, F. H. Bauer, J. R. Carpenter, E. Davis, G. Davis, L. Jackson. "Preliminary Results of the GPS Flight Experiment on the High Earth Orbit AMSAT-OSCAR 40 Spacecraft", AAS 02-004, AAS Guidance, Navigation and Control Conference, Breckenridge, Colorado, United States, February 2002.

M. Moreau, E. Davis, J. R. Carpenter, G. Davis, L. Jackson, P. Axelrad. "Results from the GPS Flight Experiment on the High Earth Orbit AMSAT OSCAR (AO-40) Spacecraft", Proceedings of the ION GPS 2002 Conference, Portland, Oregon, United States. 2002.

W. Enderle, R. A. Walker, Y. Feng, W. Kellar, "New Dimension for GEO and GTO AOCS Applications Using GPS- and Galileo Measurements", ION GPS 2002, Portland, Oregon, United States, 24–27 September 2002.

F. H. Bauer, M. C. Moreau, M. E. Dahle-Melsaether, W. P. Petrofski, B. J. Stanton, S. Thomason, G. A. Harris, R. P. Sena, L. Parker Temple III. "The GPS Space Service Volume", ION GNSS, September 2006.

W. Enderle, H. Fiedler, S. de Florio, F. Jochim, S. d'Amico, W. Kellar, S. Dawson, "Next Generation GNSS for Navigation of Future SAR Constellations", International Astronautical Congress, Valencia, Spain, 2–6 October 2006.

Frank van Graas, "Use of GNSS for Future Space Operations and Science Missions", Sixth Meeting of the National Space Based Positioning, Navigation, and Timing Advisory Board, November 2009. http://www.gps.gov/governance/advisory/meetings/2009-11/vangraas.pdf

James J. Miller, "Enabling a Fully Interoperable GNSS Space Service Volume", International Committee for GNSS WG-B Meeting, Tokyo, September 2011. http://www.unoosa.org/pdf/icg/2011/icg-6/wgB/8.pdf

James J. Miller, Michael Moreau, "Space Service Volume", Ninth Meeting of the GNSS Providers' Forum, Beijing, November 2012. http://www.unoosa.org/pdf/icg/2012/pf9/pf-2.pdf

Badri Younes, "ICG: Achieving GNSS Interoperability and Robustness", International Committee for GNSS Eighth Meeting, Dubai, United Arab Emirates, November 2013. http://www.unoosa.org/pdf/icg/2013/icg-8/1a.pdf

Frank Bauer, "GNSS Space Service Volume and Space User Data Update", Thirteenth Meeting of the GNSS Providers' Forum, Dubai, United Arab Emirates, November 2013. http://www.unoosa.org/pdf/icg/2014/PF-13/pf13_01.pdf

V. Kosenko, A. Grechkoseev, M. Sanzharov, "Application of GNSS for the High Orbit Spacecraft Navigation", International Committee for GNSS WG-B Meeting, Dubai, United Arab Emirates, November 2013. http://www.unoosa.org/pdf/icg/2013/icg-8/wgB/B3.pdf

Frank Bauer, Stephan Esterhuizen, "GNSS Space Service Volume Update", International Committee for GNSS WG-B Meeting, Dubai, United Arab Emirates, November 2013. http://www.unoosa.org/pdf/icg/2013/icg-8/wgB/B1.pdf

V. Kosenko, A. Grechkoseev, M. Sanzharov, "Application of GNSS for the high orbit spacecraft navigation", ICG-8 WG-B, Dubai, United Arab Emirates, November 2013.

X. Zhan, S. Jing, X.Wang, "Beidou space service volume parameters and performance", Eighth meeting of International Committee on GNSS, WG-B, Dubai, United Arab Emirates, November 2013.

Frank Bauer, "GNSS Space Service Volume Update", Eleventh Meeting of the GNSS Providers' Forum, Prague, November 2014. http://www.unoosa.org/pdf/icg/2013/PF-11/1.pdf

V. Kosenko "GLONASS Space Service Volume", International Committee for GNSS WG-B Meeting, Prague, November 2014. http://www.unoosa.org/pdf/icg/2014/wg/wgb04.pdf.

Dee Ann Divis, "Space – The Next GPS Frontier", Inside GNSS, November to December 2014. http://www.insidegnss.com/node/4278

V. Kosenko, A. Grechkoseev, M. Sanzharov, "GLONASS space service volume", ICG-9 WG-B, Prague, November 2014.

X. Zhan, S. Jing, H. Yang, X. Chang, "Space service volume characteristics of BDS", Ninth meeting of International Committee on GNSS, WG-B, Prague, November 2014.

María Manzano-Jurado, Julia Alegre-Rubio, Andrea Pellacani, Gonzalo Seco-Granados, Jose A. López-Salcedo, Enrique Guerrero, Alberto García-Rodríguez, "Use of Weak GNSS Signals in a Mission to the Moon", 2014 IEEE, 978-1-4799-6529-8/14.

Frank H. Bauer, "GNSS Space Service Volume and Space User Data Update", International Committee for GNSS Tenth Meeting, Boulder, Colorado, United States, November 2015. http://www.unoosa.org/pdf/icg/2015/icg10/wg/wga08.pdf

Frank H. Bauer, "GNSS Space Service Volume and Space User Data Update", Fifteenth Meeting of the GNSS Providers' Forum, November 2015. http://www.unoosa.org/pdf/icg/2015/icg10/03pf.pdf

S. Wallner, "Galileo's Contribution to Interoperable GNSS SSV", International Committee for GNSS Tenth Meeting, Boulder, Colorado, United States, November 2015.

X. Chang, X. Mei, H. Yang, "Space service volume performance of BDS", Tenth meeting of International Committee on GNSS, WG-B, Boulder, Colorado, United States, November 2015.

Willard A. Marquis, Daniel L. Reigh, "The GPS Block IIR and IIR-M Broadcast L-band Antenna Panel: Its Pattern and Performance", *Navigation*, Journal of the Institute of Navigation, Vol. 62, No. 4, winter 2015.

Frank H. Bauer, James J. Miller, A. J. Oria, Joel Parker, "Achieving GNSS Compatibility and Interoperability to Support Space Users", AAS 16-71, American Astronautical Society, February 2016.

J. Parker, J. Valdez, F. Bauer, M. Moreau, "Use and Protection of GPS Sidelobe Signals for Enhanced Navigation Performance in High Earth Orbit", AAS 16-72, American Astronautical Society, February 2016.

Luke Winternitz, Bill Bamford, Sam Price, Anne Long, Mitra Farahmand, Russel Carpenter, "GPS Navigation above 76,000 km for the MMS Mission", AAS 15-76, Thirty-ninth Annual AAS Guidance, Navigation and Control Conference. February 2016. https://ntrs.nasa.gov/archive/nasa/casi.ntrs.nasa.gov/20160001693.pdf

Werner Enderle, “Space Service Volume – using GNSS beyond GEO”, ESA Space Technology Workshop, the Netherlands, April 2016.

Frank H. Bauer, “GNSS Space Service Volume Update”, Sixteenth Providers’ Forum, International Committee for GNSS Intersessional Meeting, Vienna, 6 June 2016. http://www.unoosa.org/pdf/icg/2016/pf-16/2.pdf

W. Enderle, “ICG SSV – Simulation Phase 2 Link Budget Setup”, ICG 2016 Preparation Meeting, Vienna, 7 June 2016.

X. Chang, P. Li, “Interoperable GNSS space service volume simulation configuration”, Interim meeting of ICG-11, Vienna, June 2016.

James J. Miller, Frank H. Bauer, A. J. Oria, Scott Pace, Joel K. Parker, “Achieving GNSS Compatibility and Interoperability to Support Space Users”, Institute of Navigation (ION) GNSS+ 2016, September 2016.

Frank H. Bauer, “Space Service Volume Update”, Seventeenth Providers’ Forum. International Committee for GNSS, Sochi, Russian Federation, November 2016. http://www.unoosa.org/pdf/icg/2016/icg11/pf17/pf2.pdf

Alexander Grechkoseev, Maxim Sanzharov and Dmitry Marareskul, “Space Service Volume and Russian GEO Satellites PNT”, Seventeenth Providers’ Forum, International Committee for GNSS, Sochi, Russian Federation, November 2016. http://www.unoosa.org/pdf/icg/2016/icg11/pf17/pf1.pdf.

James J. Miller, Frank H. Bauer, Jennifer E. Donaldson, A. J. Oria, Scott Pace, Joel J. K. Parker, Bryan Welch, “Navigation in Space: Taking GNSS to New Heights”, *Inside GNSS*, November to December 2016. http://www.insidegnss.com/node/5196

D. Marareskul, “Russian Federation view on further stages of SSV simulation”, Working Group Meeting, 8–10 November 2016, Sochi, Russian Federation.

D. Marareskul, “Space users’ navigation equipment: development, classification and unification principles”, Working Group Meeting, 8–10 November 2016, Sochi, Russian Federation.

W. Enderle, “ESA Activities related to GNSS Space Service Volume”, Presentation to the GPS PNT Advisory Board, Redondo Beach, California, United States, 8 December 2016.

Frank H. Bauer, Joel J. K. Parker, Bryan Welch, Werner Enderle, “Developing a Robust, Interoperable GNSS Space Service Volume (SSV) for the Global Space User Community”, ION International Technical Meeting, Monterey, California, United States, January 2017.

W. Enderle (on behalf of the ICG WG-B), “Status of Activities on Interoperable GNSS Space Service Volume”, Munich Satellite Navigation Summit 2017, Munich, Germany, 16 March 2017.

M. Paonni, M. Manteiga Bautista, "Galileo Programme SSV Actions", Munich Satellite Navigation Summit 2017, Munich, Germany, 16 March 2017.

W. Enderle, E. Schoenemann, "GNSS Space Service Volume – User Perspective", Munich Satellite Navigation Summit 2017, Munich, Germany, 16 March 2017.

Stephen Winkler, Graeme Ramsey, Charles Frey, Jim Chapel, Donald Chu, Douglas Freesland, Alexander Krimchansky, Marcho Concha, "GPS Receiver On-Orbit Performance for the GOES-R Spacecraft", Tenth International ESA Conference on Guidance, Navigation & Control Systems, Salzburg, Austria, 29 May–2 June 2017. https://ntrs.nasa.gov/search.jsp?R=20170004849

X. Chang, H. Yang, "Navigation satellite system space service volume and its applications", the Eighth China Satellite Navigation Conference, Shanghai, China, May 2017.

Reference tables of GNSS-utilizing missions

The International Operations Advisory Group (IOAG) is working to identify current and future space missions relying on GNSS signals for PNT and science applications. The IOAG provides a forum for space agencies to identify common needs across multiple international agencies and to coordinate space communications policy, high-level procedures, technical interfaces, and other matters related to interoperability and space communications. IOAG members currently include the Agenzia Spaziale Italiana, Canadian Space Agency, Centre National d'Études Spatiales, Deutsches Zentrum für Luft- und Raumfahrt, European Space Agency, Japan Aerospace Exploration Agency, NASA, and the Russian Federal Space Agency. Observer members include the China National Space Administration, Indian Space Research Organisation, Korea Aerospace Research Institute, South African National Space Agency, and the United Kingdom Space Agency.

These reference tables are updated annually and, in turn, have been used by the ICG in its work to develop interoperable capabilities to support space users.

IOAG Website: www.ioag.org

J. Parker, "NASA GNSS Activities", Twelfth Meeting of the International Committee for GNSS, Kyoto, Japan, 2–7 December 2017, pp. 53-57. http://www.unoosa.org/oosa/en/ourwork/icg/meetings/ICG-2017.html

Abbreviations and acronyms

ACE	NASA GPS Antenna Characterization Experiment
AEP	Architecture Evolution Plan
AFSPC	Air Force Space Command
AOP	Argument of Perigee
BDS	Beidou Navigation Satellite System
BPSK	Binary Phase Shift Keying modulation
C/No	Carrier-to-Noise Ratio
CAO	Cabinet Office, Government of Japan
CAST	China Academy of Space Technology
CBOC	Composite Binary Offset Carrier
CDMA	Code Division Multiple Access
CS	Commercial Service
ESA	European Space Agency
FDMA	Frequency Division Multiple Access
FOC	Full Operational Capability
GCS	Ground Control Segment
GEO	Geostationary Orbit
GLONASS	Global Navigation Satellite System
GNSS	Global Navigation Satellite System
GOES	Geostationary Operational Environmental Satellite-R series
GPS	Global Positioning System
GRC	NASA Glenn Research Center
GSFC	NASA Goddard Space Flight Center
GSO	Geosynchronous Orbit
GTO	Geo Transfer Orbit
HEO	Highly Elliptical Orbit
ICD	Interface Control Document
ICG	International Committee on GNSS
IF	Intermediate Frequency
IGSO	Inclined Geosynchronous Orbit
IOAG	International Operations Advisory Group
IOV	In-Orbit Validation
IRNSS	Indian Regional Navigation Satellite System
IS	Interface Specification
ISAC	ISRO Satellite Centre
ISRO	Indian Space Research Organisation
JAXA	Japan Aerospace Exploration Agency
JPL	Jet Propulsion Laboratory
LEO	Low Earth Orbit
LNA	Low Noise Amplifier
LoS	Line of Sight
MEO	Medium Earth Orbit

MMS	Magnetospheric Multi-Scale
MOD	Maximum Outage Duration
MRTP	Minimum Radiated Transmit Power
NASA	United States National Aeronautics and Space Administration
NavIC	Navigation with Indian Constellation
NEC	Nippon Electric Company
NOAA	National Oceanic and Atmospheric Administration
OCS	Operational Control Segment
OCX	Next Generation Operational Control System
OOSA	United Nations Office for Outer Space Affairs
OS	Open Service
PNT	Positioning, Navigation and Timing
POD	Precise Orbit Determination
PRN	Pseudo-Random Noise
PRS	Public Regulated Service
PVT	Position, Velocity, Time
QZS	Quasi-Zenith Satellite
QZSS	Quasi-Zenith Satellite System
RAAN	Right Ascension of the Ascending Node
RCP	Right-hand Circular Polarised
RE	Earth Radius
RF	Radio Frequency
RS	Restricted Service
SIS	Signal in Space
SJTU	Shanghai Jiao Tong University
SMC	Space and Missile Systems Center
SPS	Standard Positioning Service
SSV	Space Service Volume
SV	Space Vehicle
TCM	Trajectory Correction Manoeuvres
TLI	Trans-Lunar Injection
TTC	Telemetry, Tracking and Command station
URE	User Range Error
UTC	Universal Time (Coordinated)
WG-B	ICG Working Group B

Acknowledgements

This booklet was published by the United Nations Office for Outer Space Affairs in its capacity as executive secretariat of ICG and its Providers' Forum. Sincere thanks to all who have helped, and who recognize the in-space advantages of the SSV specification and provide leadership in developing an SSV specification for the GNSS constellations.

- United States Air Force Space and Missile Systems Center (SMC) GPS Directorate (GP)
- Air Force Space Command (AFSPC)
- United States National Space-Based Positioning, Navigation, and Timing (PNT) Advisory Board
- NASA Magnetospheric Multi-Scale (MMS) Mission Team
- NASA/NOAA GOES-R Team
- NASA GPS Antenna Characterization Experiment (ACE) Team

Benjamin W. Ashman, NASA GSFC

Frank Bauer, F Bauer Aerospace Consulting Services, retired NASA

Daniel Blonski, ESA

Alexey Bolkunov, PNT Center, TSNIIMASH, State Space Corporation Roscosmos

Henno Boomkamp, ESA

Xinuo Chang, CAST

Nilesh M. Desai, ISAC/ISRO

Jennifer Donaldson, NASA GSFC

Werner Enderle, ESA

Claudia Flohrer, ESA

Dale Force, NASA GRC

Ghanshyam, ISAC/ISRO

Francesco Gini, ESA

Motohisa Kishimoto, JAXA

Mick N. Koch, NASA GRC

Satoshi Kogure, CAO

Dmitry Marareskul, ISS-Reshetnev Company

Jules McNeff, Overlook Systems Technologies, Inc.

James J. Miller, NASA Space Communications and Navigation (SCaN)

Michael Moreau, NASA GSFC

Mruthyunjaya L., ISAC/ISRO

Yoshiyuki Murai, NEC

Koji Nakaitani, CAO

Yu Nakajima, JAXA

A. J. Oria, Overlook Systems Technologies, Inc.

Scott Pace, George Washington University

K. S. Parikh, SAC/ISRO

Joel J. K. Parker, NASA GSFC

G. Ramarao, ISAC/ISRO

Ramasubramanian R., ISAC/ISRO

John Rush, Retired NASA

O. Scott Sands, NASA GRC

P. V. B. Shilpa, ISAC/ISRO

Erik Schönemann, ESA

Vishwanath Tirlapur, ISAC/ISRO

Stefan Wallner, ESA

Bryan Welch, NASA GRC

Hui Yang, CAST

Lawrence Young, NASA JPL

René Zandbergen, ESA

Xingqun Zhan, SJTU

Essential TEFL

Grammar, Lesson Plans and 300 Activities to Make You a Confident Teacher

by James Jenkin and Emma Foers

i-to-i UK
Woodside House
261 Low Lane
Leeds LS18 5NY
United Kingdom

www.onlinetefl.com

First published in the UK in 2011

978-0-9568063-0-7

Edited by Andrew Jack.
Designed by Dee Grismond.

Contents

Introduction

Here at i-to-i we believe that teaching English overseas can be extremely challenging (especially if it's your first time in the classroom), but with the right guidance and information your lessons can be fun and effective for both you and your students.

One of our missions as an organisation is to try to make TEFL as accessible as possible. We often hear stories of new teachers taking a suitcase full of grammar, methodology and activity books with them to their first job, only for them to still struggle to know what to do! So, we thought we'd do the sensible thing and put everything you really need into one simple format, and here it is: Essential TEFL.

We hope this book will be an ongoing support and inspiration to you. It aims to bring together everything a teacher needs to plan and deliver effective and engaging lessons: teaching techniques, lesson plans, activities, and instant grammar help.

Think of this book as your new best friend… it'll always be on hand with the information you need to survive the stickiest of teaching situations and ensure you walk into any classroom with total confidence.

Good luck with your teaching adventure ahead!

How to Use this Book

We want this book to empower you - you're the one in the classroom! While we can show you what works from our experience, you should use this to supplement your own individual teaching style.

We recommend you select and tweak the activities and lesson ideas in here as you see fit – every classroom is different and you should try to adapt activities to suit the needs, dynamics and abilities of your classes. Try the activities, then shape and refine them each time you use them (make notes in this book so you can remember them for next time!).

For sheer teaching inspiration, you'll want to head to 'Activities' (section 2) and 'Lesson Plans' (section 3) – you'll find hundreds of ideas to bring your English lessons to life. Alternatively, if you want a reminder of the teaching basics, then take a look through 'How to Teach' (section 1), particularly if you need to adapt your materials for a specific type of class.

Finally, if you're unsure about a specific grammar point, or need to remind yourself of the meaning of a specific TEFL term, then head to 'Grammar' (section 4) or the Appendix for instant and easy-to-digest explanations.

Remember that there are no hard and fast rules in teaching. As teachers we're dealing with people - what's really important is to put the rule book aside, be receptive to what our students need and adjust our lessons accordingly!

About the Authors

James Jenkin – i-to-i Academic Director

James Jenkin has been teaching English as a foreign language since 1994, and has managed English language programmes in Vietnam, China and Australia. His classroom career has included teaching Sudanese refugees, Vietnamese government ministers and Chinese airline pilots.

James' particular love is teacher training, having been an accredited Cambridge CELTA tutor since 1998 and an i-to-i classroom TEFL tutor since 2006. He understands the needs, worries and dreams of people entering the TEFL world.

James has a Cambridge CELTA, a Bachelor of Arts Degree in Russian and Latvian, and a Master's Degree in Applied Linguistics.

What James has to say

New teachers can often feel overwhelmed. There seems to be so much to get on top of: grammar, planning, designing activities, knowing how to teach, and that's before you've even entered the classroom! Essential TEFL is designed to make English teaching manageable. We want everything you need to be easily accessible, so you can put planning to one side, and focus on your students.

Emma Foers – i-to-i TEFL Tutor

Emma Foers obtained her CELTA qualification straight after University, before jetting off on a round-the-world teaching adventure taking in Japan, Italy, Portugal and Egypt. She now works as a TEFL tutor for i-to-i and is the author of Kick Start Your TEFL Career: 20 Classroom Activities for Elementary Learners.

Emma loves being in the classroom and creating relevant, fun materials to keep her students engaged and entertained.

What Emma has to say

I've been teaching for a long time now, but I can still remember how challenging my first experience in the classroom was. Planning was tough, especially working out how to make my lessons as clear and fun as possible. There are so many TEFL books out there, each specialising in different areas, that I didn't know which ones to choose: that's why we decided to put everything you need into one book. I certainly wish this book had been out there to help me when I first started teaching!

How to Teach

Introduction

This is a section we hope you'll return to again and again for inspiration. We wanted it to be refreshing (and slightly provocative). We know teachers love to take time – when they have the chance – to reflect on what good teaching is all about.

1. Top 10 Dos and Don'ts

These are the principles that make a TEFL class great (or not so great). It's easy to forget them – and a good idea to come back to them now and again.

2. Using Coursebooks and Resources

Here we look at the types and range of materials available to you, and how you can get the most from them.

3. Teaching Specialised Classes

Here we examine how you can make different types of classes work – whether you're tutoring one-to-one or you're in charge of 50 students.

1

1.1 Top 10 Dos and Don'ts

These are the basic principles that make a TEFL class flop or fly. They're easy to forget, so refresh your memory (and your teaching!) by coming back to them now and again.

Top 10 Dos

✔ 1. Go for it!

Throw yourself into teaching, give it your all, and have fun (your students are much more nervous than you are! Remember what it was like when you learned a language?). Encourage and motivate your learners. Make your classroom a positive and enjoyable experience.

✔ 2. Have an aim

The most important thing is that your lesson needs an aim. Your students should walk away from a lesson feeling 'Today, I learnt how to do X'. Otherwise, the lesson can seem like a waste of time. (Remember classes you've been to where you thought, 'What was the point of that?').

✔ 3. Be organised

Familiarise yourself with any new content you're going to teach. Make a running sheet. Have your materials ready to go, in plastic pockets in a file. Make sure the equipment works. Take a spare whiteboard marker. You'll feel confident, so you can relax and enjoy the time with your students.

✔ 4. Get students talking – to each other

This isn't just about making a lesson lively and fun (although that's a big plus). Learning English is a skill, like learning to swim or cook. Your students need to practise English, not just learn about English. And the best way to make sure students get lots of practice is if they talk to each other, in pairs and groups, or mingling as a whole class. (Don't make all the practice through you, or only one student gets to talk at a time).

✔ 5. Start a lesson with a warmer

A warmer is a simple activity, preferably something active and fun, where students talk to each other. As you know, it's easy to feel awkward and shy with a big group of people. A warmer removes that anxiety. It helps students feel relaxed and confident to speak for the rest of the lesson. In a warmer, students should be interacting in small groups or mingling – definitely not talking one at a time to the whole class. That's the opposite of a warmer!

✔ 6. Use variety

As you know, there's nothing worse than a boring class. But making a class interesting isn't about playing games all the time, and avoiding anything 'heavy'. Rather, it's about variety. Vary the skills your students are practising (listening, speaking, reading, writing) as well as the pace and physical activity (sitting, standing, mingling, running). Keep the expression 'light and shade' in mind – follow a quiet and serious activity with something fun and high-energy.

✔ 7. Instruct clearly

We've said it's important to have students practising in a variety of activities. The one risk that creates – unlike in a traditional class, where students just have to sit and listen – is that they won't know what to do. This is especially so since English is their second language. The result will be chaos! Instruct simply and clearly, and support your instructions with an example or demonstration.

8. Elicit

Eliciting means asking the students to tell you, rather than you always telling the students (which is the unfortunate dynamic in many classrooms around the world). Turn everything into a question. Rather than drawing a picture on the board and saying 'This is a car', ask, 'What's this?'. Always give your students a chance to tell you things. They'll find it very empowering and engaging.

9. Work on pronunciation constantly

Pronunciation seems to be the last thing many teachers think about. But if you can't understand someone's pronunciation, it doesn't matter how good their grammar or vocabulary is! Whenever you teach anything new – grammar or vocabulary or functional language – you should teach students how to pronounce it as well, and give them a chance to practise it orally.

10. Correct students (in a nice way!)

Numerous studies show language students want much more correction than they get (we've found that some TEFL teachers are possibly too kind-hearted!). Students like correction because they feel you're listening to them and trying to help them. It's how you do it, of course – correction should be gentle and encouraging, not a reprimand. Bear in mind that correction is not always appropriate. If students are involved in a discussion, don't interrupt, or you'll inhibit them. Wait until afterwards to discuss any errors you heard.

Top 10 Don'ts

1. Don't talk all the time

We mentioned that language learning is a skill that students need to practise. Well, the more time you talk, the less time your students get to practise. (Imagine a driving lesson where you just listen to an instructor talk about driving). Teachers talk a lot with the best possible intentions: usually to explain things, or to give students listening practice. Unfortunately, it's generally counterproductive. Lengthy explanations are confusing, and listening to one person for a long time is boring.

2. Don't use foreigner talk

Foreigner talk is a linguistic term meaning the weird language we use with language learners: YOU SIT, PAIRS, NOW, OKAY? Students can sense it. Even if they don't feel patronised, they'll have the impression that classroom English is far removed from real-world English. Having said that, it's good to be careful with the language you use. Reflect on how you'll say something before you say it, and choose words you know your students will understand.

3. Don't echo

Echoing means repeating what a student says.

Teacher: What's your favourite food?
Student: Italian.
Teacher: Oh, Italian! Great.

We do it all the time. Our aim is to encourage students, but in fact it stops them talking. (In the example the student only got to say one word, as opposed to the teacher's seven!). Ask for more information instead.

Teacher: What's your favourite food?
Student: Italian.
Teacher: Tell us more.
Student: Well, I like pizza and I often go...

✗ 4. Don't teach a non-standard variety of English

This doesn't mean 'put on a British accent'. It means develop an awareness of what is accepted as standard, international English around the world. It applies equally whether you're British or American or Australian or a non-native speaker. Avoid teaching vocabulary or pronunciation that is distinctly local and might be unintelligible to most speakers (unless your students need to know how people speak in a certain place, and you make it clear how it differs from standard English).

✗ 5. Don't underchallenge your students

Teachers generally don't want to overburden their students. That's a good thing. But we often go too far down the 'gentle' path. Our students want their English to improve, so they need to be challenged. They want to feel they're getting their money's worth coming to class. Judge what they can realistically cope with and push them to achieve their goals. And, incidentally, give students plenty to do in practice activities so they're not twiddling their thumbs.

✗ 6. Don't 'overurge' your students

Teachers, with the best intentions, often stand over students and urge them to speak. 'What do you think Alex? Go on, tell us. What do you think? What's your opinion about the Olympics?' Sadly, it has the opposite effect – it makes students feel under pressure and stops them talking. When you ask a question, don't stare. Look away slightly, and give your student time to think and respond. Encourage students to ask each other questions. Then, while students are talking to each other, sit down discretely nearby so you can hear. In other words, give people space.

✗ 7. Don't be unprofessional

Being professional doesn't mean being boring – rather, it means taking your job and your students' needs seriously. Follow the obvious protocols wherever you're working. If you're not sure, ask. (The way a teacher dresses, for example, is incredibly important in some cultures).

✗ 8. Don't criticise your students' country and culture

It seems obvious. But out of frustration, or tiredness, or culture shock, teachers sometimes let their guard down. You might hear locals criticising the power blackouts or bad traffic, but stay out of it. It can seriously damage your relationship with your students. Tell students what you like about their country instead. Be careful with jokes – make sure they can't be misinterpreted as making fun of your students' culture.

✗ 9. Don't preach

We're teaching language, not politics or religion. Try to avoid using the opportunity of a captive audience to preach. Your proselytising might annoy students (who may not have the linguistic resources to argue back) or even get you into serious trouble. Besides, every minute you're telling your students about the world, it's one less minute when they can practise. Ask them to tell you what they think instead!

✗ 10. Don't have favourite students

It's very important to students how they get on with their teacher. Try to build an equally warm and positive relationship with all the students. Learn everyone's names (so you're not always asking the same three students whose names you remember!). Never compare students. You'll create a great learning environment!

.2 Using Coursebooks and Resources

Your coursebook can sometimes feel like more of a hindrance than a help, but used properly it can really cut down your preparation time. This section looks at how to get the most from your coursebook, as well as the many other resources you have at your fingertips.

Introduction

How to Teach

Activities

Lesson Plans

Grammar

Appendix

Coursebooks

The situation

- Most TEFL students in classrooms around the world have a coursebook, which forms the basis of lessons.
- You can apply some simple principles to bring a coursebook to life.
- You can apply some simple principles to extract much more activity from a coursebook, reducing your preparation time, and avoiding the risk of running out of material.

Will I have a coursebook?

In most schools (in particular, larger well-organised schools) students will have a coursebook that you'll need to follow. Students and the school will see this as the foundation of the course. Often, where a syllabus exists at all, it's based on a coursebook.

However in some schools, there may be no materials (or just materials in the local language), so you'll need to consider whether to buy resources and/or create your own. We'll look at that in the next section.

What are coursebooks like?

At any specialist TEFL bookshop you can also search for some well-known TEFL coursebooks such as:

- *New Headway English Course* (John and Liz Soars, Oxford University Press)
- *New English File* (Clive Oxenden, Christina Latham-Koenig and Paul Seligson, Oxford University Press)
- *New Cutting Edge* (Sarah Cunningham and Peter Moor, Pearson Longman)
- *Language In Use* (Adrian Doff and Christopher Jones, Cambridge University Press)
- *Reward* (Simon Greenall, Macmillan)

These sorts of coursebooks are colourful and engaging and, in theory, contain everything a student needs. They generally:

- come in different levels from Elementary to Advanced
- teach a balance of the four skills (reading, writing, listening and speaking)
- provide grammar and vocabulary activities
- have units based on topics (e.g. food, travel, education)

A coursebook series usually also contains a teacher's book (with suggestions for the classroom, and grammar explanations), a workbook (for homework), audio CDs, and possibly DVDs and links to online resources.

Are coursebooks enough?

Yes and no. It's good to be aware of how you can supplement a coursebook if students need more practice in a particular area, or if the coursebook doesn't interest them.

However, many learners like having a coursebook because it makes the course feel organised. They can see a plan, and go forwards and backwards as they need to.

A number of teachers avoid coursebooks and try to use tailored material for the majority of their lessons. However, while this is admirable, it can be counterproductive. Too many handouts can seem chaotic.

Nevertheless, using a book shouldn't just be 'Turn to page 60 and do exercise 3'. There are some techniques you can use to get much more from a coursebook, both to make it more interesting for students, and also to extract a lot more useful practice. In our experience, many new teachers tend to fly through material and then run out – and resort to playing hangman for the last twenty minutes of the lesson!

How can you get the most from a coursebook?

We suggest there are three simple techniques for bringing any coursebook to life, and making it more effective.

- **Get students interested in the topic.**
- **Have students work together.**
- **Personalise everything.**

• Get students interested in the topic.

Imagine the coursebook unit you're teaching is about food. Generate interest by bringing in visuals of food, or real ingredients and cookbooks. Engage students' senses, let them taste the food, and browse through the cookbooks. Ask students to talk to each other about them. Students could discuss what they could make with the ingredients, or which recipes they like and don't like.

• Get students to work together.

Whenever possible, put students in pairs and groups. If students are in pairs, tell one student to put their book away, so they have to work together. Similarly, if you're giving handouts to groups of students, only give one to each group. It's not only a livelier dynamic, but you're also dramatically increasing the amount of student practice.

• Personalise everything.

Make everything meaningful and directly relevant to your students. After they complete a grammar exercise, have them rewrite the sentences to make them true about themselves (or their partner). After they read a story, have them write a new ending. After they practise a dialogue, have them change the script so people in your class become the characters.

Should you take coursebooks abroad?

It's generally wise not to take any coursebooks – certainly not a class set – because you may not know what level you'll be teaching, nor what your students and school will require.

It's also generally not necessary; you can buy mainstream TEFL coursebooks in most cities.

Supplementary Print Resources

The situation

- There's a great range of TEFL materials available
- Knowing where to get good resources will reduce your preparation time dramatically
- You should supplement a coursebook in an organised and restrained way

Why would I supplement a coursebook?

Many coursebooks are engaging and self-contained. However, you may find that you're stuck with one that's dull or irrelevant to your students. You might also find that students need additional practice with a particular language point or skill. Therefore you may decide to supplement a coursebook with activities.

Just be aware, using supplementary materials can seem messy unless it's planned well (and it can be expensive!). In general, make sure any supplementary material matches the content of the coursebook, for example, by addressing the same topic and learning objectives as in the unit. A little bit of additional material goes a long way. Use it less frequently but to more memorable effect.

Also, in practical terms, it pays not to be dependent on photocopied activities. Some schools don't allow teachers to photocopy at all, impose a copy quota, or require you to submit anything for copying well in advance. You may end up paying for photocopying outside of the school from your own pocket!

What sorts of publications are available?

There are lots of great print resources in TEFL because it's such a huge market. There's great variety, and the approach is often cutting-edge, with a focus on enjoyment and interaction in the classroom.

Some of the available types of materials are:

- spiral-bound, photocopiable activity books, which contain handouts, sets of picture cards, board games etc (e.g. *Communication Games*, Jill Hadfield, Longman Pearson)
- skills practice books (e.g. *Impact Listening*, Kenton Harsch and Kate Wolf-Quintero, Longman Pearson)
- handbooks for teachers, which describe activity ideas, and may include photocopiable handouts (e.g. *Teaching Large Multilevel Classes*, Natalie Hess, Cambridge University Press)

You can also introduce students to materials they can use on their own, in particular:

- grammar practice for students, available at different levels (e.g. *English Grammar In Use*, Raymond Murphy, Cambridge University Press)
- learner's dictionaries, also available at different levels (e.g. *Oxford Learner's Dictionary*)

These are just examples of materials you might want to use, not particular recommendations; choice of materials is a very individual thing and should be tailored to the needs and interests of your students. Spend time browsing and looking at the range available.

Should you take supplementary resources abroad?

Most cities will have one bookshop with at least a limited range of TEFL resources. However, it's worth taking any activity books you particularly like as they might not be available.

Online Resources and Activities

The situation

- The web has revolutionised language learning.
- There is a lot of free material on the Internet, for teachers and students.
- Students love using the Web.
- Students can practise all four skills online, including speaking.

How has the Internet changed language teaching?

The Web provides a rich source of material for teaching and learning, in particular:
- resources for you as a teacher (lesson plans, worksheets and ideas for activities)
- TEFL websites where students can practise grammar, vocabulary etc
- authentic reading and listening
- Web applications for creating and sharing content

The Web as a teaching tool has a range of particular benefits:
- It motivates students by bringing the 'real world' into the classroom.
- It caters to students' individual interests, learning styles and language levels as they can pursue what appeals to them and at their own pace.
- It helps shy students express themselves.

Many larger schools have computer laboratories you can book for regular classes.

Despite the opportunities for interaction on-line, students still want face-to-face interaction. Think of computers as just one, important, component of a language course.

• Resources for you as a teacher

There are many ready-made lesson plans and resources available for free (some require you to share your own activities to access others).

• Language learning websites

There are lots of controlled practice quizzes, and interactive speaking and listening activities, that can be used as extensions to the target structure in a coursebook, or be chosen by an individual student as an area for development.

To take full advantage of the autonomy the Web provides, give students room to explore. Let them find quizzes that not only meet their individual language needs, but that they also find enjoyable. Ask them to share what they find with other students.

• Websites for authentic reading and listening

As we know, authentic materials generally motivate students because they bring the real world into the classroom. The Internet is the perfect resource for a wealth of authentic language.

You can use a website as you would any reading or listening text in order to develop receptive skills:

- skimming (e.g. find a section)
- scanning (e.g. find information, such as names or numbers)
- reading for detailed understanding
- reading for inference

You can also use the Internet for information searches. Students can do research for a class project, or work in pairs to compete to find answers to a quiz first.

Finally, exploit the authenticity of the Internet. Ask your students what they like to do on the Web, in English or in their first language. Keep this principle in mind and get students to look for things they want to find. Also suggest activities you often do – e.g. reading the news, looking for recipes, trying to find things on eBay.

• Websites for creating and sharing content

Have students become active participants on the English-speaking Web, through:

- Exchanging emails
- Contributing to discussions and forums
- Collaborating on a class project or wiki
- Creating and uploading material such as videos, animated movies, posters and comic strips

Risks

The main problem a teacher faces is preconceptions (from students, parents and school administration) that an Internet session is just a chance for students to mess around and waste time. A teacher needs to:

- have an explicit aim. Without this, students will lose interest, and probably not learn much.
- state and enforce expectations. Students need to be on task and using English.
- be well-prepared for the task and confident with the technology.

You also need to be aware of school policy regarding risky behaviour on the Internet, and know how you will help students avoid:

- coming across inappropriate content
- compromising their privacy and personal information
- infringing intellectual property

One final issue to consider is unreliable technology. This is a fact of life in many teaching environments! Keep in mind that the Internet is just a means to an end – one potential component of a course that helps students to achieve language outcomes. Don't let the success of the course depend on unreliable technology.

Authentic Materials

The situation

- Authentic materials are any real reading or listening texts, not created for language teaching.
- Using authentic materials motivates students.
- You can use authentic materials with any level of learners.

What are authentic materials?

They are any text (written or spoken) used by English speakers in the real world, not designed or modified for language learning. These include:

- magazine articles
- brochures
- advertisements
- food labels
- maps
- songs
- TV programmes

and so on. In fact, the list is endless.

Why should you use authentic materials?

Authentic materials motivate students because they bridge the gap between the classroom and the real world. Many students report that they understand what's in the coursebook, but not what 'real' English speakers use.

Imagine, if you were learning a language, how different it would feel to read and understand a real menu from a restaurant as opposed to an obviously fake restaurant menu in a coursebook. As a language learner, whenever you understand an authentic text you feel you've made genuine progress.

How do you use authentic materials?

Many teachers' first reaction to authentic materials is that they're 'too hard', especially for lower levels. Certainly, you need to be selective. An elementary class would cope much better with a map or a timetable than a business report. However, consider 'grading the task not the text'. That is, give your low-level students a difficult text (e.g. a newspaper), but ask them to do something simple with it (e.g. locate different sections).

Of course we need to choose texts that our students want to read or listen to, whether related to their everyday life, their interests, or their future careers. We should also consider a wide range of text types, not just newspaper articles (which seem to be the authentic text of choice, for some reason!).

Aim to use authentic materials in class, as you would in real life. For example, get students to use a menu in a restaurant role play, rather than answering true/false questions. Have students use a map to plan a real excursion.

Should you take authentic materials abroad?

If you can predict what your students will be interested in, certainly. Many school environments feel very far removed from the English-speaking world; the more interest you can generate with real brochures, magazines, posters etc the better. Once you're teaching abroad, you may ask someone back home to ship you materials that will interest your students.

1.3 Teaching Specialised Classes

Each teaching situation you face, be it teaching kids or businessmen, comes with its own unique challenges and rewards. This section explains how to approach a wide range of different students and situations – and what to do when things go wrong.

Young Learners

The situation

- Parents worldwide see English as central to their child's education. From South America to Europe to Asia there are countless opportunities for teaching younger learners.
- The term generally refers to children between six and twelve, although more and more parents are sending their pre-schoolers to English classes.

Plusses	Challenges
• A very rewarding experience – children can be enthusiastic and fast learners.	• Can be a struggle controlling the class.
• Lots of fun.	• Can be tiring if you're not used to running around.
• Never boring.	• Young learners need lots of variety, so you need activities up your sleeve.
• If you're a sporty person you'll love the activity.	

Suggestions:

Keep the class active (without it getting out of hand!)

Movement not only keeps students alert and involved, but also assists learning by stimulating different connections in the brain.

1. Get students moving. They should be out of their seats at least several times per class.
2. Use different types of movement. Get students to stand, sit, pass things, move different parts of their body, and interact physically with each other.
3. Have as many students as possible practising at once. While this is a principle for all TEFL teaching, children more than adults will quickly get bored if only one student at a time is speaking.
4. Vary seating. Move students frequently into different pairs and groups.
5. Use light and shade. Follow a fun, mingling speaking activity with a pair writing activity. Have a dedicated time for quiet individual reading.

Encourage your students

This is a crucial time in your students' development. One comment can affect students for the rest of their lives. Positive comments will help them love English, and help them become intrinsically motivated.

1. Encourage your students often.
2. Encourage students equally (you can even mark it on the roll each time you praise a student to keep track). It's very demotivating not to be praised when others are, or to feel the teacher has a favourite.
3. Praise students for genuine achievement. Be on the lookout for things that students do well. Insincere praise is obvious, and can waste all your effort to encourage.
4. Making mistakes and subsequent correction has particular stigma, so do praise a student if they get something right after you correct them. Make sure that your students see you're correcting errors to help them improve, not to point out something wrong.
5. Have rewards, but don't overdo them. Too many may seem like buying the children's affection. Make rewards special – but make sure in the end all children are rewarded equally.
6. Encourage students to praise each other. For example, finish a class with students telling their partner what they did well today.
7. Praise your students in front of their parents and other teachers.
8. Let students show off their work. If they make posters, ask the principal if you can post them in the corridor or the library.

Engage your students' senses

1. Bring in real objects. If you're teaching food, bring in food they can see, smell, touch and taste.
2. Use music and songs to teach language.

Keep it fun

1. Use visual humour.
2. Laugh with (but never at) kids.
3. Be enthusiastic.

Use games

Children will get so absorbed in the activity itself that they forget there's a serious purpose. Many games have an element of repetition, so they work well as controlled practice.

1. Turn everything into a game or physical activity. Rather than just saying 'finished', have students jump up as well; rather than write something in their book, have them run to the whiteboard.
2. Give prizes, but don't overdo it, or they lose their impact. Younger children in particular may find the game in itself is fun enough.
3. Parents and school administrators may think games are a waste of time, so make sure you can explain their linguistic purpose. It's often better to call them 'practice activities'!

Be calm and patient

While activities need energy and pace, you as a teacher should be calm and patient. The more frenetic your class becomes, the more strength and control you need to exercise.

1. If a child cannot remember something you've taught, they're probably not being difficult on purpose. Everyone learns at different rates, so hide your frustration.
2. Learners may go through a silent period, where they appear not to learn. Don't rush them or force them to speak if they're not comfortable.
3. Children need more time than adults to adjust to finishing one activity and getting into a different headspace for a new activity.
4. Give students time to answer, to feel confident they will have a chance to contribute at their own pace.
5. If you have a discipline problem, don't rush to respond. Show you are in control. Use silence before you speak.

Use your authority effectively

Students will look up to you and do what you ask, if you have some simple ground rules and apply them consistently.

1. Be confident with your authority. You're the boss.
2. Establish class rules. Have the class agree on two or three rules – for example, not to hurt anyone, to speak English, and to do what the teacher asks. Post them on the wall and point to them when there's a problem.
3. Agree on a penalty – e.g. students stay back one minute for every minute they waste – and apply it uniformly.
4. Give individual students a first warning.
5. Use silence. Don't match students' disruptive behaviour. If students continue to talk, be silent and wait. If students are noisy, speak softly.
6. If a child throws a tantrum, ignore them. Don't give them the attention they're seeking. Move students away from them, or remove the child to a safe place.
7. Give children choices: 'Will you play by the rules, or do you want me to send you out?'
8. Use peer pressure to help manage the class. Give behaviour points to teams and deduct marks for disobeying the class rules.

Instruct clearly

Clear instructions prevent the biggest cause of chaos in a TEFL class – students not knowing what to do.

1. Stand in a particular place each time so unconsciously students know the next activity is coming.
2. Use a routine signal to get students' attention. You can use a routine that students join in – for example, counting down, or clapping a rhythm.
3. Show students what to do, rather than telling them.

Keep reviewing

Your students may not have developed autonomous study habits so you should help them with reviewing and practising language.

1. Recycle language in a range of different ways – games, repeating, singing, drawing, reading and writing.
2. Many of your students will be used to intensive classes where they associate reviewing with stressful 'tests'. Take the stigma away and have regular non-assessed quizzes. Make them fun – e.g. play bingo or a game show.

What to avoid

- Teaching abstract concepts. Teach grammar through actions and games instead.
- Spending too long on one activity.
- Too much frenetic activity. Make sure there's variety – light and shade!
- Favouring students.

Problems	Solutions
• Students misbehave.	• Find out the school rules and procedures for dealing with misbehaviour, so you're not alone. Then establish class rules from Day 1 and enforce them strictly but fairly.
• It's hard to control the class.	• Make sure one of your rules is that students pay attention when you signal. Be directive when you instruct.
• Students seem bored.	• Aim for as much variety as possible. Give all students something to do at the same time.

Problems	Solutions
• Students don't understand.	• Use demonstration and examples to instruct, rather than explaining how to do things.
• Parents interfere and tell you how to teach.	• This happens to the best teachers, so don't let it affect your confidence. Be ready to explain simply and clearly why you do things. Encourage them to come to class to see the results.
• The school supplies a local co-teacher, who does things very differently from you.	• Invite your co-teacher for lunch! Develop a good relationship with them, and when the time is right, talk about how you can work together best. Avoid any sort of conflict in front of the children.

Conversational English

The situation

- Conversational English focuses on developing spoken fluency rather than grammar or writing.
- Asia offers many opportunities for this type of teaching, mainly in private language schools.
- While the pay is generally not high, it's a good way to enter the TEFL field as it generally requires less experience than other types of teaching.

Plusses	Challenges
• Relaxed atmosphere – students may want to socialise as much as improve their English.	• Can be unrewarding if you want an intellectual challenge.
• It's interesting to hear what your students have to say.	• Not seen as high-status work.
• A great way to develop skills in lesson planning and classroom management.	• Can be difficult to get students to contribute.

Suggestions:

Treat a conversation class as seriously and professionally as any other. To improve speaking and listening is a serious language outcome for students. Also, deal with other language items as they naturally arise: functional language, pronunciation, vocabulary and grammar.

Have a structure

1. A routine makes it easy to plan, and students feel the class has direction.
2. Identify a topic for the day.

Start with a warmer

While a warmer is important for any class, it's particularly essential in a conversation class to establish a communicative atmosphere.

1. Ideally, relate the warmer to the topic of the lesson.
2. Use visuals or real objects to stimulate discussion.
3. Get students talking to each other (not to you) and discussing what they know about the topic. Perhaps set some simple discussion questions, or a quiz.

Use a text

Use a text to stimulate students' ideas and language.

1. Keep it short (so it doesn't take half the lesson!)
2. The text should be engaging, and related to your students' interests or needs. Some degree of controversy may be useful for stimulating discussion, but bear in mind your students' sensitivities and any cultural taboos.
3. Consider different types of text: for example, an advertisement, a flyer, a poster, food or medicine packets.
4. It's best to avoid too much unfamiliar language in the text, since its aim is to generate discussion. However, if students do need to learn any learn any vocabulary to understand the article, bring in dictionaries and have them teach each other.

Follow the text with a speaking activity

1. When we think of a conversation class we tend to think of a whole-class, free discussion. However, unless the topic is simple, this can be risky: students may be shy, they not have the language to discuss it cold, one or two students may dominate, and other students may simply tune out.
2. To prepare students for a discussion, start with a role play with assigned roles and opinions. This is a safe way to discuss ideas and practice the relevant language without being too exposed. They then feel much more equipped to express their real opinions.
3. Give activities structure so students need to interact. Perhaps they have to mingle and complete a questionnaire, or work together to solve a problem, or create a poster.
4. Try small-group rather than whole-class activities so each student gets as much practice as possible. You can finish with a five-minute, whole-class wrap up at the end.
5. If the topic is somehow depressing, finish the class on a high note – e.g. students discuss solutions or advice.

Have a regular finishing activity

This gives the class a sense of completion.

1. You can spend a few minutes looking at language. Dealing with language as it arises naturally is very memorable as it's in context. Ask students if they have any questions, or maybe pull something from the text.
2. You can look at one or two errors that you heard while students were talking. It's best to do it at this time rather than interrupt students during their speaking practice.
3. Students together can write any new language they learned on the whiteboard.

Problems	Solutions
• Students are unwilling to talk.	• Find out what interests your students, and what they're happy to talk about.
• Students are seemingly unable to talk about abstract topics.	• Avoid 'issues', especially with the whole class.
• Students don't take it seriously, and talk in their first language.	• Keep it fun. Inject some healthy competition in the activities, such as girls vs boys.

What to avoid

- Walking in unprepared and hoping students will maintain a conversation for an hour.
- Talking to fill the silence, if students are quiet – think of alternatives like small-group guided discussion.
- Topics that are either boring, or that make students uncomfortable.

Business English

The situation

- Employers increasingly demand that employees speak English – in fact in Europe it's a given – and Business English is the fastest growing area of TEFL.
- The term 'Business English' can range from office English to the language of business negotiations.
- As the fiield relies on medium- to large-scale businesses, most work is based in larger cities, and organised through private language schools, often on-site.

Plusses	Challenges
• You can give students very focused help. • Students can be very motivated as English is important for their success. • It's interesting being involved in real work environments. • It's challenging work that helps you grow quickly as a teacher. • Students often have very interesting experiences and opinions.	• Students have a lot of life pressures so may not attend regularly. • Students may be unmotivated if they're forced to attend. • It can feel daunting to teach a specialist area of English.

Suggestions:

People often feel they need specialist knowledge to teach Business English (or any other type of English for Specific Purposes). It's not actually true. Surveys of Business English students have shown that they want an English teacher to be a language expert, not a business expert.

The key is first to become aware of HOW your students will use English in their job, whether it's writing emails, conducting meetings, or making sales presentations. Then you have to help them to get the language right. Therefore, some preparation – finding out what your students' real-life needs are – will make your classes relevant, motivating, and will make you feel confident.

Be professional

1. Dress professionally and appropriately for the country. Appearances mean a lot in business.
2. Get business cards printed that say you're a Business English trainer or consultant.
3. Make sure all your materials look professional. Ensure any photocopies are clean and properly referenced.
4. Look organised in class. Have all resources in order in a file.
5. Choose words that are appropriate for the business world. Call your students 'participants'. Call games 'activities' or 'simulations'.
6. Use your students' titles until they ask you to call them by their given name.
7. Produce a report for each student and the employer at the end of the course.
8. Have professional expectations of your students. Insist they do homework and participate in class (and tell them this will be going in their report!).

Conduct a needs analysis

In your first session, conduct a serious needs analysis.

1. Find out what they need to do with English. Ask students or the company for real examples of specific things they do: the sort of emails they send, the documents they read, the conversations they have on the phone. Some companies, such as hotels, may have manuals which contain exact wording they want employees to use.
2. Assess students' English level in at least speaking and writing. You could run a short interview and writing activity.

Design a programme

Design a programme for your students, and give them and their employer a copy. Companies will feel their training dollar is being well spent. Include:

1. The outcomes: 'By the end of the programme participants will be able to …'. This should match the findings of your needs analysis.
2. A brief session-by-session outline. You don't need to create this content – you can select from one or several books, and give page references.
3. Regular assessment. This doesn't mean you need to spend long hours writing tests. Make the tests related to the course's real-life outcomes. If a course outcome is to write an email, assessment should require them to write an email.

Teach what's real

1. Make activities like real life. Use role plays to simulate real business interactions. Have your students make PowerPoint presentations. Set writing tasks just like the business correspondence they need to do at work. Avoid decontextualised exercises from a grammar book.
2. Focus on functional language, rather than grammar: e.g. introducing people, telephoning, describing company achievements, agreeing and disagreeing.
3. Focus on solving problems and making decisions, not just routine work interactions.
4. Since they're the business experts, take in dictionaries and get them to work out the meaning of specialist vocabulary and teach each other. You have a lot to offer by helping with the language issues: how the words are pronounced and used in a sentence.
5. Only teach grammar as it comes up. They're probably expecting English classes to be dry and unrelated to their real needs, so make sure it's the very opposite!
6. Don't hesitate to diverge from the programme as they raise particular things they need, but do return to it to show that the programme is organised and you're in control.
7. The real-life needs of Business English students are not just in their specialised field. Managers, for example, need language for travel, eating out, and making small talk.

Use role plays

You can base a whole Business English course around role plays, as these most closely simulate what they need.

1. Consider the personality of your students and their position when you assign roles. At least initially, someone may feel uncomfortable playing the boss of someone who is their manager in real life.
2. A role play (like any activity in real life) should have a purpose. It finishes naturally when something is achieved: for example, when students reach agreement, make a purchase etc.
3. Using simulated business cards is a lifelike way to assign roles.
4. You can use secret role play cards, giving each student a new personality and hidden motivations. Assigning personalities frees students up to try out things they wouldn't normally do at work. Telling students their motivation creates lifelike conflicts of interests, and propels a role play along.
5. Students need preparation time to decide what they'll say and how they'll say it. Students on one 'side' could prepare together.
6. Set up the room as realistically as possible. Have students move and position themselves as they would in real life. Make the students feel this is authentic.
7. Don't interrupt the role play. Hold back and observe. Make notes and discuss errors or relevant language afterwards.

Use good material

1. There are many excellent Business English materials (print, multimedia and online).
2. Use authentic material from your students' workplace.
3. Collect any authentic material you find that's related to your students' work.

What to avoid

- Feeling you need to be a business expert. There's no need to put yourself under that stress when the students know you're not there to teach them about business. Just show the students that you want to understand their English needs at work in order to help them.
- Any sort of unprofessional behaviour such as arriving late, dressing inappropriately, or being disrespectful.

Problems	Solutions
• Students are forced to go, so are unmotivated.	• Show students from the moment of the needs analysis that you're determined to make the class worthwhile and relevant to their needs.
• Students come irregularly because of outside pressures.	• There may not be much you can do! Be empathetic, as students will appreciate it. Avoid work that depends on attending the previous session.
• Your students have powerful positions and are intimidating.	• To start with, focus on what you're best at teaching, and use 'tried and true' activities. Your students may be more anxious than you, despite their positions.

One-to-One

The situation

- Teaching one-to-one can be quite lucrative and gives you independence as a teacher.
- A lot of teachers use a salaried position as an opportunity to make contacts in a new country and arrange private tuition (although check the legal situation!).
- One-to-one is intense, so consider encouraging your student to bring a friend along.

Plusses	Challenges
• A unique opportunity to give a student very focused and personalised help.	• It can be tiring as you are always part of the interaction.
• You may build a deep friendship.	• Your student can feel overwhelmed by the dynamic.
• A lot less confronting than teaching a large group.	• Can be a problem if there's a personality clash.

Suggestions:

Conduct a needs analysis

Treat your individual student as professionally as any class – not as an 'easy option'. They'll appreciate it, and word will soon spread. In your first session, conduct a serious needs analysis to find out:

1. What do they need to do with English? Ask to see real examples – material from their work, school, the community.
2. What is their current English level in each skill? At least conduct a quick interview and give them a short writing task.
3. What topics interest them?

Design a programme

Design a programme for your student, and give them a copy (and their employer if relevant). Include:

1. The outcomes: 'By the end of the programme you will be able to ...'. This should match what the needs analysis told you, and gives you a focus.
2. A brief session-by-session outline. Don't go into detail so you'll still have flexibility, but at least have a topic for each session.
3. Regular assessment. This can just be a simple quiz.

Personalise everything

It's a great chance to tailor your teaching to someone's needs and interests.

1. Focus activities on the student. If you're practising question forms, use them to ask about your student.
2. While it's an easy way to fill in time, don't make it all about you! With any topic, find out what your student thinks. Use open questions ('Why?', 'What do you think?') to get them to tell you more.
3. Do ask regularly if the course is providing exactly what they want. Balance this, however, by being confident in your evaluation of the students' language needs.
4. Take notes as you go. Use these to give your student feedback.
5. Help your student develop effective strategies for language learning that are appropriate for their learning style.

Vary the dynamic

Vary the interaction pattern and activity types. With just two people in a room, it's very easy for the situation to become intense.

1. A one-to-one lesson shouldn't all be speaking. Follow a speaking activity with some writing. Use a reading activity to trigger a discussion.
2. Give your student time and space. Don't tower over them or constantly 'urge' them to speak. Give them time to prepare for an activity in silence – e.g. to take notes or use a dictionary.
3. Change the seating. Avoid always being in exactly the same position.
4. Leave the room from time to time while your student reads or listens. They'll like the break.

Use the Internet

Obviously the Internet is a rich resource for engaging and useful material.

1. Find – or have your student find – interesting reading and listening material.
2. Get your student to be active on the Web, and to do what anyone might use the Internet for: contribute to forums, look up information relevant to them, and take part in collaborative activities.
3. If you are recommending sites, it's a good idea to check them beforehand to make sure they work!

Coach your student for real life

Help your student to rehearse for things they need to do in real life, whether it's a job interview or a social situation they'll need English for. There are lots of good resources for these.

1. Help them to script possible responses for a job interview. Role play and then give feedback.
2. Have your student give a formal presentation.
3. Your student may ask you for more general advice. If you don't feel qualified, just say so. Be aware of the influence you can have over someone's decisions.

Give your student the floor

1. In role plays, give yourself the role with less to say.
2. Hold back in discussions. Keep your opinions a mystery! It's your student who needs the practice.

Get out of the room

In TEFL it's always a challenge to bridge the gap between the real world and the classroom. With one-to-one teaching it's easy – you can go outside! You'll find lots of language is generated simply by the things you come across.

1. Go to places of relevance and interest to your student.
2. Rather than explaining the 'real world' to your student, ask them to explain things to you. Go to places you know nothing about.
3. Let your student give you a guided tour of their area.
4. Do real things that need language – like shopping.

Use real objects and a range of authentic texts

Of course this is a principle for any class, but working one-to-one it's easy to see and work with the material. You can use original texts without worrying about photocopying.

1. Bring in things to change the dynamic – rather than looking at each other, you can turn and focus on the object or text.
2. Give your student control. Let them handle the book or use the DVD remote.

What to avoid

- Seeing one-to-one as conversation. Use a range of activity types.
- Being constantly central to the action – it's tiring for you, and not good practice for your student! Hold back.
- Being the only source of input for the student. They should read a variety of texts to generate ideas, and listen to a range of voices in meaningful contexts.

Problems	Solutions
• Your student seems shy or unfriendly.	• Build rapport with a 'get to know each other' game.
• The sessions are intense and 'hard work'.	• Don't forget to include reading, writing and listening activities where you can take a back seat. Have your student bring someone else, so they practise with each other.
• You spend a lot of time travelling between sessions.	• Be clever with scheduling, and see if your students can come to you.
• Students cancel at short notice.	• Treat this as a serious business. Have your student (or their parents) sign a contract which includes a penalty for late cancellation.

Large Classes

The situation

- Teaching large classes is a way of life in many parts of the world where groups of fifty are not uncommon.
- As daunting as teaching a very large class sounds, there are some very effective management techniques that will help make your life easy, the class enjoyable, and the students' learning successful.

Plusses	Challenges
• High energy.	• Harder to get to know students and meet individuals' needs.
• Never boring – time flies!	• Students can feel anxious talking to the whole group.
• Activities take longer, so it's rare to run out of things to do.	• Classroom management.
• The class will always be responsive – someone will always contribute.	• Admin burden.
• Can feel you are really contributing to students' futures in a challenging environment.	

Suggestions:

Get to know your students as individuals

Learn students' names and something about them. Students will be used to teachers not giving them any attention, so if you make the effort, you'll dramatically change the atmosphere in the class.

1. Be open about the challenge – 'I'm going to remember your names, so you have to help me!'
2. You can use name tags to start with.
3. Ask students initially to stay in the same place.
4. Have students bring in photos. Put up a photoboard in class, or get them to complete a questionnaire which you keep with their photo.
5. Use activities to help everyone, including you, to remember names: questionnaires, 'find someone who', a mnemonic game (e.g. 'purple Peter').
6. Use students' names when you talk to them: 'What music do you like, Susan?'.
7. Pay equal attention to all students. Make sure students see that you like them all, and you aren't favouring the few students whose names you know.
8. Try to show each student that you're aware of their needs. It could be as simple as a commenting on their homework.

Make notes

1. Keep brief notes for each student. It helps you remember not only their names but also their needs and interests.
2. Make notes as you monitor. You can deal with common errors or other issues with the whole class after they finish an activity.

Have class rules

1. First, know the school rules – don't apply rules in your class that are in conflict!
2. On Day 1, agree on two or three principles of particular relevance to a large class – e.g. when someone speaks to the whole class, everyone will listen – and post them on the wall.
3. Be strict but fair. Have a consistent penalty for breaking the rules - e.g. everyone stays back several minutes.
4. Create teams, and have them select team leaders, who are responsible both for looking after the team members and also for maintaining order.

Instruct clearly

One of the biggest risks in a large class is that the students may not understand what to do, so you'll be running round the room trying to repair an activity!

1. Use a signal to get attention, and wait until you have eye contact with everyone before you instruct. Those few seconds will save an activity from chaos.
2. Don't shout to compete with students' noise – wait and then talk quietly.
3. Check your instructions ('Do you speak or write?').
4. Use the whiteboard to support your instructions. Write up a word prompt, a discussion question, or an example.
5. If there's an activity type your students particularly like, use it regularly. You won't need to instruct in detail each time.

Manage a large group differently

1. Talk to the whole class. Direct questions to the back as well as the front.
2. When a student speaks softly, walk away from them, not up to them, and gesture to show you want them to speak to the whole class.
3. Get students to help with handing out worksheets, setting up equipment, etc.
4. Give students something to do while materials being handed out. You can play a simple game or sing a song.
5. Give students an ongoing activity to do if they finish early, such as write in a journal.

Get all students to participate

1. Give students a mark for participation.
2. Assign roles in group work. Get one student to be the organiser, with the responsibility of making sure every student contributes.
3. Ensure students get to practise with different students, otherwise it's boring, and bad for their language development. Where possible, organise mingling and different groupings. If the classroom makes it a challenge to move students around, think of alternatives: can they turn around to make groups of four? Can you seat them differently at the start of each lesson?
4. Organise ongoing group projects that you can monitor.
5. Consider 'activity corners' where you organise different activities around the classroom, and students can move around as they want.

Be clever with time outside class

One challenge with a large class is the amount of administration and marking. However, there are ways of dealing with it. Do put school administration first – you'll be judged on handing attendance and grades in on time, probably above the standard of your teaching!

1. Set homework that has right and wrong answers, and students can correct each other's in class.
2. Don't set lots of writing homework that'll need correcting after hours. Focus on short pieces that you can work on in class.
3. Tell students that when you correct their writing you'll only be looking at the target language of that lesson (e.g. past tense).
4. Use a correction code (e.g. WW for wrong word).
5. Set a time each week outside class when students can come to you with questions.

What to avoid

- Competing with the noise of the class – do the opposite.
- Letting noise get out of hand. Enforce a rule that students must stop talking when you give a signal.
- Running around too much. Instruct clearly so you don't need to clarify misunderstood activities.

Problems	Solutions
• Students are unmotivated.	• Get to know the students' names and something about their different interests and motivations.
• It's hard to control the class.	• Do what's counterintuitive. Wait rather than compete with noise. Speak softly. Walk away from students to get their attention.
• Students don't know what to do.	• Make sure you establish eye contact with all students before instructing. Show students what to do rather than telling them. Check your instructions.
• Students are focused on an exam so they're unwilling to take part in 'fun' speaking activities.	• Accept this as their main motivation, but show how speaking will help their English level overall. Get students to discuss topics and grammar points relevant to the exam.

Multi-Level Classes

The situation

- 'Multi-level classes' refers to groups where learners are at different levels of proficiency (for example, elementary and intermediate) or at different levels in different skills (for example, some are stronger in speaking and others in writing).
- A multi-level class is the reality in many classrooms worldwide, especially when teaching adults with varying backgrounds and learning experiences.

Plusses	Challenges
• Increases your awareness of students' needs.	• You're constantly on your toes to ensure lower level students are coping and higher-level students are challenged.
• Challenging yet satisfying when it works.	• Tricky to plan activities and assessment suitable for the whole class.
• Can be an opportunity to create a very supportive classroom dynamic.	
• An opportunity to try different approaches (e.g. avoiding grammar teaching).	

Suggestions:

Ensure everyone feels equally valued

While language level obviously has nothing to do with intelligence, in a multi-level class it's easy for lower-level students to feel 'dumb'.

1. Always start and finish the class as a whole group, even if you sometimes divide the students into groups in between.
2. Ask questions to all students. Choose questions you know a lower-level student can answer, and address the student by name. Praise them in front of the class for genuine achievement.
3. Stop talking! Now's the time to conquer any excessive teacher talk problem. If only the stronger students understand you, the lower-level students will feel inadequate.

Use non-level specific activities

1. If you can, avoid a grammatical syllabus, which is by its nature level-specific.
2. Focus on functional language, where students can participate to their level.
3. Have the class complete tasks and ongoing projects - e.g. design a poster or make a video.

Use level-specific material flexibly

If you do have level-specific content to cover, either group by level, or deliberately mix levels.

1. Pair up stronger and weaker students. Strong students often enjoy the responsibility, and the lower-level students receive individualised help.
2. You can divide students by level but never call them 'weaker'! Just use their names to put them in groups. Then they can work on different, graded material.
3. You can offer students a choice of different worksheets or activities, not explicitly labelled 'easy' and 'hard'. The Internet is ideal for this.
4. Rather than talk about levels, you can ask which students have 'done' a particular language point in an earlier class. Have them move to a different part of the room to work on another activity, while you instruct the students who need to learn it.

Use ongoing activities

In a multilevel class, students will finish activities at different times. Have ongoing activities they can turn to, rather than sitting with nothing to do.

1. Have students write an ongoing journal. Give feedback in the journal to challenge and motivate higher-level students.
2. Give students a workbook or a bank of materials, rather than going ahead in the coursebook.

What to avoid

- Dividing the class into two distinct groups, 'weak' and 'strong'.
- Comparing individual students.
- Teacher-centredness. If you talk a lot in class, it will either be too difficult for the lower-level students, or not challenging enough for the higher-level students.

Problems	Solutions
• It's difficult to coordinate activities, as some students always finish first, and sit there bored.	• Have ongoing activities that stronger students can always turn to.
• Higher-level students are not challenged because you spend more time with lower-level students.	• Make sure you spend dedicated time with stronger students to motivate them. A little goes a long way with lower levels: there's no need to hover over them, or feel sorry for them, just because their proficiency is lower. Teach them a small amount of new language, get them working, and spend time with the higher-level students.
• Lower-level students can't possibly cope with the prescribed coursebook or tests.	• Speak to the school about using separate materials and assessment.

Teaching Without Resources

The situation

- A lack of coursebooks and other resources is the reality in many countries, but there are effective ways of dealing with this.
- In fact, in well-equipped school environments we can become focused on teaching materials rather than learners; developing skills to teach with fewer resources can lead to a more interactive experience for us and our students.

Plusses	Challenges
• Can make you very responsive to students, rather than focused on the coursebook and handouts.	• Tricky to prepare.
• You develop a range of new teaching techniques quickly.	• Easy for classes to become chalk and talk.
• Can feel you are really contributing to students' futures in a challenging environment.	• Easy to fall back on two or three familiar activities.

Suggestions:

Some of the best teaching can come from using what's in the classroom. Students can, in fact, see piles of photocopied worksheets as chaotic and pointless. Many of the activities in this book (see section 2) don't rely on materials.

Be prepared

If possible, take a number of activities with you. Don't take too much – in an environment where students are not used to resources, a little bit will go a long way!

1. Buy a coursebook for yourself to help plan a course. This will help you work out what topics, vocab, grammar etc to teach, and in what order.
2. Keep your own files of activities, pictures, texts etc, organised by topic or grammar point. Laminate any activities you plan to use more than once.
3. Find activities on the Internet. See the appendix for recommended sites.
4. Take one or two photocopiable activity books with you.

Investigate and experiment

1. Find out how the other teachers deal with the lack of resources.
2. Find out what topics are of interest to the students, and base lessons on those.
3. See what sort of games and activities the students enjoy outside class. Can you modify them to provide language practice? For example, if students like 'rock paper scissors', is there a vocabulary activity you could base on that?
4. Experiment! Use your creativity. Your students will appreciate your efforts. Try to avoid falling back on obvious time-fillers like hangman.

Get the students to do the work

1. You don't need to photocopy an activity in a book. Draw it on the board and ask students to copy it.
2. Use a running dictation. Post a text, cut into sections, on the wall. Students need to run up, memorise the text, then run back and dictate it to their partner (see p43).
2. Have students write comprehension questions for the other students.
3. Get students to write tests for each other based on what they've done in class.
4. Have an ongoing project, where students are responsible for producing all the content.

Use the whiteboard effectively

1. Plan to use the board effectively as this will be a central focal point for the class. Divide it into sections for vocabulary, activities and language analysis. Use different colours.
2. If you are teaching language from a real-life context, draw it on the board, eliciting as you go.
3. Write up word prompts for controlled practice.
4. Use the board to create mystery. Start drawing a picture, slowly, and ask students to work out what it is. Write up dates in your life and ask students to guess why they're important. Write apparently random words and ask students to work out how they're connected.
5. Write incomplete sentences which students discuss or complete in writing. This could be to practise a grammatical structure ('I've often... ' 'I've never...') or to generate debate ('Men are...' 'Women are...').
6. Write unconnected words that students have to put into a story, or into sentences that are true about partner.
7. Write some (controversial?) sentences related to a topic, which students discuss or formally debate.

Use drama and role play

1. Get students to write a dialogue based on the topic or using the language they've learnt, and then rehearse and perform it.
2. When characters come up in a story, in a song, or in the news, have students role play them.
3. Get students to improvise simple dialogues with no rehearsal time.

Use dictation

1. Use dictogloss: you dictate a short story and students have to try to recreate it perfectly, first individually, then in pairs, then in groups (see p43).
2. Dictate sentences with an error in each. Students have to find the error.
3. Dictate sentences about you, then have students dictate sentences about themselves to each other.

Use oral controlled practice

A lack of printed resources can have a very positive effect in ensuring there is lots of speaking practice in class.

1. Do lots of modelling and drilling: the whole class, half class, small group and individual.
2. Practise dialogues by dividing the class into two halves.
3. Use substitution drills. Students repeat a sentence, then you point to a substitute word or, better still, picture on the whiteboard: 'I'm going to the shop' (picture of school), 'I'm going to the school'.
4. Use transformation drills. Students repeat a sentence, then you point to a prompt that will change the grammar: 'She's going to the shop', (question mark) 'Is she going to the shop?'.
5. Use question and answer drills. Students answer according to a prompt: 'Are you a student?' (tick), 'Yes, I am'.
6. Use chain drills. Students stand or sit in a circle. The teacher starts off a series of questions and answers which go around the circle.

Use pictures and real objects

Visuals bring the real world into the classroom. Collect pictures from everywhere – newspapers, magazines, brochures, advertisements. Ideally they should be quite large, but you can always ask students to come up close to look at pictures (rather than you run round the class showing people).

1. Students copy the picture and label it.
2. Students describe what's happening in the picture.
3. Students write a dialogue between two people in a picture, then perform it.
4. Students write a story based on the picture.

What to avoid

- Falling back on teacher talk. Think of ways to get students interacting and listening to different sorts of input.
- Using the same few activities again and again.

Problems	**Solutions**
• Your students have no listening and reading practice.	• See what chance there is of finding a DVD player, and a local photocopy shop so you can provide some sort of reading material.
• Students can't remember new language because they have no written record.	• Post useful expressions and vocabulary around the room.
• The class lacks purpose because there is no coursebook.	• Write up daily aims on the board, and post weekly aims on the wall.

Activities

Introduction

We know language learners have to practise, so effective activities are central to their success. We do want to help you avoid relying on a handful of the same ones, so this section contains hundreds of activities, warmers and techniques to bring your lessons to life, and make preparation easy!

1. Twenty-Four Warmers You Must Know

A selection of fun, interactive speaking activities you can use for any class and any level. Almost all require no preparation.

2. Twelve Activities You Must Know

These are classic activities and techniques, organised by skill, which you can use in any lesson.

3. Activities A-Z

Here we've listed hundreds of our favourite activities alphabetically.

4. Activities A-Z: Photocopiable Materials

These are the photocopiable activities referred to and cross-referenced in Activities A-Z that you can photocopy and use in your classes!

2.1 Twenty-Four Warmers You Must Know

Get your students going with a good warmer. All the activities in this section are designed to create a relaxed classroom atmosphere, full of happy, chatty students! Pick ones that contain vocabulary and grammar your students are familiar with to boost their confidence for the class ahead.

Twenty-Four Warmers You Must Know

1. Word association

1. Stand with the students in a circle.
2. Start by saying a word (e.g. *hot*).
3. The student on your right says a word that is somehow linked (e.g. *cold* or *weather*).
4. The next student says a word linked to the student before them, and so on around the circle.

Variations:

- When students understand what to do, put them into smaller groups to increase STT.
- Increase the pressure – if a student can't say a word in three seconds, they're eliminated and have to sit down.
- Have students say words which are not connected in any way!

2. Hot seat

1. Put the students into small teams.
2. One person from each group sits with their back to the board.
3. The teacher writes a word on the board.
4. The students facing the board define the word without saying it. The students with their back to the board have to guess what it is.

Variation:

The students with their backs to the board ask their teams yes/no questions to work out the word.

3. Line-up

1. Ask the whole class, 'How long does it take you to get to school?'. Elicit several answers.
2. Tell the students to line up against the wall, from the shortest time at one end, to the longest time at the other (students will need to mingle and ask each other to work this out).
3. Once the students are lined up, divide them into pairs. Ask them to talk about a related topic (e.g. *what happened on the way?*).

Variation:

This works with any question that will let you sequence students: 'When's your birthday?', 'How long have you been in the UK?' etc. It has to be something they need to ask each other to be able to do (e.g. it doesn't work with height!).

4. The cup of knowledge

1. Design a 'cup of knowledge' (possibly a box or an ice-cream container) for your class to store newly learnt vocabulary in.
2. At the beginning of the lesson, let each student put their hand into the cup of knowledge and chose a word.
3. In pairs or groups, the students then have to describe their word, without saying the actual word. The other students have to guess what it is.

Variation:

Groups then have to use the word in as many different sentences as possible.

5. Board race

1. Divide students into two or three groups.
2. Draw lines on the board to give each team equal space to write.
3. Give students a time limit (e.g. two minutes).
4. Tell students a topic (e.g. 'colours').
5. One student at a time runs to the board, writes a word (e.g. *blue*) and then passes the pen to their next team member.
6. The group with the most words (spelt correctly!) in the allocated time wins.

6. Noughts and crosses

1. Split the class into two teams.
2. Draw a large noughts and crosses grid on the board.
3. Write a newly learnt word or collocation in each of the nine boxes.
4. Teams take it in turns to come to the board and say a sentence using the word correctly in a sentence to win the square.
5. The winning team has three in a row correct.

Variation:

With larger classes, do one example with the whole class, and then split the class into groups. You may want to appoint a stronger student as a monitor in each group.

7. Question time

1. Give students one minute to write one question they want to ask other students.
2. You can specify what language it must contain: if students have been studying superlatives, tell them it must contain a superlative (e.g. *who's the most beautiful actress in the world?*).
3. Students mingle and ask their question. Encourage them to ask follow up questions: *Why? Who is she?*.

Variation:

Instead of mingling, students stand in concentric circles, facing each other. Students ask their questions and follow up questions until the teacher claps their hands to signal that the outer circle moves one person to the right.

8. Wordbuster

1. Write a long word on the board, such as INTERNATIONAL.
2. Give students a time limit (e.g. two minutes) to find as many small words as they can in the word, e.g. *line, tire, liar*.

Variation:

You can devise a points system for the size of the word e.g. up to 3 letters = 1 point,
4 letters = 2 points etc.

9. Get moving!

1. Blu-tack pictures on the wall (e.g. action shots, places or people).
2. Students walk around the room in pairs and discuss the pictures.
3. Join pairs together. Students compare what they said about the pictures.

10. Fly swatter game

1. Write recently learned words in random places on the board.
2. Divide the class into two teams.
3. Have one person from each team stand near the board with a (clean!) fly swatter.
4. Describe one of the words. The first person to hit the correct word wins.

Variations:

- Have students take it in turns to describe one of the words instead of you.
- With larger classes, give each group a set of words on separate cards. The group spreads them out face up. One student describes one of the words, and the other students race to slap their hand on the correct word first.

11. Fruit salad

1. Students sit in a circle on chairs.
2. Name each student one of three fruits (e.g. apple, banana or orange).
3. Call out one fruit (e.g. 'Banana!'). All the 'bananas' also call out 'banana!' and swap seats.
4. Repeat with all three fruits at random.
5. Every so often call out 'fruit salad!'. All the students call out 'fruit salad!' and swap seats. However, take one of the chairs away, so one student is left standing. Give them a penalty: for example, they have to talk about themselves for thirty seconds.

12. Ball game

1. Students stand in a circle.
2. Students throw a ball (or stuffed toy) to each other. When they throw the ball, they have to ask a question. The person catching the ball answers.

Variation:

Once students know what to do, put them in smaller groups, each with a ball.

13. Jumbled dictation

1. Dictate a word students have recently studied, but with the letters in the wrong order (e.g. E – L – B – T – A for 'table').
2. Students race to write the correct word on the board.
3. Students continue the activity in small groups, taking it in turns to dictate.

Variation:

You can also dictate sentences with the words in the wrong order.

14. Picture back!

1. Stick a picture of a person, animal or thing on each student's back (without them seeing).
2. Students have to walk around the room asking yes/no questions to find out what's on their back.

15. Four circles

1. Ask students to draw three circles.
2. Instruct students to write words related to a topic in each circle: for example 'In the first circle, write three items that are in your fridge'.
3. Put students into groups and they ask each other about what they wrote to find out more.

16. Picture board

1. Put students in groups.
2. Give each group a series of (related or seemingly unrelated) pictures.
3. Students have to make up a story together from them.

17. Translating

For students of one nationality.

1. Put students into pairs.
2. Give each pair a different short text in English appropriate for the level of the class.
3. Students translate the text into their first language.
4. Join pairs together. Each pair reads out the translation. The other pair writes down an English translation.
5. The pairs compare the new English version with the original.

18. Interruption game

1. Choose or write a short story, suitable for the level of your students, in which there'll be lots of potential questions.
2. Demonstrate the game. The teacher begins the story, and students have to interrupt where possible with questions. For example:

Teacher: 'Once upon a time, there was a girl ...'
Students: 'When was this?'
Teacher: 'This was a long time ago. The girl's name was Amelia ...'
Students: 'Why was she called Amelia?'
Teacher: 'Her sister liked the name ...'

3. Students repeat the activity in groups, with one student reading.

19. True/false game

1. Ask students to write three statements about themselves: two true and one false (you can do an example for yourself where one is obviously false).
2. Put students into groups (or get them to mingle). They tell each other their three statements, and the others have to guess which are true and which is false.

20. Picture mingle

1. Each student draws three things in their life on a piece of paper (e.g. their house, dog and neighbour).
2. Students tape the picture to their front or their arm.
3. Play some music. Students dance around the room.
4. When the music stops they talk to the person nearest them about their pictures.
5. Play the music again and repeat several times.

21. Correct or incorrect?

1. On the board, write some sentences, some correct and some containing errors (related to what students have been studying).
2. Divide students into teams.
3. Teams say whether they think a sentence is correct or incorrect (1 point), and if incorrect, what the error is (1 point).

22. How can you use it?

1. Put students into groups.
2. Tell the class the name of an object (e.g. *a piece of paper*).
3. Each group brainstorms as many possible ways of using the object as possible (encourage them to be creative: show you can fold it up and put under the leg of a wonky table!).

23. Pictures on the board

1. Divide the class into two or three teams.
2. One person from each team comes to the board.
3. Show the students at the board a word, without showing the others. They have to draw the word.
4. The team that guesses the word first wins a point.

Variation

This time, in their groups, get the students to position their chairs so that only one person can see the board. The teacher either shows a picture or writes a word/phrase on the board. The student who can see the board must draw a picture and the other students have to guess what the student is drawing. The first group to get the correct word wins.

24. I admire...

1. Students mingle and tell each person something they admire about them (and you can finish by telling students something sincere you admire about them as a class).

2.2 Twelve Activities You Must Know

Consider these the aces you should have tucked up your sleeve in every lesson. Sorted by skill (speaking/ listening/reading/writing), these simple, effective activities can be whipped out to boost student participation, regardless of your lesson's focus.

Speaking

1. Matching halves

Have you ever been	to France?
Have you ever eaten	French food?

Chop up words, phrases or sentences in half. Students have to match the two halves. Possibilities:
- Students mingle to find the other half (don't let them show anyone their card – they have to speak), then they Blu-tack their matches on the board.
- Students play concentration, with the cards face down.
- Students play snap.
- Divide students into groups of three or more. Give one student in each group half of the cards. Spread the other cards face up on the desk. The student reads out one card at a time, and the other students race to find the other half.

2. Questionnaires and forms

	Favourite band?	Favourite song?	Play an instrument?	Sing?

Design a questionnaire. You don't need to photocopy it – students can copy from the board. Students mingle and take notes. Afterwards:
- Students discuss any surprising results.
- Students summarise the survey results in a report.

Or design a form. Students interview each other and fill them out. Then:
- If they don't have names on them – randomly distribute them, or Blu-tack them to the wall, and students work out who they describe.
- Students write about each other.

3. Information gap

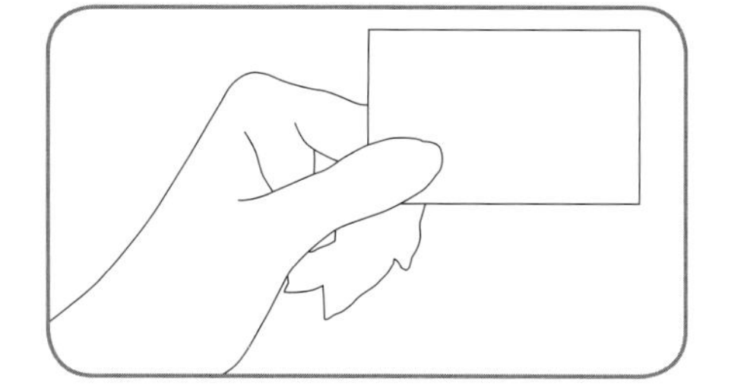

More a principle that underpins hundreds of possible activities, this is where you give different students different information, so they have to communicate to achieve something. If students are given cards with information, make sure they don't show other students!

- One student thinks of (or picks up a card describing) a person, place or thing, and the other students have to work it out by asking yes/no questions.
- In pairs, one student has a picture (or thinks of something they know like their house), and describes it to their partner, who draws it.
- In a role play, give each person different information (e.g. in a restaurant, only the waiter knows the prices, so the customer has to ask).
- In a role play, give each student a card that describes their real-life motivation (e.g. in a shop, the shop assistant will get a commission, but the customer has a tight budget). It works well if there are conflicting interests, just like in real life.

Listening

4. Complete a table/diagram/map

NUCLEAR POWER	
Pros	Cons

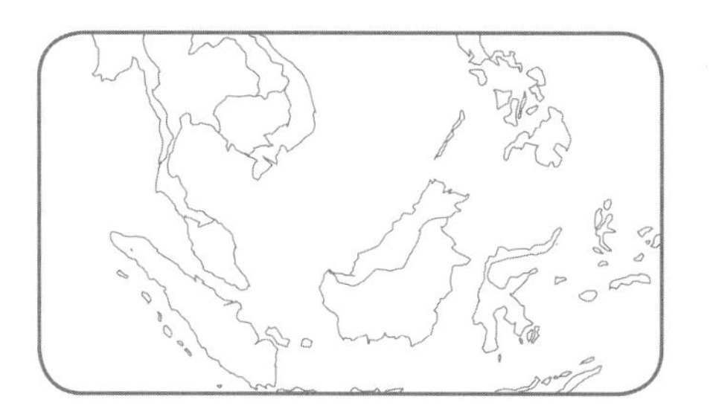

Students listen and have to:

- fill in a simple table (you don't need to photocopy the table – draw the table on the whiteboard and have students copy it)
- fill in a flow chart
- draw a diagram
- draw a route on a map
- draw a picture

Students could be listening to you, an audio recording, or each other.

5. What's next?

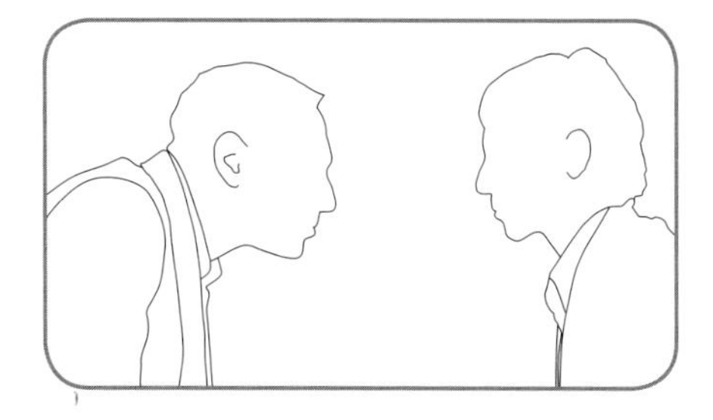

Student listen to or watch part of a story, which suddenly stops. They have to predict what will happen next.

Again, students could be listening to you, an audio recording, or each other, or watching a DVD.

As a controlled activity, divide the class into groups, and give out a text with words gapped to one student in each group; they read the text and stop at the gap, and students have a set of words to choose from.

6. Bingo

Play bingo with the class, or get students to do it in groups.
Variations:

- Use words, not numbers.
- Create cards with words. Produce a sound, and students have to find a word with that sound.
- Create cards with pictures. Say the word, and students have to find the picture.

B	I	N	G	O
7	25	44	57	62
15	22	40	50	70
11	30	FREE SPACE	46	74
2	28	37	55	68
10	27	39	59	75

Reading

7. Information exchange

A
Xxxx xxxx xxx xxxx xxx xx xxxx xx xxx xx xxxx x xxxx x
xxxx xxxx xxx xxxx xxx xx xxxx xx xxx xx xxxx x xxxx x
Xxxx xxxx xxx xxxx xxx xx xxxx xx xxx xx xxxx x xxxx x
xxxx xxxx xxx xxxx xxx xx xxxx xx xxx xx xxxx x xxxx x

B
Xxxx xxxx xxx xxxx xxx xx xxxx xx xxx xx xxxx x xxxx x
xxxx xxxx xxx xxxx xxx xx xxxx xx xxx xx xxxx x xxxx x
Xxxx xxxx xxx xxxx xxx xx xxxx xx xxx xx xxxx x xxxx x
xxxx xxxx xxx xxxx xxx xx xxxx xx xxx xx xxxx x xxxx x

Divide the class into groups. Give each group a different part of the text. They read their section, and then share the information with the other groups. Rather than just saying 'Read it' (you have no way of knowing if students have understood), you could:

- give out questions.
- use a simple-to-prepare activity, like a grid:

	Food	Drink
Japan		
Germany		

- get students to underline a distinctive feature of the text (e.g. all the numbers, or people's names) and work out what they mean or what they did.
- get students to write their own questions for their part of the grid, which they then give to another group to answer.

8. Mind map or flow chart

If a text has interconnected ideas, or some sort of sequence (even a story), students can create a mind-map to show relationships between ideas, or summarise the stages in a flowchart.

9. Following instructions

Give students some sort of instructions or directions – say a recipe or mobile phone manual – and have them actually cook, or find unusual features on their phone.

Writing

10. Dictation

1 Xx xxx xx xxx xx.	2 Xx xxx xx xxx xx.	3 Xx xxx xx xxx xx.

There are engaging variations on the classic dictation. One thing remains – students must aim to finish with something 100% correct.

Get students moving in a running dictation!
- Divide students into pairs. One will write, and one will run.
- Blu-tac parts of a text around the walls.
- The runner has to go to each text and memorise it – no writing! They then run back to their partner, who writes it down.
- The writer cannot leave their chair.
- The winning pair has the text sequenced correctly and 100% accurate, including spelling and punctuation.

Students can collaborate in a type of dictation called a 'dictogloss'.
- Read the class a short passage, twice, the first time at natural speed, the second a little slower.
- Each student tries to write down as much as possible.
- Students then work in pairs, then fours, to try to recreate the text.
- Read out the text again.
- Students again in their fours correct what they've written.
- Finally, get all students around the whiteboard to write up the text.
- Challenge the class to work together to get the text completely correct.

Have students dictate to each other. It could be a whole text, or sentences, or words and numbers they've been practising.

11. Cooperative writing

Have students work together and be creative! This could be to produce stories, posters, web-based posters, or video.
- Have a clear concrete goal, with specifications.
- Students can write the first line of a story in pairs. They pass on their first line to the next group, and each group writes the second line, etc, until it finishes back with the original pair.
- Make sure students get to show off their work – for example, you may be able to display posters in a school corridor.

12. Real-life writing

Have students write something useful in real life. For example, if they're considering study options, they can write emails to different universities to find out real information.

2.3 Activities A-Z

Consider this section your activity bible: no matter what grammar point, language function or topic you're trying to teach, you'll find something useful here. Just look up the first letter of whatever you want your students to practise, then use the activities listed to get them talking. Even if you don't find something that fits your students exactly, there's plenty to inspire.

Most of the activities can be adapted to different age groups and language levels (for example by using more advanced vocabulary in a word game), so we haven't specified who they're for. You're the best person to judge which activities will work with your students. Your primary students may be completely focused on preparing for a test, while your adult students may love running round the room. You need to make the call!

So you and your students can get the most out of each activity, we've also suggested variations (both in terms of procedure and the language being practised) and extensions (for example, how students can follow up on an activity for homework).

A Adjectives : GRAMMAR

1. Adjective mime

No preparation

1. Demonstrate the activity. Ask 'What adjective is this?'. Mime/use your hands to convey the adjective (e.g. *small*). Students call out the correct answer.
2. Students practise in small groups.

Variations:

- You could specify adjectives covered in a recent unit.
- This activity could also be used to practise nouns, verbs and prepositions.

2. What am I like?

No preparation

1. Each student completes the following statements with an adjective: *I'm... I'm not... I wish I was...*
2. Students mingle and share their statements.
3. The aim is to find another student with the same answers.

Variation:

- Students could complete three statements on any topic you've recently covered in class.

3. Chain collocations

No preparation

1. Students stand in a circle.
2. One student says an adjective (e.g. *green*). The next student has to say a noun that matches (e.g. tree). The next student says an adjective that matches the previous noun (e.g. *tall*). The next student says a noun that is relevant to their new adjective (e.g. *man*) etc.
3. If a student gets it wrong they leave the circle. The winner is the last remaining student.

Variations:

- You can have students clap a rhythm to add pressure.
- After students know what to do, break into smaller circles to maximise practice.

4. Synonyms and antonyms

Before class: Note down ten adjectives, along with one synonym and antonym for each (e.g. *big*: *large, small*), plus three unrelated adjectives (e.g. *square*).

1. Ask students to copy this table from the board (with ten empty rows):

adjective	synonym	antonym

2 Read out your words in random order. Students have to fill in every square in the table, and remove the three words that don't fit (of course students may fill out the table in a different order).

Variation:

- Students could do this in groups, with one student in each group reading out the words.

5. New combinations

Before class: Note down ten adjective + noun combinations students have recently learnt (e.g. *tall man, modern building*).

1. Write the adjective + noun combinations on the board (ask the students to guess what noun you're going to write next, in order to keep them engaged).
2. Divide students into groups. They need to write as many different combinations as they can in three minutes.
3. Discuss what combinations are and are not possible.

Extension:

- Students write a story using some of their new combinations.

6. Collocation snap

Before class: Photocopy & cut up sets of 10+ adjective + noun collocations (e.g. heavy + rain). Each word should be on a separate card. Create one set for each group of around 4-5 students.

1. Divide students into groups. Give one set of cards to each group.
2. Students match the adjectives and nouns face up on the table.
3. Ask groups to choose a group leader. The leader picks up the cards and shuffles them.
4. The leader places one card at a time face up in a stack on the table. All students say the word as its revealed.
5. When two consecutive cards make a collocation, the first student to bang their hand on top of the cards and say 'snap!' wins a point.
6. The winner is the first to five points.

Variation:

- You could use snap for any sort of vocab matching (e.g. synonyms, antonyms, picture + word).

7. Collocations from a text

Before class: Use a text students are going to read for understanding first. Make sure it has a number of relatively strong and useful collocations (e.g. *weak tea* rather than *expensive tea*).

1. First have students read the text for gist and detailed understanding.
2. Draw students' attention to an adjective + noun collocation in the text.
3. Ask them to underline any other adjective + noun collocations they can find.
4. Have students write a short text related to the topic using the collocations they found.

Variations:

- Students could find different types of collocation (e.g. noun + noun, adverb + adjective, verb + adverb etc).
- Students could use learner's dictionaries to do this – you could highlight the fact the dictionaries list common collocations.

Adverbs in -ly : GRAMMAR

1. Adverb role play

No preparation

1. In groups, students write a script for a role play based on functional language they have been studying.
2. They need to give directions to the performers in brackets, e.g. 'Customer (angrily): I want my money back!'
3. Students practise and perform their role play.

2. Adverb mime

see page 191 - Describing actions: -ly

Before class: Photocopy and cut up the cards on page 98. Create one set for each group of around 4-5 students. Keep A (action) and B (adverb) separate.

1. Divide students into groups.
2. In turn, each student takes two cards, one from A and one from B. They mime the action in the way the adverb says. The other students have to guess the complete sentence, e.g. 'You're playing pool badly'.
3. The student who guesses correctly keeps both cards.
4. The student with the most cards is the winner.

Adverbs of frequency : GRAMMAR

1. 'How often' survey

see page 195 - Adverbs of frequency

No preparation

1. As revision, elicit the adverbs of frequency on a cline (*never – hardly ever – sometimes* etc).
2. Draw a table on the board based on a topic students have been studying, for example:

name	eat fast food	eat out	cook

3. Ask students to copy the table in their notebook.
4. Students mingle, ask each other about the topics (*How often do you eat fast food? What do you have? How much does it cost?* etc), and take notes.
5. As a whole class, students share interesting information they found.

Extension:

- Students could write up a report on class preferences and see how they compare with the class average.

2. Interesting habits

see page 195 - Adverbs of Frequency

Before class: Note down ten more unusual habits you know some of your students will have (e.g. play a musical instrument, speak to someone from overseas, make something artistic).

1. Ask the class, 'How often do you do these?' and write the activities on the board.
2. Ask students to write five sentences about people in the class, and to include how often they think they do these things. Give an example: *Susan plays the drums every day.*
3. Students mingle and show their sentences to the people they wrote about, and check if they were correct.
4. As a whole class, students share the interesting information they found.

Giving advice : FUNCTION

1. Study advice

Before class: Photocopy and cut up the cards on page 99. Create one set for each group of around 4-5 students.

1. Tell the class they're going to help each other with their English.
2. Divide students into groups.
3. In turn, one student takes one card and tells the group their problem with learning English. Every other student has to give the best possible advice they can.
4. The student with the problem decides who gave the best advice, and gives that person the card.
5. The student with the most cards is the winner.

Extension:

- Follow this with a personalised discussion about solving students' real problems with learning English.

Variation:

- You could create a set of problem cards on any topic. Instead of using cards, students could write down three genuine problems they have; the person in their group giving the best advice wins a point.

2. Web advice

Before class: Find a website where people post questions looking for advice (e.g. answers.yahoo.com or wiki.answers.com). Students will need access to computers.

1. Show the class the website. Elicit its purpose (people ask questions and get answers).
2. Have students write their own questions.
3. Students share their questions (either on a class wiki, or on paper) and write advice for each other.

Extension:

- Students could actually post to a website and revisit to see what advice people have given.

Agreeing and disagreeing : FUNCTION

1. Number discussion

No preparation

If you want to ensure all students take part in any discussion (e.g. from the coursebook) this technique will make it more like a 'game' for students who feel awkward expressing their opinions.

1. Divide students into small groups.
2. Allocate each student a number.
3. Students must state an opinion in turn, according to their number, and say whether they agree or disagree with the earlier opinions.

2. Heads I agree, tails I disagree

Before class: Photocopy and cut up the cards on page 99. Create one set for each group of around 4-5 students. You also need a coin for each group.

1. Elicit the language for giving opinions, and agreeing and disagreeing.
2. Tell the class you want to know their opinions.
3. Divide students into groups. Give each group one set of cards face down.
4. One student picks up a card and gives an opinion on the topic.
5. Next, the student on their right tosses the coin. If it's heads the student must agree, and if it's tails they must disagree, even if it's not their real opinion. They also need to give a convincing reason why.
6. The group votes on whether it was a good answer. If so, the student keeps the card.
7. The student with the most cards is the winner.

3. Budget proposal

Before class: Find multiple copies of a shop catalogue related to vocabulary students have been studying (e.g. furniture, computers). You also need A3 paper and markers.

1. Tell students they are going to (for example) furnish the student lounge, or equip a new office for a business.
2. Divide students into groups.
3. Give each group a catalogue and one sheet of A3. Specify a budget they need to meet.
4. Students choose what they'll buy from the catalogue and draw their proposal on the sheet of A3. Students within each group need to agree.
5. Groups present their proposals.

4. Ranking ideas

Before class: Prepare a list of activities that students can do in a language class (e.g. listen to CDs, discuss in groups). Photocopy and cut up. Create one set for each group of around 4-5 students.

1. Tell students you want to know what they like doing in their English class.
2. Divide students into groups. Give each group a set of the activities.
3. Students need to rank the activities from most to least favourite. People within each group need to agree.
4. Have students move around the class and look at the other groups' rankings.
5. As a whole class, discuss people's ideas.

Variation:

- This could be any topic students have been studying

Animals : TOPIC

1. Animal discussions

Before class: Note down discussion questions.

1. Divide the class into groups of around 4-5.
2. Write up two or three discussion questions. Depending on the level and preferences of the group, these could be general and personalised:

   - What animals do you like? Why?
   - What would be your ideal pet? Why?
   - If you could be any animal, what would it be? Why?

   or more provocative:

   - Do people in your country treat animals well?
   - 'Meat is murder.' Do you agree?
   - Which animals would you never eat? Why?

3. Students discuss the questions in their groups.
4. Finish with whole-class feedback where students share some ideas their groups discussed.

Variations:

- Students may find it difficult to share opinions, either because they are not used to discussing controversial topics, or simply because they are not familiar enough with the language. Students can initially participate in a role play, e.g. as journalists interviewing factory farmers and as animal rights activists, before students give their real opinions (taking on a role is 'safer', and gives students a chance to practise with the language first).
- This could lead to a whole-class debate.

2. Animal talk

Before class: Note down the names of animals you're going to use.

1. Tell students they're going to learn animal sounds in English.
2. Say an animal sound (e.g. *Woof!*). Elicit the name of the animal and write (or draw) it on the board.
3. Point to the name of the animal. Students make the sound.
4. Put students in groups. Ask them to brainstorm the equivalents in their first language.
5. Finish with the whole class. Ask students to teach you their animal sounds.

Extension:

- Students could write and perform short dialogues between animals using a mixture of English and the animal sounds.

3. Amazing animals

Before class: Prepare a random montage of pictures of animals, their names, and interesting facts (without naming the animal, for example '*They can run at 100 kilometres per hour*'). Make one copy for each group of around 4-5 students.

1. Divide the class into groups of around 4-5.
2. Students have to match the elements in the montage.
3. Students share what else they know about the animals.
4. As a whole class, students share the interesting information they heard.

Variation:

- This could be any topic that interests students.

4. What animal am I?

No preparation

1. Divide students into groups. Have them brainstorm all the names of animals they've learned. Ask representatives from each group to write their list on the board.
2. Point to one animal. Elicit statements about the animal: *I'm (small)/I've got (sharp teeth)/I can (see at night)*. Elicit the questions: *Are you .../ Have you got .../Can you ...?*
3. Demonstrate the activity: 'I'm thinking of an animal. Ask me questions to find out what I am. I can only say yes or no'.
4. Students do the activity in groups.

Variation:

- This could be adapted for other vocabulary.

Apologising : FUNCTION

1. One-minute apologies

Before class: Photocopy and cut up the cards on page 100. Create one set for each group of around 4-5 students.

1. Demonstrate with a student to the whole class. Pick up a card that describes something bad you did (e.g. broke a window - role play the situation after the event, not the event itself). Students need to guess what happened. For example:

   Teacher: That glass will cost me fifty pounds!
   Student: I'm sorry. I'll pay for it.
   Teacher: What happened?

2. Students do the activity in groups.

2. Rewrite history

Before class: Think of a difficult dispute you had with someone.

1. Describe to the whole class the dispute you had. Ask for suggestions regarding what you should have done to make it better.
2. Act out the suggestions with a volunteer.
3. Divide students into pairs. Students discuss difficult situations they've been in and act out possible scenarios that would have made the situation better.

Making appointments and arrangements : FUNCTION

1. Practice partners

see page 235 - Present continuous 4

No preparation

1. Tell students they're going to find two other people to practise English with outside class.
2. Have students write out their diary for the week. Do an example on the board, including some free time:

	Mon	Tue
9.00	Class	Free
10.00	Class	See sister

3. Students mingle and find two other people who are free at the same time.
4. They should also arrange where they're going to meet, and what they're going to do.

2. Problem appointment

Before class: Think of a problem appointment you've experienced (e.g. the doctor didn't come).

1. Tell the class about your problem appointment.
2. Divide students into pairs. Ask them to share similar experiences.
3. Ask each pair to script one of the experiences.
4. Students practise and perform for the class.

Articles : GRAMMAR

1. A/the chain

see page 164 - Articles 1: a vs the

Each pair needs one blank piece of paper.

1. Divide students into pairs.
2. Give each pair one sheet of paper. Instruct one student to write.
3. Ask each pair to copy a short sentence containing 'a' (e.g. *I met a man*).
4. Students pass their sheet to the next pair. They write a follow-up sentence with '*the*' and '*a*' (e.g. *The man had a job*).
5. Elicit the reason why they're using '*the*' and '*a*' ('a' is new information, 'the' is something we've mentioned before).
6. This continues (*The job was in a bank...*) until it comes back to the original writers.

2. Recreate a text

see page 164 - Articles 1: a vs the

No preparation

1. Divide students into pairs.
2. Ask each pair to find a short section they liked in the coursebook.
3. Instruct one student to write. They should copy out the text, but change or omit five of the articles.
4. Pairs swap their texts and try to correct the articles.
5. Join the pairs together to discuss what they did.

3. Postcard

see page 164 - Articles 3: place names

Before class: Find a map that contains place names but without articles (e.g. *Thames*). Make one copy for each pair.

1. Review rules for '*the*' with place names (you could use the map to elicit these).
2. Tell students they're going on a trip. They need to write a postcard (or email) back home.
3. Instruct one student to write. They need to talk about where they've been to, and what the places were like. Remind them to use 'the' correctly.

Extension:

- Students talk about where they really want to travel to, and why.

The arts : TOPIC

1. I know what I like

Before class: Find examples of varied and controversial visual art. Create one handout for each group of around 4-5 students.

1. Divide the students into groups.
2. Students describe the artwork. Elicit the language and write it on the board.

3. Students then discuss what they like and don't like, and why.
4. Students share ideas as a whole class.

Extension:
- Students role play art dealers and buyers and try to sell the works.

2. Posters

Before class: Bring A3 paper, coloured markers, paint etc. Find a public place where you'll be able to display students' work.

1. Divide the students into small groups.
2. Students create an artistic poster on a topic they've covered in class.
3. Display the posters somewhere prominent.

B Business : VARIOUS

1. Business cards

Before class: Bring one blank or completed card for each student.

1. You (or the students) create business cards with imaginary names and titles.
2. You can use these for role plays throughout your Business English course.

Adopting a role reduces students' inhibitions (especially if your students work for the same company!).

2. Dream interview

No preparation

1. Ask the class what would make a dream job interview. Elicit/suggest 'it would be ideal if you could write your own questions'.
2. Ask students to write down five things they're very proud of in their professional lives (e.g. *10 years' experience, fluent in three languages*).
3. Ask students to write the questions that match these answers on a different sheet of paper (e.g. *How much experience do you have? How many languages do you speak?*).
4. Divide the students into pairs. Students swap their questions, and ask and answer them.
5. Students share interesting information they found out about their partner with the whole class.

3. Find your replacement

Before class: Find an advertisement and a short position description for a job in the same field as your students.

1. Analyse the language and vocabulary in the advertisement and position descriptions.
2. Tell students they're being sent to a different country for a year, and they have to write an advertisement with a position description for someone to replace them.
3. Students write an advertisement and a position description for their own job.
4. Post the ads and job descriptions on the walls.
5. Students find a job that interests them (make sure someone chooses every job) and write an application.
6. Each student then reads the application and interviews the candidate for their position.

4. Presenting the company

Before class: Note down presentation guidelines and criteria for audience feedback. If possible have access to a data projector.

1. Tell students they need to present their company at an important meeting.
2. Divide students into pairs. They should either work for the same company, or have to invent one. Each pair needs to think of a real-life meeting where they would need to present, and what sort of presentation they would need to give (e.g. an overview of the company, or a proposal for maximising profit).
3. Students script their presentation and create visuals and props.
4. Students do their presentation. The audience should make notes on the performance according to criteria (e.g. pronunciation, clarity, interest, non-verbal language).

Extension:
- You can video the presentation and have presenters evaluate their own performance using the same criteria.

5. Business meetings

Before class: Photocopy enough role play cards on page 101 for each student.

1. Divide students into groups of around five. Four students will be employees, and one student will be the company director.

2. Tell the class that each employee will propose an idea to motivate staff. The director will run the meeting.
3. Give each student a role play card. Employees have four minutes to prepare their argument. The director should note down phrases they can use to facilitate the meeting.
4. Ask the directors to begin. Each meeting will need to come to a decision.
5. Students share their decisions with the whole class.

6. In the office

No preparation

1. Elicit several annoying things that workmates can do: 'not wash up cups', 'talk loudly on the phone' etc.
2. Divide students into pairs. Ask each pair to brainstorm a list of other bad habits.
3. Elicit polite phrases to ask people to change their behaviour, e.g. *I wonder if you could, I'm sorry to mention this, but...*
4. Students prepare and practise model exchanges as examples of how to change someone's behaviour.
5. Have pairs perform for the class.

7. Inappropriate behaviour

Before class: If possible, find images or video of appropriate and inappropriate non-verbal language or dress.

1. Introduce topic of appropriate and inappropriate behaviour in the business world (with visuals if possible). Elicit several ideas as a whole class.
2. Divide students into groups. Students brainstorm dos and don'ts.
3. Each group plans a short simulation of a business meeting. Each student will display one of the behaviours they discussed.
4. Groups perform for the whole class. The audience identifies the inappropriate behaviour.
5. Finish with a whole-class discussion to summarise advice for businesspeople.

Extension:

- In the whole-class discussion, address issues of appropriateness in different cultures, and whether norms are changing.

1. Urban solutions

Before class: Note down words, or find pictures of urban problems (e.g. *overcrowding, crime*). Bring one sheet of A3 for each group.

1. Elicit some examples of urban problems.
2. Divide students into groups. Give one sheet of A3 to each group.
3. Draw a grid on the board for students to copy:

Problem	Solution

4. Students work together to make notes in the table.
5. Ask a representative from each group to fill in the grid on the board.
6. Finish with a whole-class discussion. Will these solutions happen? Do the advantages of city life outweigh the disadvantages?

Extension:

- Students can write up the ideas for homework.

2. Design a city

Before class: If possible find pictures of attractive and innovative cities. Bring one sheet of A3 for each group.

1. As a class, discuss what makes a city a good place to live (use visuals if possible).
2. Divide students into groups. Give one sheet of A3 to each group.
3. Students need to draw a model city. They should label their diagram to explain the purpose of different features.
4. Put the designs on the wall and let students look at them. Groups should be prepared to explain their ideas. Possibly get students to vote on the best design.

Variation:

- This activity can be used for any sort of design ideas.

Clothes : TOPIC

1. Fashion magazines

Before class: Bring one fashion magazine for each group.

1. Show the class one of the magazines. Ask whether they like fashion (if you have reluctant students you can reassure them they will be able to express their opinion).
2. Divide the class into males and females, and then into smaller groups. Give each group a magazine.
3. Ask them to discuss what they like and don't like.
4. After, join one group of males with one group of females.
5. Ask them to exchange their ideas.
6. Finish with a whole-class discussion on whether males and females had similar or different opinions.

2. Find your twin

Before class: Find two copies of the same clothes catalogue or fashion magazine. From one magazine, cut out pictures of people (if possible, the people you choose should be wearing similar clothing to increase the challenge in the activity). In the other magazine, cut out the same pictures so you have a matching pair for each. Paste each picture onto an individual piece of card.

1. Elicit language used to describe clothing and people's appearances.
2. Give one card to each student. Tell them not to show anyone.
3. Tell students they need to find their twin by describing the person only.
4. Students mingle and find their matching picture.
5. After, each pair sit down together and write a description of their person.

3. Body swap

Before class: Find a clothes catalogue or old fashion magazine. Cut out pictures of people so there is one for each student (the pictures should be as large as possible). Paste each picture onto an individual piece of card. Cut each piece of card in three to make a head, a torso and a pair of legs.

1. Give each student an un-matching set of head, torso and legs. Tell them not to show anyone.
2. Tell students something has gone terribly wrong, and they need to find a body that matches.
3. Students mingle and swap parts of their body, but only by describing them.

4. Clothes and culture

No preparation

1. Have students give a presentation on clothing in their country (this can engage male students who might be reluctant to talk about fashion).
2. Stress to students that you are very interested to find out about your students' culture(s).

Comparative adjectives : GRAMMAR

1. Comparing countries

see page 183 - Comparatives: -er and more

Preparation: Bring scrap blank paper.

1. Put students into pairs.
2. Get students to compare their own country with another: *Italy has better food than Spain, Italians are more romantic than English people* etc. The more controversial, the better. Ask them to make notes as they go along (e.g. *better food*).
3. Give a piece of paper to each pair. Ask them to write AGREE in big letters on one side and DISAGREE on the other.
4. Join up pairs to make groups of four. Ask them to share their opinions. The other pair has to listen and hold up their paper showing either AGREE or DISAGREE.

2. Adjectives in a bag

see page 183 - Comparatives: -er and more

Before class: Write adjectives on separate pieces of paper. Make one bag of adjectives for each group.

1. Divide students into groups. Give each group a bag of adjectives.
2. Students have to put their hand into the bag and make comparisons using the adjective, e.g. *'pretty'→ Christina Aguilera is prettier than Britney Spears.*

4. Comparative brainstorm

see page 183 - Comparatives: -er and more

Before class: Bring a range of realia or visuals.

1. Divide the class into pairs.
2. Instruct one person in each pair to write.
3. Show the class two unrelated visuals or objects (e.g. a banana and a calculator). Tell the class they have three minutes to write down as many comparisons as possible (e.g. *a banana's cheaper than a calculator*).
4. The pair with the most valid comparisons gets a point.
5. The winning pair is the first to three points.

Variations:
- Allow students to choose the objects, and to score. Have students check each other's sentences and challenge any suggestions.
- Do this on the board – split the class into two teams facing the board, hold up the two objects and one member from each team has to run to the board and write a sentence (they get a point if it's correct; the winning team is the first to get five points).

Complaining : FUNCTION

1. Neighbours

No preparation

1. Draw a block of apartments on the board. Highlight apartments 10A and 10B. Elicit who they are ('neighbours') and what problems they might have.
2. Divide the class in two. Half are from Apartment 10A: they are professional musicians who practise at home. Half are from Apartment 10B: they have a dog locked up in the apartment who barks a lot.
3. Put the musicians together to brainstorm why they need to practise at home. Put the other people together to brainstorm why it is their right to have a dog in the apartment
4. Set up the students in two concentric circles: 10A in the middle facing out, and 10B on the outside facing in.
5. Students need to talk to the person facing them. Use a bell (or something similar) to signal two minutes when the outer circle moves around one.
6. Students must complain to each other and try to reach an agreement.
7. As a whole-class discuss who found the best solution and why.

2. Win-win

No preparation

1. Start with a whole-class discussion on complaining. *When did you last complain? What types of businesses are particularly bad? What happens when you complain?*
2. Divide students into pairs.
3. Each pair needs to think of a typical situation where people complain (or use one from experience) and write a dialogue. The dialogue needs to have a happy ending for both people.
4. Students practise and present their dialogue.

Conditionals : GRAMMAR

1. Facts

see page 256 - zero conditional

No preparation

1. Divide students into pairs.
2. Each pair writes down five scientific facts starting with 'If you ...' (e.g. *If you heat water to one hundred degrees, it boils*. Four statements should be true, and one false.
3. Students mingle. They have to listen to other pairs' statements and work out which one is false.

2. Decisions, decisions

see page 257 -1st conditional

Before class: Note down as many options as you can think of for an event for the class (e.g. *park/beach, bus/train, picnic/restaurant*).

1. Tell the students they're going to plan for an upcoming event (e.g. an end-of-term party or excursion).
2. Write the options for the event on the board (*park/beach* etc).
3. Divide students into groups.
4. Students have to discuss the options using 'If we...' (e.g. *If we go to the park we'll be cold*).
5. Finish with a whole-class discussion and make concrete plans for the event.

3. Imagination chain

see page 259 - 2nd conditional

No preparation

1. Ask students to stand in a circle.
2. Elicit an ending to the statement *If I won a million dollars I'd...* (e.g. buy a new house).
3. Elicit a follow-on statement e.g. *If I bought a new house I'd live there*.
4. Have students continue the chain in the same way (*If I lived there...*).
5. Once students are confident with the structure, divide them into smaller groups to continue.

Variation:

- When in smaller groups, students can set a time limit for each response to increase the competitive edge.

4. What would I do?

see page 259 - 2nd conditional

Before class: Note down a list of prompts for hypothetical situations (e.g. *have a million dollars, be the opposite sex*).

1. Write up prompts for amazing and unlikely events (have a million dollars etc).
2. Elicit an example of what someone would do if it was true (using second conditional) e.g. *If I had a million dollars I'd have parties every night*.
3. Ask students to note down two opinions for each prompt, one what they really think, and one which is a false opinion (e.g. *give it to charity, buy a sports car*).
4. Put students into groups. Each person says their two sentences, and the other students guess which is their real opinion.

5. If I was famous

see page 259 - 2nd conditional

Before class: Bring a visual of a famous living person.

1. Show students the picture of the person. Elicit who they are and what they do.
2. Divide students into groups. Students discuss what they'd do if they were this person.

Variation:

- Bring in visuals of a number of famous people.

6. Fixing regrets

see page 260 - 3rd conditional

Before class: Note down a regret you can use as an example.

1. Introduce the topic of regrets, but also how bad decisions can turn out well (for example, 'I dropped out of medicine and my parents were really unhappy. But then someone told me to become an English teacher. That was the best decision I ever made').
2. Have students write down three regrets. Model an example using the third conditional (e.g. *If I'd stayed at home I would have got married*).
3. Ask students to memorise their sentences and close their books.
4. Students mingle and share their regrets. Other students have to give them good advice to make sure it will turn out well.

Conjunctions : GRAMMAR

1. Complete the sentence

see page 208 - Conjunctions and prepositions

Before class: Note down conjunctions you want students to practise (e.g. *and, but, despite the fact that*).

1. Put students in pairs.
2. Ask each pair to write ten short sentences that are true about them (e.g. *Van likes music*).
3. Instruct each pair to swap their sentences with another pair.
4. Write conjunctions on the board you would like students to practise.
5. Each pair needs to add a second half to each sentence, using one of the conjunctions on the board (e.g. *Van likes music and he plays in a band*).
6. Join the pairs together. Students discuss whether the sentences are correct, and also whether the information is true.

2. As long as possible

see page 208 - Conjunctions and prepositions

Before class: Note down short sentences for the board. Bring blank paper.

1. Divide students into pairs. Give each pair a piece of paper. Instruct one student to write.

2. Write a short sentence on the board (e.g. *Milk is white*) and ask students to copy it.
3. Ask students to add anything they like on the end of the sentence (e.g. ...*and I like it*).
4. Students pass their paper to the next pair. They need to add something else to the end to continue the sentence (e.g. ...*but it's expensive*). Continue until the paper reaches the original writers.
5. Finally as a whole class discuss how it could be broken into natural sentences.

Maintaining a conversation : FUNCTION

1. Start a sentence

Before class: Copy and cut up one set of cards on page 102 for each group.

1. Divide students into small groups. Tell students they are going to have a conversation.
2. In turn they pick up a card and have to use it to start the next sentence.
3. The challenge is to maintain the conversation as long as possible.

Variation:
- You can also specify the topic students must talk about.

2. Don't make me laugh

Before class: Note down a list of potentially funny or embarrassing topics.

1. Divide students into pairs.
2. Write one topic on the board.
3. Tell students they need to talk about the topic and to try to make the other student laugh. The person who manages not to laugh gets a point.
4. Write up a new topic from time to time.
5. The winner in each pair has the most points.

Extension:
- To finish, have several confident and high-scoring students try in front of the whole class.

Countries : TOPIC

1. How's your geography?

Before class: Prepare a map of a country or continent, but blank out the names (you could also add facts and figures and blank out key information). Make one copy for each group.

1. Show the map to the whole class. Elicit one or two facts.
2. Divide students into groups. Give one handout to each group.
3. Have students work out the obscured names and information.
4. Join groups together, or allow students to move about freely, to check what they're not sure about.
5. Discuss as a whole class.

2. Country presentation

No preparation

Have students research and give a presentation on a country of their choice.

The countryside : TOPIC

1. City and country lives

Before class: Bring visuals of a city and a country person.

1. Use the pictures to elicit how the two people's lives are different.
2. Divide students into pairs.
3. Students discuss and draw the daily routines of each person.
4. Join pairs together. Students show each other their drawings and describe their routines.

Variation:
- This could be used to describe routines of different jobs, people in different countries etc.

2. New life

Before class: Photocopy and cut up one set of statements on page 102 for each group.

1. Discuss as a group some of the advantages and disadvantages of living in the city and the country.
2. Divide students into groups.
3. Give one set of statements to each group.

4. In turn, one student picks up a statement card and reads it out (e.g. *I might move to the country to have a quiet life*). The other students have to convince them either to do it or not to do it.
5. The student gives the card to the student with the strongest argument.
6. The winner is the student with the most cards.

Variation:

- This activity could be used for practising opinion language on any topic.

Crime : TOPIC

1. The punishment fits the crime

Before class: Photocopy and cut up one set of crime cards on page 103 for each group.

1. Discuss as a group what sorts or punishments are often given for what crimes. Elicit key vocabulary and write it on the board.
2. Divide students into groups.
3. Give one set of crime cards to each group.
4. Students rank the crimes from least to most serious.
5. Ask students to decide what punishments are appropriate for each crime.
6. Finish with a whole-class discussion.

2. Crime controversies

Before class: Note down statements for discussion.

1. Divide the class into groups of around 4-5.
2. Write up two or three statements for discussion. Choose statements that are appropriate and interesting for your group (some students may be uncomfortable with the topic). They might include:

   - An eye for an eye.
   - Crime is the product of an unjust society.
   - It's not the penalty, it's the chance of being caught.

3. Finish with whole-class feedback where students share some ideas their groups discussed.

3. Crime fiction

Before class: Find pictures related to crime that are potentially related. Paste the pictures in random order onto a sheet of A4. Make one copy for each pair.

1. Divide students into pairs.
2. Students need to write a story that connects all the pictures.
3. Join pairs together. As one pair tells their story, the other pair has to point to the relevant picture.
4. Ask students to share any good ideas with the whole class.

Extension:

- Students may like to act out their stories in front of the class with a narrator.

Culture : TOPIC

1. Gesture discussion

Before class: Bring pictures of gestures in different cultures.

1. Use visuals to introduce the topic. Elicit 'gestures'.
2. Divide students into groups.
3. Ask groups to discuss gestures in their own country, and what they know about other cultures.
4. Finish with a whole-class discussion.

Extension:

- Students could write a guide (e.g. in poster form) to gestures in their own country and English-speaking countries.

2. Culture discussion

Before class: Note down discussion questions.

1. Divide the class into groups.
2. Write up two or three discussion questions. Depending on the level and preferences of the group, these could be general and uncontroversial:

   - What are the differences between two cultures you know about?
   - Should we protect traditional cultures?
   - What are the most important things a visitor to another country should think about?

or more provocative:

- Are some cultures simply wrong?
- Will some cultures never get on?
- Should immigrants change their beliefs?

Some topics are very sensitive so exercise caution.

3. Finish with whole-class feedback where students share some ideas their groups discussed.

D Asking for directions : FUNCTION

1. Around my city

Before class: Create two versions of a map of the students' city, 'A' and 'B'. 'A' should have certain destinations marked (e.g. 'bookshop', 'supermarket') while 'B' does not. Copy 'A' for half the students in the class and 'B' for the other half.

1. Divide the class into pairs. One student is 'A' and one 'B'.
2. Tell the class student B has to get to certain destinations around the city. Write the places from Map A on the board.
3. Hand out the maps. Tell students not to show anyone. Student B needs to ask student A for directions and mark the route on the map.
4. Student A and B compare maps to see if student B found the places correctly.

2. Classroom city

Before class: Write city features (e.g. 'library', 'shop') on sheets of A4. Bring blindfolds for half the students.

This activity needs movable desks or tables, and students who are happy to be blindfolded!

1. Show the class your signs. Say, 'You're going to visit these places in a new city'.
2. Move the desks to set up the room like a city, with 'streets' between the desks.
3. Divide students into pairs. One student in each pair needs to put on a blindfold.
4. Attach the signs around the walls.
5. The blindfolded student needs to ask their partner for directions and try to reach the destinations.
6. After a while, swap blindfolds, and change the location of the destinations.

Using drama : VARIOUS

1. Invented language

Before class: Note down a dramatic scenario for each pair to act out (e.g. 'bank robbery', 'love at first sight').

1. Put students in pairs. Quietly tell them the scenario which they are going to act out.
2. Tell students to use an invented language, not English. They still need to use English stress and intonation as if they were really talking to one another.
3. The rest of the class watch and guess what the situation is.

Extension:
- Students could write out the real dialogue in English for one of the scenes.

2. Future you

No preparation

1. Elicit some personal information questions from the whole class (e.g. *What do you do? How old are you? What are your hobbies?*). Write them on the board.
2. Announce it's now 20 years in the future.
3. Students mingle and ask each other the questions, answering as if it is twenty years in the future.

3. Character swap

Before class: Note down ideas for character swaps.

1. Ask students to stand up. Divide them into pairs.
2. Tell students to have a conversation about what they do in their free time.
3. Every couple of minutes, make a signal (e.g. ring a bell) and write a new character on the board, e.g. 'teachers', 'children', 'opposite sex'. Students have to maintain the conversation but pretend to be the new character.

E Education : TOPIC

1. School system

Before class: It's worth having some knowledge of your own and your students' school systems in case students ask questions. Bring one sheet of A3 for each group.

This works best with students of one nationality.

1. Draw a chart of your country's education system on the board. Try to elicit the details. For example:

18+		
12-17	secondary	
5-11	primary	
3-4	kindergarten	play games, ...

2. Divide students into groups. Give each group a sheet of A3.
3. Students draw an equivalent chart of education in their own country.
4. Students look at other groups' charts and discuss any disagreements.

2. Education controversies

Before class: Note down statements for discussion.

1. Divide the class into groups of around 4-5.
2. Write up two or three statements for discussion. Choose statements that are appropriate and interesting for your group. They might include:

   - Students should choose a course that will make them money.
   - Traditional teaching works best.
   - Private schools are unfair.

3. Students discuss the statements in their groups.
4. Finish with whole-class feedback where students share some ideas their groups discussed.

3. Successful student

Before class: Photocopy and cut up one set of statements on page 103 for each group.

1. Discuss as a group different strategies for learning English.
2. Divide students into groups.
3. Give one set of statement cards to each group.
4. Students rank the strategies from most to least effective.
5. Finish with a whole-class discussion. Suggest that different strategies might suit different students, and ask students to identify techniques that might suit them best.

Environment : TOPIC

1. Causes and effects

Before class: Create a random montage of pictures of environmental causes and effects (e.g. polluted river, factory etc). Make one copy for each group.

1. Divide students into groups.
2. Give one handout to each group.
3. Students discuss the causes and effects in the pictures.
4. Finish with a small-group or whole-class brainstorm of possible solutions to the problems.

Extension:

- Students could write about the causes, effects and solutions.

2. Development role play

Before class: Find a colour picture of an unspoilt natural environment. Draw a map of an island showing major towns and natural features (rivers, forests etc). Draw pictures of factories/houses etc over an environmentally significant area to show where development is being planned. Make enough copies for every third student.

1. Show the class the picture. Ask, 'Where's this?'. Announce 'it's the Island of...' (Give your made-up island a name).
2. Divide the class into three. Explain the groups are developers, environmentalists, and the island government. The developers want to make changes to the island; the environmentalists want to stop them. They have to persuade the government because in the end, the government members will vote.
3. Divide each of the three groups into smaller groups. Give each small group a copy of the map.

4. The developers and the environmentalists need to prepare their arguments. The government members need to brainstorm both advantages and disadvantages of any development on the island in preparation for discussions.
5. Set up the room with meeting areas and separate the class into smaller groups to discuss.
6. In the end bring all the government members together to vote on the proposal.

3. Nature controversies

Before class: Note down statements for discussion.

1. Divide the class into groups of around 4-5.
2. Write up two or three statements for discussion. Choose statements that are appropriate and interesting for your group. They might include:

   - People's needs come first.
   - We need to change our lifestyles or the earth will die.
   - Developing countries have the right to pollute, just as the West does.

3. Finish with whole-class feedback where students share some ideas their groups discussed.

F Family : TOPIC

1. Different language, different family

Before class: Prepare a generic family tree (complexity depends on level). You need one A4 sheet for each pair.

This works best with students of one nationality who have different kinship terms in their own language to English.

1. Draw the family tree on the board. For example:

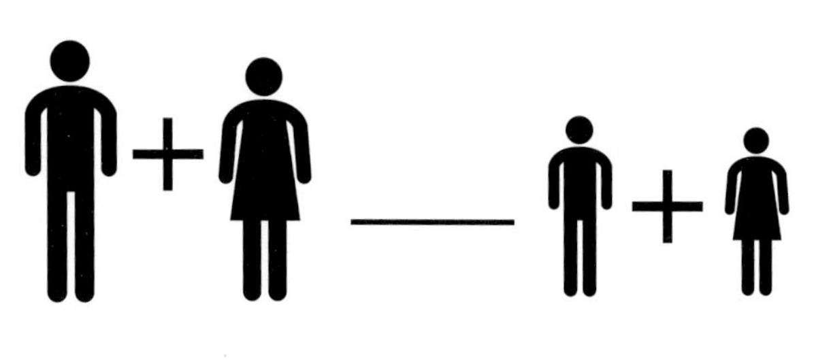

2. As you draw, elicit the English words. For example: 'Kate is Bill's what?'. Write '*daughter-in-law*' on the side of the board.
3. Divide students into pairs. Give each pair a sheet of A4 paper.
4. Ask students to draw a family tree in their own country, and list their relationship to each member down the side of the page.
5. Invite a couple of students to use the board to explain the relationships to you and the class.

Variation:

- This method can be used whenever the students' first language significantly differs from English, such as in terms of addresses or colours.

2. Family opinions

Before class: Note down the start of sentences you plan to use.

1. Write up the start of an opinion about families, e.g. 'Parents should always...'. Elicit some ideas from the class.
2. Write up some more half sentences, for example:

   - Children should always...
   - I think grandparents...
   - In my opinion, arranged marriages...

3. Ask students to finish the sentences individually, according to their own opinion.
4. Form groups of 4-5 people.
5. Students share and give reasons for their opinions.
6. Finish with the whole class. What did most students agree about? What created the most discussion?

Famous people : TOPIC

1. Interview a famous person

Before class: Bring pictures of several pop stars/ politicians/actors that your students will know. Consider preparing a table for students to take notes on, for example:

name	childhood	now	other details

1. Elicit the names of some famous people using the visuals.

2. Tell students, 'Today some of you will be famous. And some of you are journalists. You're going to interview them'.
3. Put students into groups of 2-3 and assign the role of interviewers and a famous person.
4. Give the students time to prepare their interview questions and answers.
5. Once they have finished they have to conduct the interview in front of the class. Get the other students in the class to make notes (possibly using a pre-prepared sheet).

Variation:

- Consider raising the challenge for students as they listen by choosing people at random to ask follow-up questions.

2. Famous pairs

Before class: Bring pictures of famous people or characters who are associated with another person (e.g. Victoria Beckham, married to David Beckham; Batman, with partner Robin). Think of an unlikely scenario for the writing activity (e.g. 'buying groceries at the supermarket').

1. Use the visuals to elicit the names of the people or characters.
2. Ask students to tell you who their partners are.
3. Divide students into groups of two or three. Students discuss what they know about the famous pairs.
4. Students write a dialogue between the two people in an unusual scenario (e.g. going shopping, meeting the other's parents for the first time etc).
5. Choose one or two pairs to act out their dialogue for the class.

3. Celebrity heads

Before class: Note down names of famous people your students will know.

1. Divide the class into two teams.
2. Place two chairs at the front of the class facing away from the board.
3. Ask one student from each team to sit on one of the chairs.
4. Write the name of one famous person in large letters on the board.
5. In turn, the two students ask their team a yes/no question (e.g. '*Am I male? Am I an actor?*'). If the answer to their question is 'yes' they can ask another question.
6. The team scores a point when they get a correct answer. The team losing the point has to replace their student at the front.
7. The team to score five points is the winner.

Variation:

- You can create 'hats' from paper with the names written on them. The students in the chairs wear them and guess two different names.

Practising fluency : VARIOUS

1. Simon says

No preparation

1. Teach the class the rules of 'Simon says' (Students must do what you tell them to do, but only if you begin with 'Simon says'. Anyone who gets it wrong drops out of the game).
2. Demonstrate with examples more challenging than 'Touch your...': for example, '*Sing a song in English*' or '*Pick up something that belongs to someone else*'.
3. After students show they understand what to do, divide the class into groups. Students take it in turns being Simon.

2. Topic cards

Before class: Create a list of topics appropriate for your class (e.g. MUSIC/PETS ... or GLOBALISATION/ CORRUPTION ...). Copy and cut up a set for each group.

1. Divide the class into groups.
2. Give each group a set of topic cards face down.
3. Appoint a time keeper in each group. In turn, one student picks up a card and needs to talk about it for thirty seconds without pausing.
4. If a student is unsuccessful they drop out of the game.
5. The winner is the last remaining student.

3. Topic corners

Before class: Write four topics of interest to students on four pieces of card.

1. Post one topic on the wall in each corner of the room.
2. Tell students when they're in a corner they have to discuss the topic on the wall.

3. Invite students to go to the corner of their choice and start talking.
4. Every few minutes, give a signal (e.g. ring a bell or play music). Students have to move to a different corner and talk about the new topic.

4. Topic board game

Before class: Create a board game on any topic students have been doing that looks like this:

Make one copy for each group. Bring die and counters.

1. In turn each student throws the die and moves their counter.
2. When they land on a topic, they have to talk about it for thirty seconds without pausing (another student should keep time).
3. If they are unsuccessful, they go back to their previous position. The winner is the first student to reach the end.

Variations

- To keep the other students involved when people speak, they could ask a follow-up question. If the student can't answer, they have to go back to their previous position.

5. Don't say it

Before class: Copy and cut up one set of words on page 104 for each group.

1. Divide students into groups.
2. Give each group a set of cards, face down.
3. In turn each student must turn over a card. The other students have to ask the person questions to try to make them say the word.

6. Ripple questions

Before class: Note down 10 sorts of questions students have been practising in class.

1. Ask students to form a circle. Stand as part of the circle.
2. Ask the student on your right your first question. They answer, then ask the person on their right, and so on, as the question moves around the circle.
3. After a few seconds, ask the student on your right your second question. Then repeat with your other questions.
4. When you've finished your list of questions, move out of the circle.

Variation:

- This can work equally well if students are seated in rows. You'll need to ask the questions to the student at the end of each row.

7. Words on the board

No preparation

1. Ask students for a topic they'd like to talk about (e.g. 'sport'). Write it on the board.
2. Elicit ten words related to the topic (e.g. 'football', 'Olympics'). Write them on the board.
3. Divide the students into groups.
4. Students need to maintain a conversation for as long as possible, but they must use one of the words on the board in every sentence.

Food : TOPIC

1. Food quiz

Before class: Create a food quiz. While it should be challenging, it should be related to food your students will know something about. Write ten multiple choice questions with the answer highlighted. For example:

1. Where do tacos come from?
a Colombia.
b Mexico. ✔
c Spain.

Also, create another version of the quiz for the final. Make one copy of the first quiz for each group, and just one copy of the final quiz for the whole class.

1. Divide students into groups.
2. Ask each group to choose a quizmaster or quizmistress. They will ask the questions and keep score.
3. Give the quizmaster/mistress the quiz. They then read out one question and all three possible answers to their group. The first student to say the right answer gets a point.

4. Bring the winner from each group out to the front of the class. Ask for two volunteers to be quizmaster/mistress and scorer. Use the final version of the quiz to determine the class winner.

Variation:
- This activity could be used for any topic.

2. Guess from pictures

Before class: Make a montage of different types of food. Copy one for each group.

1. Divide the students into groups.
2. Give one handout to each group.
3. In turn, one student thinks of one type of food from the montage. The other students, in any order, ask questions to find out what food the student is thinking of (e.g. *Are you a vegetable? How much do you cost?*). However, they cannot mention the name (e.g. *Are you a banana?*) or ask directly (e.g. *What are you?*).
4. When a student is ready to guess, they have to say 'You're (a banana)'. If they're correct, they score a point. If they're wrong, they forfeit the round.
5. The winner is the first person with five points.

Variation:
Bring in lots of real food (consider fruit, vegetables, packets, tins, sauces, spices etc). If the class is small enough, have all the students gather round one table.

3. Good food guide

No preparation; just decide what format the guide will be in, and ask for permission if necessary.

1. Tell students they're going to create a good food guide to the local area (it could be a poster, a booklet or a webpage).
2. Ask the class to decide who the primary market will be (e.g. cheap eating places for students? Traditional food for tourists?).
3. Divide students into groups. Ask each group to choose one person to write. Groups brainstorm what they know about local places to eat.
4. Ask the class for a volunteer to be chief editor (or several volunteers to share the role).
5. The chief editor (with your support) organises the design, writing and production process. You may want to dedicate a set time each day or week for work on the guide.
6. Make sure the final product is displayed or distributed.

Variation:
- You could organise a similar project producing a cookbook of students' recipes.

4. Plan a party

No preparation; you may need permission from the school.

1. Plan a real-life party (or picnic or barbeque) with your students.
2. Agree on a budget and plan the dishes and what ingredients students will need to buy.
3. Appoint several people as organisers – when having the party, ask them to ensure people use English throughout.

Free time : TOPIC

1. Find someone who ...

Before class: Design a handout which reflects what you know your students do (copy one for each student). For example:

> Find someone who ...
> - practises guitar after school ___________
> - always wins online games ___________
>
> etc.

1. Tell students 'You're going to find out what your classmates do outside class'.
2. Elicit an example of a question students will need to ask ('*Do you practise guitar after school?*').
3. Students mingle and ask the questions. When someone answers 'yes' they write their name in the space. They can ask follow-up questions, but don't need to make notes.
4. Finally, bring students together as a class. Ask them to share anything interesting they heard.

2. Pie chart interview

Before class: Note down a rough pie chart of your typical day (e.g. showing '33% working', '5% watching TV', 30% sleeping' etc).

1. Introduce the topic by drawing your pie chart on the board. Elicit as you go: 'What do you think this is?' etc.
2. Divide the student into pairs.
3. Students ask each other about their days and draw each other's pie charts.
4. Finish with a whole-class discussion about what people have in common or not in common.

Future forms : GRAMMAR

1. Prediction pyramid

see page 249 - will *future 2*

Before class: Bring pictures showing predictions of the future.

1. Use visuals to introduce the topic (this could be 'the future' in general, or limited to cities, the economy, the environment etc).
2. Divide students into groups.
3. Each group discusses and writes down five predictions for the future.
4. Join groups together. They have to agree on the five most likely predictions between them.
5. Finally have representatives of the groups come to the board. The whole class has to agree on the five most probable predictions.

2. The restaurant with no food

see page 247 - will *future 1*

Before class: Find or create a simple restaurant menu. You'll need one copy for each group of three.

1. Use visuals on the board to elicit the fact we're in a restaurant.
2. Ask, 'How do you ask for something in a restaurant? Elicit the formulas *'I'll have...'* and *'I think I'll have...'*
3. Set up the room with restaurant tables.
4. Divide students into groups of 3. Assign one person in each group to be the waiter; the others are customers.
5. Ask the waiters to follow you out of the room (so the customers cannot overhear). Give each waiter a menu. Tell the waiters that the manager wants to promote a certain dish (choose an unlikely dish on the menu) and will give you a 100 dollar/pound/euro etc bonus if the customers order it. They have to try to make customers choose it.
6. Return to the room. Ask the waiters to take customers to their table and take their order.
7. To finish, reveal to customers why their waiter was being difficult!

3. Shared plans

see page 250 - going to *future 1*

No preparation

1. Ask students why they're studying English. After some free discussion, ask 'How do I say I have a plan?'. Elicit the structure *I'm going to (work in the US)*.
2. Ask students to write down two plans they have after their English course starting with *'I'm going to...'* They should then memorise the two sentences and close their books.
3. Ask students to stand up and find someone with similar plans to yours.
4. When most students have found a partner, get students to sit with them (you can pair up the other students) and discuss what they'll do to achieve their plans. Can they offer any good advice?
5. To finish, students share some ideas with the class.

4. DVD drama

see page 252 - going to *future 2*

Before class: Arrange a DVD player and screen. Bring a DVD with obvious dramatic moments (a TV soap opera is ideal).

1. Divide the students into pairs.
2. Play a short extract of the DVD, and pause it just before the drama reaches a climax (for lower-level students you can do this with the volume off).
3. Ask pairs to discuss what happens next: *She's going to shout at him.*
4. Press play to show students if they were right.
5. Repeat with other sections of the DVD.

Variation:

- Instead of using a DVD, pairs can plan and then act situations, pausing just before the dramatic climax.

G Greeting people and saying goodbye : FUNCTION

1. New roles

Before class: Note down situations you're going to use.

This is for practising different levels of formality.

1. Set the scene for a party (e.g. draw balloons and a table of drinks on the board; play some music).
2. Tell students, 'You're at the party. Go and say hello to people'. Students mingle and greet each other.
3. Assign students as A or B. Say, 'Student A, you're students. Student B, you're teachers. Go and say hello'. Students need to change the formality of their language.
4. Repeat with different variations (an elderly person, a child, the King or Queen etc).
5. Discuss with the whole class any language difficulties that arose during the activity.
6. Have students try again in pairs.

Variation:
- Once students are used to the activity, you can write the variations on the board rather than interrupting students. You can also change the situation as well as the participants.

2. Musical goodbyes

Before class: Bring an upbeat music CD.

1. Instruct students to walk around the room and talk to people when the music plays.
2. Play the music. Students mingle.
3. Pause the music. Tell students they have to end the conversation politely, and move on.
4. Repeat several times.
5. Discuss with the whole class any difficulties students had breaking off a conversation.
6. Have students try again in pairs.

3. Hello teacher

No preparation

This works best with a class of one nationality.

1. Divide students into pairs. Each pair writes a dialogue, in their first language, meeting a teacher in the corridor.
2. Ask students to translate their dialogue, word for word, into English.
3. Join pairs together to share their translations.
4. Discuss with the whole class what translates well, and not well, into English (greetings may have very different social conventions, e.g. whether a title is necessary or not).
5. If they wish, students can rewrite and practise their dialogues (just make sure students do not get the impression you think their first language is 'wrong'. If you speak the students' first language, you may like to act out their first dialogue with them, and ask for feedback on your performance).

Variation:
- This can be used to raise students' awareness of any type of social language.

H Health : TOPIC

1. Health questionnaire

Before class: Prepare the topics for the questionnaire.

1. Draw a table on the board, for example:

name	use traditional medicine	use Western medicine	eat special food

2. Ask students to copy the table in their notebook.
3. Students mingle, ask each other about the topics (*How often do you use traditional medicine? What do you do?* etc), and take notes.
4. Divide students into groups. Students summarise the information they found and present it to the class.

Extension:
- Students could write up the results in a report. Students may also like to make recommendations.

2. Health quiz

Before class: Bring sheets of A4.

Everyone has opinions about health! Students shouldn't need additional information to create this quiz.

1. Tell students they're going to test each other's knowledge about health.
2. Do an example question on the board (e.g. 'What is one food that contains Vitamin C?' Elicit possible answers: oranges, peppers...).
3. Divide students into pairs. Ask them to write ten questions about health.
4. Pairs swap quizzes and write their answers to the other pair's questions.
5. Join the pairs together for scoring and discussion.

Variation:

- This activity is suitable for any challenging topic. Students can research the topic to write the quiz, but there needs to be a possibility other students can answer the questions!

3. Healthy living poster

Before class: Bring two visuals representing healthy and unhealthy living (e.g. riding a bike and smoking). You'll also need A3 paper and coloured markers. Find out where students will be able to display their posters.

1. Introduce the topic by showing the two visuals. Elicit a few ideas regarding healthy and unhealthy activities.
2. Divide students into groups. Instruct one student to write. Students brainstorm healthy and unhealthy activities in two columns.
3. Tell students they're going to produce posters to give other students useful information about their health. Do a very simple example on the board, for example:

GET HEALTHY!
Do:
- Ride a bike. It makes your heart stronger.
Don't:
- Smoke. Smokers die young.

4. Students plan what they'll write. Monitor to help ensure it's accurate before it goes on the final poster.
5. Distribute the A3 paper and pens. Students produce their posters.
6. Display the students' work.

Variation:

- This activity could be used for any topic. If possible, students could research the topic first; help them put text into their own words rather than plagiarising!

History : TOPIC

1. Dates on the board

Before class: Make sure you know several dates in history (e.g. 1949 Chinese Revolution, 2008 Beijing Olympics).

1. Test the class with several historical dates (e.g. write *Chinese Revolution* and ask 'When was the Chinese Revolution?'). Elicit and write up the answer *(1949)*.
2. Tell the class, 'Now it's your turn. You're going to test me'.
3. Divide students into groups. Ask them to write five similar questions.
4. Students ask you from the floor.

Variation:

- You could make this a competition between you and the students: have five questions already prepared. If you're unsure about dates, you could ask several students to join you when trying to answer the questions.

2. Who am I?

Before class: Find short biographies (e.g. on the Web or in an encyclopaedia) of historical figures your students will know something about.

1. Divide students into teams.
2. Ask the students, 'Who am I?'. Start reading the person's life history (e.g. *I was born in* ...).
3. When a team guesses correctly, they win a point.
4. The winning team is the first to three.

Variation:

- You could also have students read out the biographies.

Holidays : TOPIC

1. Holiday plans

Before class: Bring multiple copies of different types of information to help plan one trip (accommodation, transport, things to see and do etc). You can get these from a tourist information centre. You'll need one copy of everything for each group.

1. Tell students they're going to plan a class trip.
2. Elicit information we'll need to find out (e.g. dates, travel, accommodation, things to do, cost). Write these on the board.
3. Divide students into groups. Hand out the brochures etc.
4. Students present their proposal to the class.

Variation:

- Students could do this to plan a real-life excursion.

2. Travel agent

Before class: Find brochures or web pages of unappealing travel destinations (just make sure they're not in your students' countries!). You'll need enough different ones to give to half the students in the class.

1. Tell the class they're going to plan a trip abroad.
2. Divide the class into two groups: travel agents and customers. Move the two groups apart.
3. Give the travel agents the destination information. Tell them they have to think of all the good points about their destinations. They also need to decide on a price.
4. Tell the customers to note down what sort of holiday they want, and how much they can afford.
5. Set up the room like travel agents. Travel agents should be seated, and customers have to approach them.
6. Tell customers they have ten minutes to find their dream holiday.
7. As a whole class, students share their experiences. Who's the cleverest customer? Who's the best salesperson?

3. Travel advice brochure

Before class: Bring A4 paper and coloured markers.

1. Introduce the topic of travellers coming to your students' country (if you're teaching in that country), or your students' experiences coming to an English-speaking country. Elicit some of the difficulties people have when they first arrive.
2. Divide students into groups. Instruct one student to write. Students brainstorm what a visitor must, must not, should, and should not do.
3. Tell students they're going to produce a brochure to advise visitors. Do a simple example on the board:

BRAZIL TRAVEL ADVICE

DO...	DON'T...

4. In groups, students first write the text. Help students correct errors before they make the final brochure.
5. Distribute the A4 paper. Students create their brochures.
6. Students show each other their work.

Variation:

- This could be directed at a particular market: students, business people etc. Rather than a brochure, students could produce a webpage that they upload.

Homes : TOPIC

1. My partner's home

No preparation

1. Ask students to draw their home (either a floor plan, or the front of the building) without showing anyone.
2. Divide the students into pairs.
3. In turn, one student has to describe their home, and the other student has to draw it.
4. Compare the two pictures.

2. Ideal home

Before class: If possible, bring 'Home Beautiful' type magazines (in any language). You'll need one sheet of A3 paper and markers for each group.

1. Divide the students into groups.
2. Hand out the magazines (if applicable), and ask students to discuss what sort of home they like and don't like.

3. Tell students they're going to design a dream home. Hand out the A3 paper and markers. Say they should write notes on the poster explaining features of the home.
4. Students present their designs to the class.

Variation:

- Students could cut out pictures from the magazine to create their dream home.

3. Make the sale

Before class: Bring real estate flyers or 'Home Beautiful' type magazines (in any language).

1. Divide the class into groups and give each group a number of flyers (or a magazine). Ask them to find photographs of three different homes and describe them.
2. Students look at the magazines and make notes. Deal with any vocabulary queries.
3. Divide the class into two groups: real estate agents and customers.
4. Set up the room like estate agents' offices. Ask estate agents to sit in the 'offices' and give each one a magazine. Ask customers to stand at one end of the room.
5. Tell customers they need to go to different agents and find the best place to live.
6. As a whole class, students share their experiences.

I Using idioms : FUNCTION

1. Idiom halves

Before class: Create a list of sentences containing idioms (e.g. *I learnt the song off by heart*). You'll need one for every two students. Cut them in half:

I LEARNT THE SONG	OFF BY HEART

You'll also need Blu-tack or tape.

1. Shuffle the sentence halves.
2. Give one to each student. Ask them not to show anyone.
3. Students mingle and tell each other their half sentence in order to find their match.
4. Once students find the other half, they Blu-tack them together on the board.
5. Stay with the students around the board and elicit what the idioms mean.

2. Proverbs

No preparation

1. Give an example of a proverb in English, for example 'Too many cooks spoil the broth'. Elicit what it means.
2. Ask students for several examples in their first language.
3. Divide students into groups.
4. Students write down five proverbs in their first language and translate them into English.
5. Bring students together, and ask each group to present their favourite to the class. Tell students if there is an existing equivalent in English.
6. Discuss how common proverbs are in your students' first language. You might suggest English speakers use them rarely, and generally only say the first half (e.g. 'Too many cooks!').

3. New similes

Before class: Prepare five similes you'll use as examples.

1. Ask students to come to the board.
2. On the left of the board write five beginnings of similes (e.g. as brave as, as cool as), and on the right write the endings in different order (e.g. a cucumber, a lion).
3. Students draw lines to try to connect the halves.
4. Rub off the endings. Tell students you want them to create new similes.
5. Elicit ideas from students. Have them practise a conversation on a topic relevant to the lesson using the similes they created.

Introducing people : FUNCTION

1. One-minute party

Before class: Bring a music CD – if possible, other party or function paraphernalia.

1. Set up the room like a party/business function etc.
2. Divide students into pairs.
3. Tell students you'll play the music for one minute only. In this time, with their partner they need to meet another pair. Everyone needs to be introduced in one minute.
4. After one minute, stop the music. Students move to another pair. Play the music again.

2. Identity swap

Before class: Bring a music CD – if possible, other party or function paraphernalia.

1. Set up the room like a party/business function etc.
2. Divide students into pairs.
3. Tell students they have swapped identities. You are now your partner. You have their name, and you act like them.
4. Play the music. Students mingle and introduce their partner.

Invitations and suggestions : FUNCTION

1. The perfect invitation

Before class: Photocopy the invitation cards on page 104 so there is one for each student (several students can have a copy of the same card).

1. Elicit places you can invite a friend to (e.g. *restaurant, cinema* etc).
2. Give one card to each student. Ask them not to show anyone.
3. Tell students they have free time tonight and want to do something fun. They can ask anyone to do anything. However, you can only say 'yes' if you are invited to the place on your card (students can give clues: if they have theatre, they can say they like ballet and drama).
4. Students mingle and invite each other.

2. Go on, you'll love it!

Before class: Create cards that list unusual or unappealing activities (e.g. *visit a factory, eat rats*). You'll need one set for each group.

1. Divide students into groups.
2. In turn, one student picks up a card and tries to convince the group to do the activity together.

3. Class get-together

Before class: Bring copies of 'things to do' in the local area (preferably in English). You need one for each group.

1. Divide students into groups.
2. Hand out the copies of 'things to do'. Students have to agree on a score for each one: 1 = interesting, 5 = boring.
3. Tell students they're going to plan a class get-together.
4. Join groups together to discuss their opinions about the options in the 'things to do'. They can also suggest other ideas.
5. Bring the class together. Decide what the class will do.

L Asking about language : FUNCTION

1. How do you say …?

No preparation

This works best if all students are of the same nationality.

1. Revise expressions for asking about language: *How do you say … in English? How do you say/spell/write/pronounce …?*
2. Divide students into groups.
3. Tell students their first language is English, and they are in their first class, learning their real first language (e.g. in Korean).
4. Ask each group to choose a teacher.
5. The teacher teaches the students some basic conversational language (e.g. Korean). The students can ask any questions they like (but in English, of course!).

Variation:

- If you have students of different nationalities, they can mingle and teach each other expressions in their first language.

2. What does *bugaloo* mean?

No preparation

1. Ask students to write down five true statements about themselves. Suggest the information should be interesting that other people might not know.
2. Tell students to replace one word in each sentence with a nonsense word, e.g. 'bugaloo'.
3. Students mingle and share their personal information. Students will need to ask questions when they don't understand.

Talking about likes and dislikes : FUNCTION

1. Snakes and ladders

Before class: Create a snakes and ladders game tailored to topics of interest to your students:

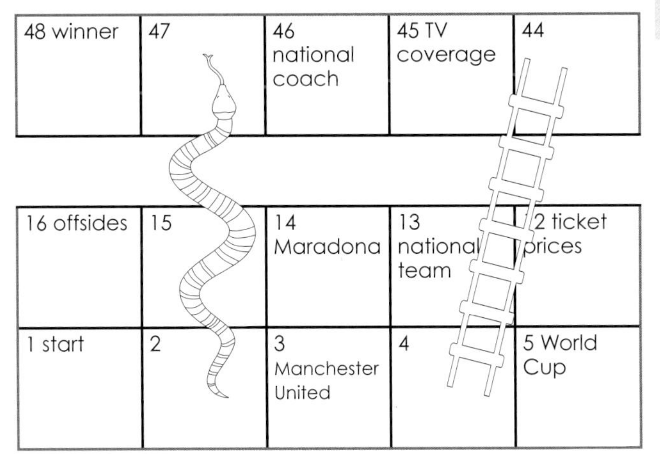

Make one copy for each group. Bring dice and counters.

1. In turn each student throws the die and moves their counter.
2. If they land at the bottom of a ladder, they go up; if they land at the top of a snake, they go down.
3. When they land on a topic, they have to say whether they like it or not, and give reasons why.
4. If the other students decide the reasons are unsatisfactory, the player goes back to their previous position. The winner is the first student to reach the final square.

Variation:

- You can add sneaky variations that give players advantages, send opponents backwards etc.

2. Partner's likes and dislikes

No preparation

1. Divide students into pairs.
2. Ask students to write down three things they think their partner likes, and three things they dislike.
3. Students share their guesses with their partner.

M Men and women : TOPIC

1. Relationship flowchart

Before class: Bring pictures/real objects associated with love and marriage: a ring, a Valentine's Day card etc. You'll need one sheet of A3.

This works best if all students are of the same nationality.

1. Use the pictures and realia to introduce the topic of love and marriage. Challenge students to tell you everything they know about the topic!
2. Draw a simple flowchart to elicit common stages in a relationship in your country:

Write any key vocab on the board.

3. Tell students you want to know what happens in their country.
4. Divide students into groups. Give each group a sheet of A3. Ask them to draw a flowchart. Say you want to know all the details.
5. Groups present their flowcharts to the class.

2. Ideal spouse

Before class: Bring two pictures, one of a man and one of a woman. Photocopy one 'ideal husband' or 'ideal wife' quiz sheet on page 105 for each group.

This works best with a substantial number of both male and female students.

1. Show students the pictures and ask, 'What makes an ideal husband? What makes an ideal wife?'. Elicit some ideas.
2. Separate the male and female students. Further divide the males and females into smaller groups. Give each group of females the 'ideal husband' sheet, and each group of males the 'ideal wife' sheet.
3. Tell groups to rank the characteristics from 1 to 10.
4. Join each group of males with a group of females. Ask groups to compare their responses.
5. Finish with whole-class discussion.

Modals : GRAMMAR

1. Signs

see page 264 - Modals 1: obligation

Before class: Plan what signs you're going to use as examples (if possible find ready-made visuals).

1. Draw some signs on the board and elicit what they mean. They should represent at least *have to* (obligation to do something) and *can't* (obligation not to do something). For example: (*You have to stop*)   (*You can't enter*)

2. Divide students into pairs. Ask them to draw six signs.
3. Join pairs together. They quiz each other on what their signs mean.

2. Invent a game

see page 264 - Modals 1: obligation

Before class: Bring at least two inflatable balls

1. As a class, brainstorm rules for a ball game (e.g. volleyball). Write important rules on the board using *You have to …, You can …* and *You can't …*
2. Divide students into groups.
3. Ask students to invent a new ball game and to write down the rules.
4. Join groups together. Ask them to teach the other group the rules, using language and demonstration. The other group should be able to try out the game.

3. Mysterious things

see page 268 - Modals 2: deduction

Before class: Bring mysterious pictures or recordings of sounds (you need to know what they are!).

1. Show or play one example. Elicit guesses using modals of deduction (e.g. *It must be a house because I can see a door*).
2. Divide students into pairs.
3. Students write down their guess for each item. Monitor and help students with language.
4. Reveal the answers!

4. Crime scene investigation

see page 268 - Modals 2: deduction

Before class: Find a picture of a crime scene (make sure it's large enough (or you have copies) so students can see the detail). Plan an engaging background to the scene – consider using pictures for anyone involved. You might like to have a surprise explanation that you reveal at the end.

1. Show students the picture. Engage them with the situation (*Tony had had a long day at work… Something seemed strange when he entered the building…*).
2. Divide students into groups. Tell them they are investigating the scene, and must produce a police report (you can specify the language students must use, e.g. 2x *must have*, 2x *could/ might have*, 2x *can't have*).
3. Join groups together. Have them discuss their deductions and come to an agreement.
4. Have groups share their ideas with the class.
5. You might want to give the class an actual answer!

Money : TOPIC

1. Country priorities

No preparation

1. Start by asking what students know about spending habits in your country. Avoid anything that seems superior; be prepared to be critical of your own country (e.g. *the average person is in debt with no savings*).
2. Tell students you want to know how people spend money in their country.
3. Divide students into pairs.
4. Ask students to think about the average adult's salary, and what percentage goes on what (students could complete a pie chart for this).
5. Pairs share their opinions with the class.

2. Money expert

No preparation

1. Elicit typical money problems with the whole class (e.g. my salary is too low/I owe my friend money).

2. Ask for suggestions how to solve these problems.
3. Divide students into groups.
4. Instruct one person to write. Ask the groups to list ten problems people have with money.
5. In turn, one student reads out one of the problems. Each other student has to propose a solution. The first student decides which is the best solution, and that person wins a point.
6. The winner is the student with the most points.
7. As a class, ask students to share good ideas they came up with.

Movies : TOPIC

1. Movie review

No preparation

After students see a movie, ask them to review it (and award it stars).

2. Movie controversies

Before class: Note down statements for discussion.

1. Divide the class into groups of around 4-5.
2. Write up two or three statements for discussion. Choose statements that are appropriate and interesting for your group. They might include:

   - Romantic movies are boring.
   - We need to protect the local movie industry from Hollywood.
   - Historical movies should be 100% accurate.

3. Finish with whole-class feedback where students share some ideas their groups discussed.

3. Re-enactment

No preparation

1. With the class, brainstorm famous movies all the students know.
2. Divide the students into pairs or groups.
3. Ask students to script a very short scene from a famous movie.
4. Students perform for the class. Students guess which movie it is.

Music : TOPIC

1. Emotional responses

Before class: Bring several very different pieces of music.

1. Ask students. 'How do you feel when you hear this?'. Play an extract from one piece of music.
2. Elicit some responses.
3. Tell students you now have more questions. Write these questions on the board:

   - How do you feel?
   - What adjective best describes it?
   - Do you like it?
   - What sort of person would like it?

4. Play the same piece of music again. When it finishes, ask students to discuss the questions with their partner. Elicit some responses to the whole class.
5. Play the other pieces of music. Each time, ask students to discuss their response with their partner.
6. Finish with whole-class discussion. Which pieces did students agree and disagree about?

2. Music mimes

No preparation

1. Demonstrate the activity. Ask, 'What am I playing?'. Mime playing a piano to elicit 'piano'.
2. Students practise in small groups.

Alternatives: You could increase the challenge with other wh- questions (Where am I? How am I playing?).

3. Opinions

Before class: Note down the starts of sentences you plan to use.

1. Write up the start of an opinion about music, e.g. 'Jazz is'. Elicit some responses (e.g. 'interesting', 'strange', 'boring').
2. Write up some more half sentences, for example:

   - Traditional music makes me feel...
   - Classical music is...
   - The best musicians can...

3. Ask students to finish the sentences individually in writing, according to what they believe.
4. Form students into small groups.
5. Students share and give reasons for their opinions.
6. Finish with the whole class. Ask students to share interesting opinions they heard.

Using music : VARIOUS

1. Relaxed atmosphere

Play music when students are first coming into class, and during the break. It helps create a very relaxed and friendly atmosphere.

2. Light and shade

Play soft background music during slower, more peaceful activities, such as project work or writing. It's a good contrast with the more high-energy activities in the lesson. It conveys the message that students have as much time as they need to complete a more reflective task.

3. Music mingle

You can use instrumental music for any mingling activity. When the music plays, students have to walk or dance randomly around the room. When the music stops, they have to talk to the nearest person.

4. Learn a song

Songs are an effective and enjoyable way to teach – especially when teaching children! There are pre-prepared songs you can find for free or buy on the Web.

5. Write your own song

Students can write their own songs, or write new lyrics to existing songs.

N Numbers, dates and times : TOPIC

1. Number dictation

Before class: Have numbers students are familiar with ready to dictate.

1. Dictate a series of numbers. Students copy them down in their notebook.
2. Ask students to compare answers. If there's a lot of disagreement, read the numbers again.
3. Have several students at once write the answers on the board.
4. Divide students into pairs.
5. Students write lists of numbers which they then dictate to each other.

Variation:

- While initially students will find the above activity satisfying simply because of their achievement, consider an additional mental challenge (e.g. including arithmetic).

2. Important dates

Before class: Note down three important dates in your life.

1. Write up important dates in your life (e.g. *1980, 1992, March 1 2008*).
2. Ask students to guess why they're important (*you were born in 1980?*). Let them know the answers.
3. Divide students into groups.
4. Students do the same activity with each other.
5. Ask a couple of volunteers to do the activity with the whole class.

3. Culture timeline

Before class: Be familiar with several key dates in your country's history. You'll need one sheet of A3 for each group. If possible, take in a reference that lists key dates in the history of your students' country.

This works best with students of the same nationality.

1. Ask the class what they know about your country's history. Draw a simple timeline with several dates on the board. Ask students to tell you what happened on these dates. Add the information to the timeline.
2. Divide students into groups. Give each group a sheet of A3.
3. Ask students to create a timeline of their own country's history (it's useful to have a reference handy that students can consult).
4. Students share their timelines with the class.

4. Taboo number

No preparation

1. Divide the class into two teams.
2. Ask each student to write a number between 1 and 10. They can show their number to people on their own team, but not the other team.
3. In turn, one person from each team asks someone on the other team three questions, with the aim of getting to say your number. For example, with the hidden number three:

A: How many times a day do you brush your teeth?
B: Two.
A: What's six minus three?
B: The same as two plus one.
A: What does F-O-U-R spell?
B: Four. Oh no!

4. A student that succeeds in getting someone to say their number wins a point.
5. The first team to score five points wins.

Giving opinions : FUNCTION

1. Debates

Before class: Note down useful language and topics you intend to use.

Debates are a great way to get students talking, and to help push their language beyond everyday topics.

1. Pre-teach/revise useful expressions (e.g. *I believe; I suppose; To my mind; If you ask me; I'm convinced that; The way I see it*).
2. Set topics your students will be interested in. Think about the ages and interests of your students, and any cultural sensitivities. Students need to feel comfortable to give their opinion, so steer away from choosing topics that might cause friction. (It's often advisable to stay away from religion or politics!).

Teenagers
- School uniforms should be worn at school
- Violent video games should be banned
- Girls are more intelligent than boys (you could get the boys to argue that girls are more intelligent and the girls to argue that boys are more intelligent!)
- Homework is good for you
- Television is a bad influence

Adults
- Money buys happiness
- Human cloning should be allowed
- Violence should be banned in films
- Women are worse drivers than men
- The legal age for drinking should be 16

3. Assign roles. Often students may agree (or pretend to in order to be polite) and the debate can finish quite quickly. Therefore it's a good idea to ask students to be devil's advocate, or specify which side they need to support.
4. Give students time to prepare (in particular lower levels). You can group students who'll be arguing the same point to brainstorm ideas.

Variation:
- Recreate a real-life debating activity in the classroom (e.g. a parliament or a TV discussion panel). Set up the room as realistically as possible. Assign roles. You can create role play cards (e.g. 'You're an environmentalist. You believe industry should pay additional tax to fund environmental programs.').

2. Room in the balloon

Before class: Note down which famous identities you'll use.

1. Draw a picture on the board to elicit that we're in a hot air balloon.
2. Put students into groups, and give each person a famous identity.
3. Tell the students the balloon is too heavy and one person must jump/be pushed out of the balloon. They need to present an opinion on why they should stay in the balloon, then debate!

3. Reporting opinions

No preparation

This works for any discussion on any topic.

1. Have students mingle and exchange views with a number of students.
2. When they're ready, ask students to sit down.
3. Ask students to write down three opinions they heard that impressed them. For example: Deng thinks it's important to live in another country to learn a language.

4. Students then show their sentences to the students who gave the opinions, to check they've expressed their ideas accurately.
5. Finally as a whole class students share the ideas they wrote down and tell the class who said them.

4. Time to think

No preparation

We often force students to come up with opinions when they may need time to consider the options. This works for any discussion topic.

1. Students respond to a statement by writing down two reasons why they agree, and two reasons why they disagree.
2. They then mingle and discuss the four points they wrote down, and ask for people's reactions.
3. Students return to their seats and tick the two points they've decided are strongest. They still may not have formed an opinion.
4. In small groups, students discuss their current thinking about the topic.

P Passives : GRAMMAR

1. Passives board game

see page 245 - Past simple passive

Before class: Photocopy one board game on page 106 for each group. You'll need counters and a die for each team.

1. As revision, elicit some examples of present and past passive in their singular and plural forms (e.g. present *This car's driven by Madonna* (singular) and *These cars are driven by Madonna* (plural); past *This car was driven by Madonna* (singular) and *These cars were driven by Madonna* (plural)).
2. Put students into groups of three or four. Give one student the answers (tell them not to show the answers to anyone) and ask them to be the judicator.
3. Students throw the die and move their counter. When students land on a square they have to complete the sentence with the correct passive form. If they get the answer correct, they can stay on the square. For all wrong answers they have to move back a space.
4. The first to finish wins!

Answers:	
1) are grown	14) was bought
2) is played	15) were cut
3) was made	16) was won
4) are sold	17) was paid
5) was invented	18) was flown
6) was stolen	19) was taken
7) are delivered	20) is given
8) was broken	21) was shot
9) are brought	22) was spent
10) were built	23) were left
11) are usually increased	24) are released
12) was written	25) was worn
13) is added	

2. Active passives

see page 245 - Past simple passive

No preparation

1. Write up 10 passive sentences spread out on the board, eight with errors. Tell students only two are correct.
2. Divide the class into two teams, and give each team a different coloured pen.
3. Give a signal to start the activity.
4. One student from each team runs to the board and corrects a sentence. They then run back and give the pen to another student. They then run to the board to correct another sentence, and so on. Students can correct a change made by the other team.
5. When one team is satisfied they call out 'Finished!'.
6. The winning team is the one who has corrected the most errors.

Variation:
- This activity can be used for other types of errors.

3. Describe a process

see page 244 - Present simple passive

Before class: Bring visuals of a process of relevance to your students (e.g. manufacturing/politics/education etc). You'll also need Blu-tack or tape.

1. Ask students to come to the board.
2. Blu-tack the visuals to the board (in the wrong order) and elicit what the process is.

3. Ask students to move the visuals to put them in the correct order.
4. Hand out board markers and ask students to write key words next to the pictures.
5. Elicit a description of one stage of the process using a passive (e.g. the votes are counted).
6. Divide students into pairs. Students write the description of the process.
7. Invite several students at one time to write the stages up on the board.

4. Crime report

see page 245 - Past simple passive

Before class: Copy a short newspaper article report of a crime. Make sure it has several examples of passives (e.g. The girl **was seen** leaving her house at 9am / Her bag **was found** in the local park).

1. Students look at the headline and visuals to predict the main idea of the article. Write two or three predictions (including the correct one) on the board.
2. Students read quickly to see which prediction was correct.
3. Students make bulleted notes of the key events in the article (*See girl house* 9). They then share their notes with a partner.
4. Ask students to underline examples of the passive. Check the meaning (e.g. *Do we know who did it? No.*) and revise the structure on the board (*The girl* (subject) *was* (auxiliary 'be') *seen* (past participle)).
5. Students write their own crime report, either individually or in a pair. They should use passives, and can model their story on the article.
6. Students share their stories with other students.

Past continuous : GRAMMAR

1. Guess your partner's whereabouts

see page 238 - Past continuous 2

No preparation

1. Write yesterday's date on the board. Elicit yesterday. Write five times (e.g. *8am, 10am, midday, 4.11pm, 10.30pm*).
2. Ask students, '*Do you know anything about my day yesterday?*'. Point to the first time. Elicit a statement in the past continuous (for example: '*you were driving to work*').
3. Divide students into pairs.
4. Tell students to guess their partner's activity at each of the five times yesterday. They need to write it in a complete sentence.
5. Students show their partner the sentences to see if their guesses are correct, and need to change any wrong information.
6. Students share anything interesting they found out with the class.

2. Alibi

see page 238 - Past continuous 2

Before class: If possible bring visuals to set the scene for the story.

1. Tell students, 'A woman went missing in your town at 8pm last night (look serious!) and she had a list of your names in her bag. The police want to question you'.
2. Elicit one or two questions the police might ask (e.g. *What were you doing at 8pm?*).
3. Put students into pairs. Give students five minutes to think of other questions the police might ask.
4. Ask, 'What do you call it when you tell police *I didn't do it, I was in this other place when it happened?*'. Elicit alibi. Give students 5-10 minutes to make up their alibi together (At 8pm we were...).
5. Put students into groups of 4. Two are the police. They need to interview the suspects. They should do this separately and then compare notes to see if the alibis are different. Great fun!
6. Students swap roles.

3. Tell a story

see page 237 - Past continuous 1

Before class: Bring a number of visuals of different places (e.g. a street, a café, a park).

1. Tell students they're going to write a story. Explain it will happen in one of the places in the pictures.
2. Ask, 'How do you describe the situation?' Elicit past continuous (e.g. *I was walking in the park*). Ask, 'How do you then say the things that happened?' Elicit past simple (e.g. *I saw a man...*).

3. Tell students they need to write a four line story
   - first describing the situation
   - then three things that happened.
4. Students write their story on their own.
5. Put students in small groups. Instruct students to tell their story, but stop before the last line. The other students have to guess the ending.

Extension:

- Students can write a number of stories based on all the pictures.

Past perfect : GRAMMAR

1. Guess your partner's day

see page 229 - Past perfect simple

No preparation

1. Put students in pairs.
2. Tell students they're going to guess what their partner did yesterday, and get all the details right. However, they have to start at the end of the day, and work backwards. They can use past perfect to show something that happened before something else. For example:

   A: You went to bed at 11.
   B: No, 10.
   A: Before that you did your homework?
   B: That's right.
   A: So you went to bed at 10 after you'd done your homework?
   B: That's right.

3. Choose several students to show their impressive memory to the class.

2. Picture reorder

see page 229 - Past perfect simple

Before class: Find a series of four or six pictures that tell a story (many coursebooks have these; alternatively you could ask students to draw a story as part of another activity and ask for their permission to use it). Copy and cut the pictures up. You'll need one set for each pair of students.

1. Divide the students into pairs.
2. Give one set of pictures to each pair.
3. Ask students to put the pictures in order to make a story (there is not necessarily a correct order – different pairs can create very different stories).
4. Students write the story. Tell students the story should be in past simple, and that every two pictures should be connected. For example, 'He turned off the computer and left'.
5. Ask students to reverse each pair of pictures and have students rewrite the sentences using past perfect. For example, 'He left after he'd turned off the computer'.
6. Once students have practised the structure you might want to let them change some back into past simple (since it's unusual to repeat past perfect a number of times in a row) and make any other changes they like.
7. Join pairs together. Students jumble their cards. One pair reads out their story, and the other pair has to put the pictures in the order they actually happened (not the order they're mentioned).

Variation:

- You can use a picture story just to practise past simple, or past continuous and past simple combined. To maximise practice, give pairs different sets of pictures, which they pass on to the next pair to write a new story.

Past simple : GRAMMAR

1. Past mime

see page 221 - Past simple 1

No preparation

1. Divide the students into pairs.
2. One student mimes a complete scene to show what they did this morning/at the weekend/last week/on their last holiday etc.
3. The other student tries to work it out, and commentates (*You made a cup of tea? Or coffee?*).
4. Ask a couple of students to mime for the whole class. Students call out what they think happened.

2. Yesterday I blanked

see page 221 - Past simple 1

No preparation

1. Ask students to write down five interesting things they did yesterday (e.g. *I bought some new clothes*).
2. Divide students into groups.

3. Students tell the group their sentences, but use the word 'blank' to replace any verb (*I **blanked** some new clothes*). Students have to work out what the verb should be.

3. Past chain

see page 221 - Past simple 1

No preparation

1. Divide class into groups.
2. Select one group to help demonstrate the rules. You start by saying *Yesterday, **I got up***. The student on the teacher's right follows by saying *Yesterday, I got up and then **I had** my breakfast*. The next student continues by saying *Yesterday, I got up and then **I had** my breakfast and then **I watched TV***.
3. The activity continues with each student repeating what previous students have said, and adding one more action each time.
4. Start students off in their groups.

Variation:

- This is a good game to practise all kinds of verb tenses and vocabulary!

4. Board race

see page 221 - Past simple 1

Before class: Make a list of 20 short sentences containing regular and irregular verbs.

1. Divide the class into two teams facing the board.
2. Draw a line down the middle of the board, giving half to each team.
3. Read out a short sentence containing present simple (e.g. *I know him*). One student from each team runs to the board and writes the sentence in past simple (*I knew him*). Students can just write the verb, but using language in a sentence is always more meaningful.
4. The team that writes it correctly first gets a point.
5. The winning team is the first to ten points.

Variation:

- You can read out a mixture of present and past simple sentences. If the verb's already in past simple, the students shouldn't change it.

5. Table competition

see page 221 - Past simple 1

Before class: Photocopy and cut up the sets of cards on page 107. You need one set per group.

1. Divide students into groups.
2. Give one set of cards to each group.
3. Students sort the cards into two groups: present and past.
4. Tell students to put the present tense cards in a pile, and to spread the past tense cards out on the table, face up.
5. The reader reads out a present tense card (e.g. *go*).
6. The other students compete to bang their hand down on the past tense card (*went*) and call it out. They keep the card.
7. The winner has the most cards.

Variations:

- You could use the cards in a number of ways: to play memory or snap, or as prompts to say or write sentences.
- This can be used to practise any pairs (e.g. synonyms or phrasal verb stems and particles).

6. Collaborative story

see page 221 - Past simple 1

Before class: Note down the initial sentence in the story (it's fun to make the topic about your students and their English class). You'll also need one blank A4 sheet for each pair.

1. Tell the class they're going to write stories together.
2. Divide the students into pairs. Give each pair a sheet of paper. Write the first sentence on the board (e.g. *Yesterday we came to class as usual*). Tell them to copy it down, and then write the next sentence in the story.
3. Students pass their story to the next pair on the right, and then write the next sentence in the story they just received.
4. Students continue to pass the stories around, writing the next sentence each time.
5. When the stories reach the original writers (keep track of this!) ask them to write the concluding sentence.
6. Post the stories on the wall so all students can see them.

7. DVD story

see page 221 - Past simple 1

Before class: Arrange a DVD player and screen. Bring a DVD with lots of activity that students can describe.

1. Divide the students into pairs.
2. One student in each pair faces the screen, and the other has their back to it.
3. Play a short scene of the DVD with the sound down.
4. The first student has to describe what happened. The other student writes it down.
5. Then all students watch the same scene to see if they were correct.
6. Get students to swap and show a different scene.

Personal information : TOPIC

1. Noisy names

No preparation

An uninhibited way to remember names!

1. Students sit in a circle.
2. One student points at another student and calls out their name (they can ask '*Sorry, what's your name?*' if they can't remember). The other student does the same back.
3. Get all students to do this at the same time!

Variation:

- Students can add another piece of information: *You're Sanjay, and you can paint!*

2. Getting to know you

Before class: Note down the prompts you'll write on the board.

1. Divide students into pairs. Ask 'How much do you know about your partner?'.
2. Demonstrate. Write one prompt on the board (e.g. 'hobbies'). Look at one student and say 'I think you like... let me think... computer games?'. The student should confirm whether it's true or not. Encourage them to give more details.
3. Write other prompts on the board (e.g. age, family, sport, food, future plans).
4. Students make guesses about their partner, who confirms if they are correct or not.
5. Ask students to share something interesting they found out about their partner.

Extensions:

- Students can write a short paragraph about their student, but without using their name (just he/she). Collect the profiles and distribute them randomly. Students mingle and try to work out who the profile describes.
- Students can create a profile of their partner to display on the wall. In the previous lesson, ask students to bring a photo of themselves. Students glue the photo in the middle of a blank sheet of paper and write the information they found out around the photo.

3. Hidden personality

Before class: Make one copy of the prompts on page 108 for each pair.

1. Divide students into pairs. Tell students, 'We're going to find out what out partners are really like!'
2. Give one set of prompts to each pair.
3. Students make guesses about their partner, who confirms if they are correct or not. Encourage students to find out as much as possible.
4. Students ask their partner what they would like other people to know about them.
5. Students introduce interesting information about their partner, with their consent, to the class.

4. Questions in the box

Before class: Bring a box (or a bag) for each group, and small strips of paper.

1. Divide students into groups. Give each group a box.
2. Give each student two strips of paper. Ask them to write a personal question on each (e.g. *Do you like sport?*).
3. Students fold their strips of paper and put them in the box.
4. In turn, each student pulls out a question at random and answers it. If they're uncomfortable with a question, they can choose another one.

Phrasal verbs : GRAMMAR

1. Ball synonyms

see page 270 - Phrasal verbs

Before class: Bring a ball (or stuffed toy); note down a list of phrasal verbs and non-phrasal synonyms.

1. Brainstorm some (informal) phrasal verbs and (formal) non-phrasal equivalents (e.g. *come in /enter*).
2. Ask students to form a circle.
3. One student says a phrasal verb, and throws the ball to another student, who says the non-phrasal synonym (encourage students to help each other so students don't feel on the spot).

Variations:

- This activity can be used to practise other synonyms or pairs (e.g. present and past tense, singular and plural).
- To maximise opportunities for speaking, bring in several balls and have students practise in smaller groups.

2. Phrasal mime

see page 270 - Phrasal verbs

No preparation

1. Ask each student to write five sentences containing an idiomatic phrasal verb (e.g. *The plane took off*). They shouldn't show anyone their sentences.
2. Divide students into pairs.
3. Students in turn need to mime their sentence: first what it really means (e.g. mime a plane on a runway taking off) and then the literal meaning of each word (e.g. stretch out your arms to be a 'plane'; take something with your hand; indicate something falling off a desk). The other student has to work out the sentence.

Plurals/countable & uncountable/ much & many : GRAMMAR

1. Plural noughts and crosses

see page 162 - Plurals

Before class: Note down nine singular countable nouns (or preferably more for multiple games). Make sure some are irregular.

1. List nine singular countable nouns down one side of the board.
2. Divide the students into two teams.
3. Draw a noughts and crosses on the board:

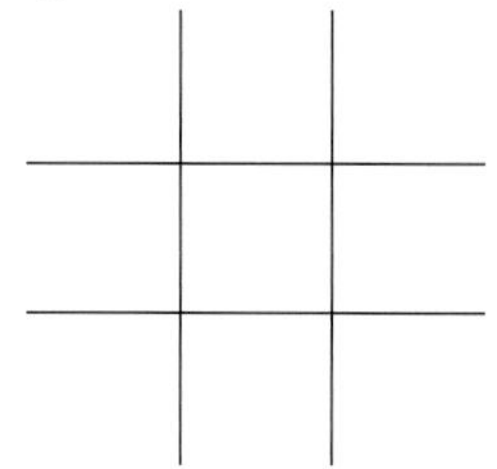

4. In turn one member of each team has to write the correct plural form of any of the nouns in one of the spaces. The first team to have three words in a row wins the game.

Variation:

- Ask students to create their own lists of nouns, either to play at the front or in groups.

2. Label the picture

see page 160 - Countable/uncountable nouns

Before class: Find a picture in the coursebook or elsewhere that contains lots of people and things.

1. Divide students into pairs.
2. Set students a time limit (e.g. three minutes) to list as many countable and uncountable nouns as they can find in the picture.

3. Board plurals

see page 162 - Plurals

Before class: Make a list of 30 singular nouns, a mixture of countable and uncountable.

1. Divide the class into three teams facing the board. Give each team a board marker.
2. Read out a noun.

3. One member of each team runs out the front and writes either the plural (if countable) or unchanged (if uncountable).
4. The team that writes it correctly first gets a point.
5. The winning team is the first to ten points.

4. Recipe swap

see page 162 - Plurals

No preparation

1. Put students in pairs.
2. Ask students to write down a recipe they know. They need to list the ingredients (eggs, flour etc), but without specifying the amount.
3. Pairs swap their recipes.
4. The new pair needs to ask how much or how many to complete the recipe (e.g. *How many eggs does it need?*).

Politics : TOPIC

Politics is a risky topic – make sure your students are comfortable with it.

1. Politics mind maps

Before class: Bring one sheet of A3 per group. Note down key topics you'll ask students to brainstorm.

1. Demonstrate the activity on the board (with a relatively uncontroversial example!). Write a key political term in the centre, and elicit associations:

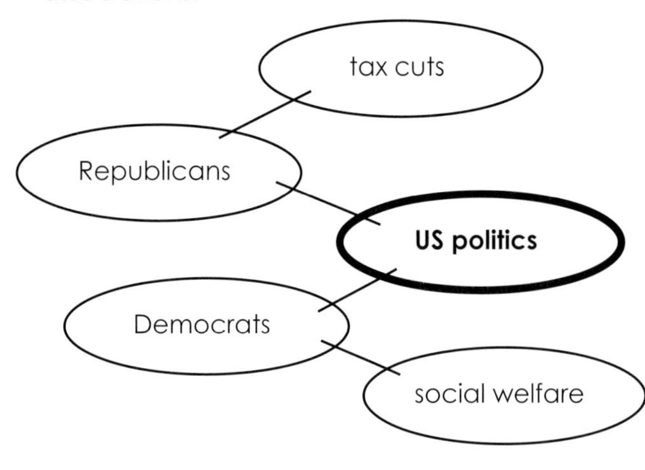

Accept students' varying viewpoints; show alternatives are possible.

2. Divide students into groups. Have groups brainstorm with other topics.

2. New laws

Before class: If possible, bring visuals of social and economic problems (crime, unemployment etc). You'll need a sheet of A3 for each group.

1. Tell students they are in government and need to create new laws to fix the country's problems.
2. As an example, elicit a problem (using a visual if you have one): for example, 'crime'.
3. Ask students to suggest laws that might address this problem. Allow students to be light-hearted (e.g. *Make all young people stay home after dark*).
4. Divide students into groups.
5. Give each group a sheet of A3. Ask students to copy this grid:

problem	proposed law

6. Students discuss and make notes.
7. As a whole class, ask students to share and discuss their proposed laws.

Prepositions : GRAMMAR

1. Map

Before class: Find a map of a city or country relevant to students' interests. Make one copy for each student and for you. On your copy draw a route.

1. Engage students with the place on the map (e.g. use visuals; have students discuss what they know about the place).
2. Give students a reason why they'll look at a map (e.g. 'I'll tell you about my trip').
3. Give one map to each student.
4. Describe the route. Students have to draw it.
5. Get students to compare answers, and then confirm the route with the whole class (e.g. on an OHP).

Extension:

- Students can repeat the activity in pairs or groups. Students draw a route themselves and then tell their partner(s).

2. Picture differences

Before class: Draw a picture related to a topic you've covered in class. Make another version with some (e.g. ten) people and/or objects in different locations. Copy one of each version for half the students in the class.

1. Divide students into pairs.
2. Give each student a different version of the same picture. They must not show it to their partner.
3. Tell students that the pictures are different. There are (e.g. ten) things in different places.
4. Students describe their picture and circle anything they think is different.
5. Students compare their pictures to see if they found all the differences.

3. Find it

No preparation

1. Divide the students into groups.
2. One student leaves the classroom while the others hide an object in the room.
3. The student returns and has to ask yes/no questions to locate the object (e.g. *Is it in a desk? Is it near the window?*).
4. When the student locates the object the next student leaves the room.

4. Three dimensions

Before class: Gather materials as per plan below.

Activities using three-dimensional objects bring prepositions to life. You need two identical sets of objects (e.g. toy blocks, or a floor plan with furniture cut from card). Have two pairs work together. Create a barrier so students can't see each other's objects. One pair arranges their objects, and instructs the other pair to replicate the arrangement. Students then remove the barrier to compare.

Present continuous : GRAMMAR

1. What am I doing, how am I feeling?

see page 231 - Present perfect continuous 1

Before class: Prepare two sets of flashcards: half with actions (e.g. *swim*) and half with emotions (e.g. *angry*).

1. Ask a student or group of students (if students are shy) to come to the front and select one card from each set.
2. Students have to mime and the class has to guess the action and the emotion (e.g. *She's swimming* and *she's angry*).
3. Correct the students as they shout out their answers.

Variation:

- Students could produce the 'feeling' as an adverb: *She's swimming angrily*.

2. Describe a scene

see page 231 - Present perfect continuous 1

Before class: Find a picture in the coursebook or elsewhere that contains a lot of activity.

1. Divide students into pairs.
2. Tell students this is a competition.
3. The first pair to write five correct sentences wins. Students must use present continuous to describe the activity in the scene.

3. Change places if...

see page 231 - Present perfect continuous 1

No preparation

You need moveable chairs and floor space for this activity.

1. Have students form a circle with their chairs.
2. Write 'Change places if you're...' on the board.
3. Stand in the centre of the circle. Say (using present continuous) '*Change places if you're...*' (e.g. *wearing black shoes*).
4. Signal that students with black shoes need to move to a different seat. Rush and sit down on an empty seat to show one student each time will lose a seat.
5. The one student left standing says '*Change places if you're...*'

Variation:

- You can use this to practise other structures (e.g. present perfect: '*Change places if you've ever...*').

4. Who is it?

see page 231 - Present perfect continuous 1

No preparation

1. Divide students into groups.
2. In turn, one student thinks of someone in the class and secretly writes down their name.
3. The other students ask yes/no questions to work out who it is. Short/temporary actions will be in present continuous (*Is she talking right now? Is she wearing a jacket*?) and states will be present simple (*Does she have long hair? Is she tall*?).

5. Complaints

see page 231 - Present perfect continuous 1

Before class: Prepare a short list of topics for discussion.

Make sure you don't criticise anything sensitive about the students' school or country in this activity!

1. Tell students in your country people often complain. Ask if it's the same in their country.
2. Write a list of topics that people complain about on the board (e.g. *transport, prices, TV*). Elicit more topics from the students.
3. Elicit an example of a complaint using present continuous (e.g. *The buses are always breaking down.*).
4. Divide students into groups.
5. Students share complaints about the topics on the board.
6. Join groups together to share their ideas, and discuss possible solutions.

Present perfect : GRAMMAR

1. Help line

see page 225 - Present perfect simple 1

Before class: Note down topics for discussion.

1. Write several topics on the board that people often have problems with (e.g. money, relationships). Ask the class to suggest what sort of problems people might have (e.g. gambling).
2. Tell students to imagine a person who has a lot of problems. Write down their name, and list five problems they have.
3. Students will simulate a role play between this person and a help line counsellor. If possible, arrange students' chairs back to back so they can't see each other when they talk on the phone.
4. In turn, one student rings up to ask for advice. The counsellor should offer helpful suggestions.
5. To finish, have students share some of the good ideas they heard.

2. Job interview

see page 228 - Present perfect continuous 3

Before class: Find a number of job advertisements of interest to your students.

1. Have students look at the different advertisements and find one that they'd like to apply for.
2. Get students to note down their relevant experience for the position.
3. Divide students into groups of three.
4. In turn, one student will be an applicant and the other two will be interviewers. Remind students to use present perfect for life experiences (*I've worked for two banks*) and for a situation that started in the past and is still true (*I've worked in my current role for two years*).
5. To finish, have students share what they found easy and difficult being interviewed for a job.

Present simple : GRAMMAR

1. What's the question?

see page 216 - Present simple 1

No preparation

Write some information about you on the board (numbers or words):

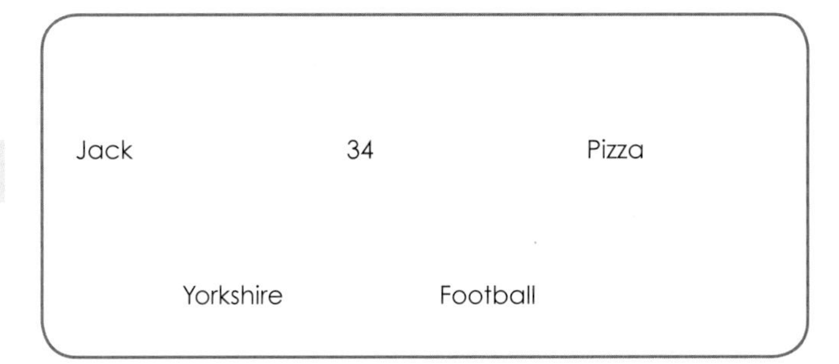

1. Ask students to make a question to find the correct answer. Model and drill the correct answer: What's your name?
2. When students produce the correct question, circle the answer.
3. Students may form a question correctly (e.g. *What's your favourite number?*) but it may not be the right question (*What's your house number?*). Tell them: Your grammar's good but it's the wrong question!
4. Write the correct questions up on the right hand side of the board.
5. Divide the students into groups.
6. Ask students to write six words and numbers about themselves, and repeat the activity in their groups.

2. Third person mystery

see page 216 - Present simple 1 & 2

No preparation

1. Divide students into groups.
2. Ask each student to write five questions to ask the people in their group about their current lives (family, work/study and free time): *Do you have any children? What do you do on the weekends?*
3. Students ask the other students in their group and make notes.
4. Tell students to stand up.
5. Students need to talk to people from other groups and describe someone from their group (using third person -s, for example *He watch**es** football every night after work*). They can only say *he* or *she*, and not mention their name. The other student has to guess who it is.

Pronouns : GRAMMAR

1. Spin the bottle pronouns

see page 174 - Subjects and objects: I or me?

Before class: Bring one empty bottle for each group.

1. Divide students into groups. Have them sit on the ground in circles.
2. Sit with one group to demonstrate. Spin the bottle so it points at another person. Say any subject pronoun (e.g. *he*). The other student says the object (*him*).
3. That student now spins the bottle and repeats the process.
4. When students are confident, instruct students to say the pronoun in a sentence (e.g. ***We** like music; They know **us***).

2. Recreate the text

see page 174 - Subjects and objects: I or me?

Before class: Find two short texts in the coursebook (or elsewhere) containing a number of pronouns.

1. Divide the students into pairs.
2. The two students look at different texts.
3. Tell students to write out the text, replacing all the pronouns with the nouns they refer to.
4. Students swap the rewritten texts and try to recreate the original.
5. Finally students compare their versions with the originals.
6. As a class, discuss any difficulties students had.

Pronunciation : FUNCTION

1. Sound bingo

Before class: Identify a pair of sounds students are having trouble with (e.g. /θ/ vs /s/). Make a list of six minimal pairs (or very similar words): e.g. *thin* and *sin*.

1. Write the list of words on the board.
2. Ask students to copy the following grid and write nine of the words in any spaces:

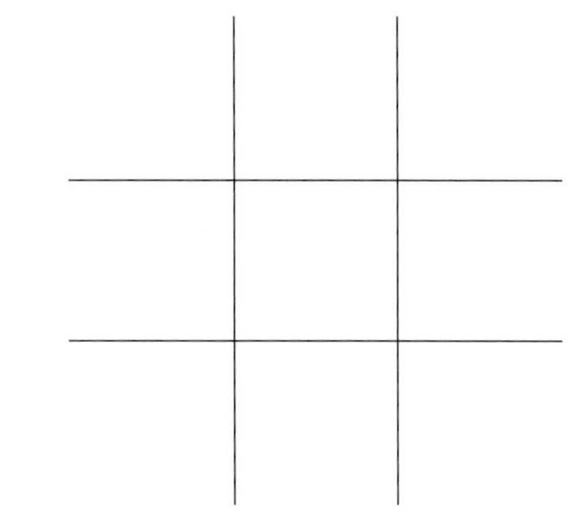

3. Rub out the list of words.
4. Read your words in a different order (number them as you go so you know what you've said).
5. Students cross them out as they hear them. When a student has crossed all nine words they shout 'BINGO!'.
6. Have students practise in groups, taking it in turns being the caller.

2. Mobile sounds

Before class: Make a list of five minimal pairs (or very similar words): e.g. thin and sin.

1.	sin	2.	thin
3.	thick	4.	sick

1. Write 1, 2, 3 ... 9 and 0 on the board. Next to each number, write one of the words:
2. Students mingle and exchange mobile numbers. However, they need to say the words, not the numbers.
3. As the first student says the words, the other student writes the numbers.
4. The first student checks the number is correct.

3. Tongue twisters

Before class: Identify one sound students are having trouble with (e.g. /r/).

1. Put students in pairs.
2. Students write their own tongue twister, containing many examples of the problem sounds.
3. Students practise and memorise their tongue twister.
4. Join pairs up. First, each pair dictates their tongue twister, and the other pair writes it down. Then students try saying the other pair's sentence.

4. Silent syllables

Before class: Find four words in the coursebook that have different stress patterns (e.g. *BU*sy, con*NECT*, *MO*nitor, intro*DUCE*).

1. Write up the four words in a row. Elicit and draw the stress pattern above each (e.g. O o o for MOnitor). Model and drill the words.
2. Ask students to copy the table into their notebook.
3. Students look back in their coursebook and add as many words as they can in five minutes.
4. Get students to compare answers.
5. Have four students at once add their words to the columns on the board. Model and drill any challenging words (not all of them!).
6. Rub off the board.
7. Get students to mingle. They have to say one of the words with the stressed syllable at high volume and the other syllables almost silent. The other student has to guess what the word is.

5. Stress pattern match

Before class: Make pairs of matching cards. They contain words with the same stress pattern, in the same order:

Dubai	Nepal
Italy	Hungary
Denmark	Thailand

1. Tell students you will give them a card that contains three words. They need to find someone with three words with the same stress pattern, by saying their words to other people. Demonstrate this with two cards that don't match, and two cards that do.
2. Give one card to each student. Make sure they don't show anyone their card.
3. Students mingle and try to find the student with the matching pair.

6. Show your feelings

Before class: Note down adjectives for the activity.

1. Revise the rule that narrow voice range suggests you're bored; wide voice range shows you're interested.
2. Have students mingle and chat (on a particular topic if you wish).
3. Give a signal (e.g. ring a bell or tap on the board). Write up the emotion that students need to show when they talk (e.g. 'bored').
4. Do this every few minutes with different adjectives ('interested', 'angry', 'amused' etc).

7. Am I asking?

Before class: Blank cards

1. Review the rules for intonation in questions.
2. Put students in groups.
3. Give each group ten cards. On each card, the group needs to write one question and answer related to topics they've studied in class. Five must be yes/no questions, and five must be wh-questions.
4. Have groups swap cards.
5. In turn, students ask their group one of the questions. Appoint someone to keep score in each group. A student wins a point for a correct answer, and the questioner wins a point for correct intonation.

8. Student teachers

No preparation

After you have modelled and drilled target language, give students the chance to be teachers. Put them in groups and let them model, drill and correct errors.

Questions : GRAMMAR

1. Textbook writers

No preparation

1. When students encounter a new text (in the coursebook or a supplementary text), have students write the comprehension questions.
2. Divide the class in half, and then in pairs. One half of the class reads the first part of the text, and writes questions. The other half reads the second part of the text and writes questions.
3. Students swap questions, and then give back their answers for correction.

2. Question and answer matches

Before class: Prepare pairs of questions and answers based on topics from class. Cut them into strips. There needs to be one strip for each student. You'll also need Blu-tack.

1. Give one question or answer at random to each student. They must not show anyone else.
2. Students mingle and ask and answer their questions to find their match.
3. Have students Blu-tack their questions and answers to the board. You can stand around the board and discuss any issues that arose.

Variation:

- Rather than feed complete sentences, you can give prompts, e.g. *What / capital / Japan?*

3. Here are the answers

No preparation

1. Write possible answers on the board (e.g. *Yes, I do. Six times a day. Never in a million years.*).
2. Elicit possible questions from the class; challenge them to be interesting and funny. Write the questions on the board.
3. Ask students to memorise the questions. Rub the questions and answers off the board.
4. Students mingle and ask each other the questions. Students can answer with anything they like.
5. As a class, ask students to share anything interesting or funny they heard.

4. Your partner's questions

No preparation

1. Ask students what they would like to ask someone their age when they meet for the first time. Elicit several interesting ideas (e.g. *What are your dreams?*).
2. Ask students to write down five questions.
3. Get students to swap their questions with a partner. They have to memorise the questions their partner gave them.
4. Have students mingle and ask each other the questions they've memorised.
5. Students sit back with their partner and report some of the interesting responses they heard.
6. Ask students to share anything interesting with the whole class.

R Relative clauses : GRAMMAR

1. Common opinions

see page 212 - Relative clauses 1: overview

Before class: Note down the beginning of sentences you are going to use.

1. Write on the board the first half of sentences starting with We..., for example: *We like films that... / We enjoy food that.../We like people who...* They need to contain a relative clause, just up to the relative pronoun.
2. Elicit several possible ways to finish the sentences.
3. Divide students into pairs.
4. Students need to find what they and their partner actually have in common.
5. Pairs report back to the class.

2. Teach the teacher

see page 212 - Relative clauses 1: overview

Before class: Note down terms and names of famous people and places from your students' country and culture, for example *guanxi, Lei Feng, Great Wall.*

This works best with students of one nationality.

1. Tell your students you want to know more about their country.
2. Write the words on the board. Elicit some ideas. Model an answer with a relative clause (e.g. *Lei Feng was a soldier who helped other people*).
3. Divide students into small groups.
4. Ask students to write explanations of the terms.
5. Have groups check each other's definitions for grammar and factual information.
6. Have several students at once write their sentences on the board.

3. Call my bluff

see page 212 - Relative clauses 1: overview

Before class: Bring in one learner's dictionary for each group.

1. Put students into small groups.
2. Give each group a dictionary.
3. Tell students they have to find three unfamiliar words. For each one, they need to write three definitions using relative clauses: one correct, and two made up.
4. Put two groups together.
5. Groups quiz each other. For every correct guess, the team gets a point. The winning group has the most points.

Reported speech : GRAMMAR

1. Class interviews

see page 210 - Reported speech

No preparation

1. Students write interview questions for each other (these could be real-life, or as part of a role play).
2. Divide students into pairs.
3. Students take turns being interviewer and interviewee.
4. Students report on the interview they conducted using reported speech (either in writing or orally to the class).

Variation:
- Students could interview someone outside class and report back.

2. Interview jigsaw

see page 210 - Reported speech

Before class: Find a transcript of an interview related to a class topic (internet news sites often have transcripts to accompany audio/video). Cut it in half. Make a copy for each student.

1. Divide the class into two groups.
2. Give each groups one half of the interview transcript.
3. Students have to write sentences to report the key information in the interview (e.g. *She said the Government was planning to raise taxes.*).
4. Match students from two groups. They report the information to each other.
5. Students then do something with this information (e.g. role play the interview, or discuss their reactions to it).

Variations:
- If you have access to two audio players, students could work from recordings rather than transcripts.
- You could use two different interviews with people on different sides of a debate; students then come together to compare the two arguments.

3. Newspaper reconstruction

see page 210 - Reported speech

Before class: Bring a newspaper or magazine article with examples of reported speech. Make one copy for each pair.

1. Divide students in pairs.
2. Ask students to underline all the examples of reported speech.
3. Students reconstruct the original words (e.g. *He announced there would be a review* → "There will be a review").

4. They weren't true

see page 210 - Reported speech

No preparation

1. Tell the class about a time you were told something that wasn't true (make sure it's easy to understand: for example, when someone said they were too sick to come to work, and then you saw them out shopping).

2. Briefly use the example to revise the difference between direct and indirect speech (He said, "I'm sick" → He said he was sick).
3. Ask students to note down three things people have said, or that they've read, that weren't true. They should write the exact words the person used (or a translation of them).
4. Put students in groups.
5. Students share their experiences.

Making requests : FUNCTION

1. Please, teacher

Before class: You need a coin for each group.

1. Elicit from the class things students always want teachers to do. Keep it light and humorous (e.g. *let us out early, give less homework*). Write as many ideas as possible on the board.
2. Divide students into groups.
3. Give each group a coin.
4. Groups will role play teacher and students. Students practise asking the requests on the board (e.g. *Could you please let us out early?*).
5. The teacher tosses a coin. If heads, the teacher agrees (e.g. *Sure, no problem*); if tails, the teacher has to refuse politely and give a reason (e.g. *Sorry, but we have too much work to do*).

Variation:

- You could use any relevant scenario for this language (e.g. parent and children, manager and employees).

2. Excuse first

No preparation

1. Write an example of a refusal for a request on the board (e.g. *Sorry but I don't have a car*). Ask students to tell you what the original request was (e.g. *Could you please give me a lift?*).
2. Ask students to write ten requests on one page in their book, and ten matching refusals on another page.
3. Divide students into pairs. Students tell each other their refusals. The other student has to guess what the request was (*the first student can guide them by saying cold, hot* etc).

4. Finish by asking for volunteers to try with the whole class on the board.

Reviewing grammar and vocabulary : VARIOUS

1. Kapunk!

Before class: Prepare cards with numbers from 10 to 1,000,000, and some with *Kapunk!*, written on them.

This is a fun way of scoring any competitive activity on the board. Put your students into teams and have one student come to the board and compete against the others to win the chance to select a points card. The team that finishes first selects a points card. If they're unlucky enough to select a Kapunk! card they lose all of their points. Tasks can range from anything from writing a correct sentence using the picture you have selected to spelling tasks using flashcards.

Variations:

- You may want to let two students come to the board at a time to make games more communicative and to build the confidence of weaker/shyer students. You can also give groups one chance to spot mistakes and help their team members (this keeps them interested in what their team members are doing!). You might want to enforce rules e.g. any team not using English loses 1000 points.

2. Categories

Before class: Choose categories based on topics students have been studying.

1. Divide students into pairs.
2. Ask each pair to copy a grid such as the following (the choice and number of categories depend on the class):

A	B
sports	
jobs	
countries	
furniture	
colours	

3. Write a letter on the board (e.g. *B*). Pairs compete to fill each square with a word starting with that letter, for example:

A	B
sports	*baseball*
jobs	*builder*
countries	*Belgium*
furniture	*bedside table*
colours	*blue*

3. Word chain

No preparation

Demonstrate with the whole class first.

1. Students stand in a circle.
2. One student says a word (e.g. *house*). The next student has to say a word starting with the final letter of the previous word (e.g. *egg*). The next student does the same (e.g. *golf*), and so on around the circle.
3. Devise a penalty if a student gets it wrong (e.g. they have to sing a song).
4. Once students understand the activity you can divide the students into smaller groups.

Variation:

- You can specify a topic area (transport, food etc). The group can challenge a student if they don't believe a word fits into the category.

4. Backs to the board

No preparation

1. Divide the class into two teams.
2. Ask each team to list ten words they've recently learnt.
3. Ask one member of each team to write their words in large letters on the board.
4. Place two chairs at the front of the class facing away from the board. Ask one student from each team to sit on one of the chairs.
5. Stand behind the students and circle one of the words.
6. In turn, the two students ask their team a yes/no question (e.g. *Can you eat it? Is it a vegetable?*). If the answer to their question is yes they can ask another question.
7. The team scores a point when they get a correct answer. The team losing the point has to replace their student at the front.
8. The first team to score ten points is the winner.

Variation:

- The teams have to mime the circled word. Any action is allowed except mouthing or spelling the word. The students in the chairs try to guess the answer.

5. Blankety blank

Before class: Prepare a list of sentences containing vocabulary or grammar you want to revise.

1. Write one word from each sentence on the board.
2. Divide the class into groups.
3. Read out your sentences to the students, replacing the word with blank e.g. 'I've just blank my lunch'.
4. Each group writes the answer in large letters and holds it up. The first correct team wins a point.

Variation:

- You can increase the challenge by not writing the missing English words on the board, or including two blanks in each sentence.

6. Grammar auction

Before class: Write at least ten sentences in large letters on separate pieces of card. Half of the sentences need to contain a grammar error. You also need toy money and Blu-tack.

1. Divide students into small teams.
2. Give each team an equal amount of money.
3. Tell students you'll show them a sentence. It could be correct or incorrect. If they think it's correct, they should bid on it. The team with the most correct sentences at the end is the winner.
4. Blu-tack a sentence to the board. Ask for bids. Give the sentence to the team with the winning bid.

7. Word cards

Before class: Print sets of small cards, each with a different word on them. Aim for a wide range of parts of speech, for example:

(pronouns) *I, he, me, we, it*
(past tense verbs) *was, sent, took, said, met, spoke, wrote*
(plural nouns) *letters, flowers, parks*
(state verbs) *remember, love*
(prepositions) *to, in*
(other useful words) *never, how*

You'll need one set for each pair.

1. Divide students into pairs.
2. Students must arrange the cards to make complete sentences, for example: *I remember how we met, He sent me flowers*. Each card can only be used once.

Variation:

- To prepare, you can write a number of complete sentences and write those words on the cards. You can then challenge students to use up all the words to make their sentences. The first pair to use all their words wins.

8. Whispers

Before class: Prepare a list of sentences containing vocabulary or grammar you want to revise.

1. Divide the class into groups of six or more, and arrange each group in a straight line.
2. Ask for a volunteer listener from each group. Take them outside of the classroom and tell them one or more sentences (depending on students' level).
3. Open the door, and let the students run to the first member of their group to whisper the message.
4. Each member passes on the message, by whispering, to their neighbour.
5. When the message reaches the end, the last person has to run to the board and write the message.
6. The winner could be either the first team to pass a legible, complete message (even if it's wrong), or the first team with a message closest to the original.

9. Story challenge

see page 221 - Past simple 1

Before class: Note down ten words you'd like students to practise. They can be completely unrelated by topic.

1. Write the words on the board.
2. Students have five minutes to write a coherent story that contains all ten words.
3. Students swap their stories and give feedback to each other.

Variation:

- You can also specify the genre (e.g. love story, newspaper report).

10. Sentence barter

Before class: Make multiple copies of sentences based on any material you've studied in class (the font needs to be quite large). Cut the sentences up into three or four parts in the spaces between words, for example:

WOULD YOU	LIKE TO	SEE A MOVIE

1. Shuffle all the cut-up sentences.
2. Divide students into pairs. Give approximately the same number of parts of sentence to each pair.
3. Set a time limit of ten minutes. Students can swap parts of sentences to make complete sentences.
4. The winning pair has the most complete and correct sentences.

Variation:

- You can cut up words rather than sentences, and students have to finish with as many words as possible.

S Science & technology : TOPIC

1. The perfect invention

No preparation

1. Elicit problems people have in their everyday lives (e.g. *traffic jams, boring jobs*).
2. Divide students into groups.
3. Students brainstorm inventions that would solve these problems.
4. Ask groups to present their best idea to the class.

2. Best and worst

Before class: Bring several pictures of 'good' and 'bad' technology (e.g. medical equipment and an atomic bomb). You also need A3 paper, coloured pens and Blu-tack.

1. Show the class the visuals and ask for other examples of good and bad technology. Write them in two columns on the whiteboard.
2. Divide students into groups.

3. Ask them to design a poster that describes the three best and worst forms of technology in history.
4. Display the posters.

Variation:
- Students could research the technologies and print pictures for their poster.

3. Nature mystery

Before class: Create a handout comprising photos of mysterious creatures or phenomena (the descriptions should not require scientific knowledge – e.g. 'the eye of a fish'. Also, you need to know what they are!). Number the photos. Make one copy for each group. Blow up one of the pictures so it's large enough for the whole class to see.

1. Show the large picture to the class. Challenge students to tell you what it is.
2. Divide students into teams. Ask them to give their team a name.
3. Give one copy of the handout to each team. Ask them to write their team name at the top. They have five minutes to try to name or describe the objects.
4. Teams swap their answers for marking.
5. Reveal the answers and find the winner.

Variation:
- This activity can be used for any topic: guessing people, places etc.

4. Science presentation

Before class: Note down presentation guidelines and criteria for audience feedback. If possible have access to a data projector.

1. Tell students they need to give a presentation on an area of science or technology. The topic should interest the class.
2. Divide students into pairs.
3. Allow students to research their presentation (either in class or as homework). They work together to script their presentation and create visuals and props.
4. Students present. The audience should make notes on the performance according to criteria (e.g. pronunciation, clarity, interest, non-verbal language).

Extension:
- You can video the presentation and have presenters evaluate their own performance using the same criteria.

Shopping : TOPIC

1. Shopping secrets

No preparation

1. Tell students you need their help. You're not sure where you should shop locally (for one type of product, e.g. food). Students call out ideas.
2. Elicit different types of shops (food, clothes, books etc). Write them on the board.
3. Divide students into groups.
4. Ask student to brainstorm their suggestions for each type of shop.
5. Bring the class together. Ask for volunteers to come to the front to present ideas. They can use the board to write the name, draw a map etc.

2. Find a bargain

No preparation

1. Divide the class in half: sellers and customers.
2. Each seller needs to choose something they have with them to sell (for example, a pen or a mobile phone). They decide what price to charge (they can change the price later).
3. Set a time limit of four minutes. Customers need to find the best bargain at the best price.
4. Swap roles.
5. As a whole class, discuss what made the most successful buyers and sellers.

3. Gift shopping

Before class: Bring blank scraps of paper, toy money.

1. Announce it's shopping time (before Christmas/ Spring Festival/Eid etc). Elicit examples of typical gifts for this occasion.
2. Divide the class into sellers and customers.
3. Give each seller six scraps of paper. They need to draw one different product on each sheet of paper.
4. Give each customer an equal amount of money. They need to write a list of four people they need to buy gifts for, and note down what they might like.
5. Set up the classroom like a shopping centre or market.
6. Start the role play. The sellers must aim to sell everything in their shop, and customers need to buy all the gifts they need.

Variation:
- Use real objects instead of drawings.

Special occasions : FUNCTION

1. Idiom race

Before class: Note down 8-10 hard-to-guess idiomatic expressions (e.g. *Break a leg! I do!*)

1. Write the expressions on the whiteboard (in large letters, randomly spaced). Ask students to suggest when you use them (*before a performance, when you get married*).
2. Divide the class into two teams.
3. Call out the situation (e.g. 'Before an exam!'). One student from each team runs to the board, puts their hand on the expression and calls it out (*Break a leg!*).
4. Finally, ask students write and perform mini dialogues.

2. Three-liners

No preparation

1. Divide students into pairs.
2. Ask students to write three-line dialogues that include the language for special occasions they've learned. They should start with the situation, for example:
   - I failed my exam.
   - I'm sorry to hear that.
   - Thanks.
3. Tell students to close their books.
4. Students mingle and practise three-line dialogues (of course students can't predict what others will say).

Sport : TOPIC

1. Sport collocation volleyball

Before class: Bring one balloon for each group.

Play sport to practise sport vocabulary!

1. Review sport-related collocations students have learnt (*kick a ball* etc).
2. Divide the class into groups, and each group into two teams.
3. The teams play 'volleyball' with the balloon. The serving team calls out the first part of a collocation (kick) and hits the balloon. The other team calls out the second part of the collocation (a ball) and hits the balloon back.
4. If a team can't respond they lose a point.
5. Unlike real volleyball, the losing team serves.

Variations:

- This could be used for any sort of word pairs (e.g. tennis – net).
- An alternative rule is that the receiving player gives the response (a ball), then must pass it to another player on their team, who calls out the first part of a new collocation (e.g. tackle).

2. Sport commentary

Before class: You need a DVD of a sports game and a DVD player.

1. Analyse one or two sentences of the commentary. Highlight the vocabulary, grammar (typically, sports commentary is in present simple: *He catches the ball …*), voice range and speed of delivery.
2. Tell students they're going to give their own commentary.
3. Play a one-minute section of the game several times with the sound down. Students note down what happens.
4. Divide students into pairs. Ask them to write a commentary and practise delivering it.
5. Ask for volunteers to come to the front and deliver their commentary along with the DVD.

3. Plan a sports day

No preparation, but you may need permission from the school.

Plan a real-life party (or picnic or barbeque) with your students. Appoint several people as organisers. Plan anything students will need to buy and agree on a budget.

Using stories : VARIOUS

1. Stand up

No preparation

1. Find a short story suitable for your students' level and age.
2. On pieces of paper write key words
3. Hand out a word to each student.
4. While you read the story, students must stand up when they hear their word and call it out (while simultaneously holding the word in the air).

2. Stories from words

No preparation

1. This activity helps activate new vocabulary and improve narrative techniques.
2. After reading a text, ask groups of students to select certain words from the text.
3. When they have finished, get groups sitting next to each other to swap lists.
4. Individual groups must then create a story containing the words from the list, either orally, or in writing. Winners can be selected by vote.

3. Draw the story

Before class: Use the story on page 109, or find another short story suitable for the level of your class (you can adapt one). Identify around 10 key images in the story. Create a version of the text with key words gapped, and make one copy for each group. You'll also need sheets of A3 and coloured markers.

1. Bring students to the front of the class. Read out the key images, and ask the students to draw them on the board.
2. Inform the students you'll tell a story and they have to listen carefully.
3. When telling the story, pause and encourage the students to identify the next word through use of the images: *Once upon a time there was a* (point to the picture of the farmer) *and he lived on a big* (point to the picture of a farm)'. Use a different tone in your voice, along with hand gestures, to emphasise adjectives and other key words.
4. Tell the story once more, encouraging the students to join in.
5. Give students the gapped copy of the story. Students have to fill in the gaps with the correct word from the story (you could focus on a grammar point here, such as the past continuous).
6. Put students into groups and give them a sheet of A3 and a pen. They have to draw the images to help them recount the story again in their groups and to the class.
7. Give them around 10-15 minutes to draw their pictures and decide who will say what part of the story.

Extension:

- After class, get students to write up the story for homework.

4. Story gestures

Before class: Find a short story suitable for the level of your class (you can adapt one). Decide on hand gestures to represent key words.

1. Tell the students you'll tell them a story and you want them to copy your hands. Hold your index finger up and ask the students to copy you. Do a few more actions and get them to copy so they understand and feel comfortable.
2. Tell the story in sections (go back to the beginning a few times, so the repetition helps them remember the actions). For example, *Once* (hold one finger up) *upon a time* (point to your watch), *there was a beautiful girl* (circle face) ...
3. Tell students you'll repeat the story, but this time, they have to do the hand actions and say the story with you (the hand actions will help them remember the story).
4. Now tell the class you won't say a word, and they have to remember the story. All you do is the hand gestures.
5. Put the students into pairs. One person has to do the hand gestures, and the other tells the story.

Variations:

- If you have an open ending to your story, students can finish off the story for homework.
- Draw a story sequence on cards, and, to create interest, get the students to predict the story first based on the drawings.

Superlatives : GRAMMAR

1. Superlative survey

see page 185 - Superlatives: the -est and the most

Before class: Prepare a survey based on what you know about your students and their interests, for example:

Who	Notes
...has met the most famous person?	
...has the least money in their pocket/purse right now?	
...is wearing the nicest pair of shoes?	

Copy one for each student.

1. Show students the survey (you could do the first item as an example on the board).
2. Tell students they'll need to ask as many people as possible to know which person is unusual. Elicit the type of question students will need to ask (*Have you met a famous person?*). They should take notes as they speak to people.
3. Students mingle and ask each other the questions.
4. Have students sit in small groups to compare notes and decide on their answers.
5. Students share their decisions with the whole class.

2. Topic superlatives

see page 185 - Superlatives: the -est and the most

Before class: Note down five nouns you're going to use.

1. Divide students into pairs.
2. Tell students this will be a race.
3. Write up five nouns on a recent topic in class (e.g. *Jaguar, Toyota, motorbike, Henry Ford, LPG*).
4. Students must write one sentence containing a superlative for each noun (e.g. Jaguars are the most attractive cars in the world).
5. The first pair with five correct sentences wins.
6. Finally, get pairs to compare their opinions.

T Tag questions : GRAMMAR

1. Find someone tag questions

see page 206 - Tag questions

Before class: Create a handout containing statement prompts, based on what you know about your students (e.g. speak/Japanese). Make one copy for each student.

1. Tell students they have to speak to other students and check information about them.
2. Write the first prompt on the board as an example. Students must use either a positive sentence with a negative tag (*You speak Japanese, don't you?*) or a negative sentence with a positive tag (*You don't speak Japanese, do you?*). Elicit the possible responses (*Yes, I do or No, I don't*).
3. Give one handout to each student.
4. Students mingle and ask each other.
5. Ask students to share any surprising information they found out.

Extension:

- You could follow with less controlled practice where students ask their own questions.

2. Group quizzes

see page 206 - Tag questions

Before class: Bring A4 paper.

1. Divide students into groups.
2. Students write ten tag questions about a topic recently covered in class. Half should be true, and half false, and they should draw a tick or cross accordingly (for example: *Rome's the capital of Italy, isn't it?*✔ / *Caesar was Greek, wasn't he?*✘).
3. Each group chooses a quizmaster or quizmistress.
4. Groups swap their statements.
5. The quizmaster or mistress reads out the statement without the question tag (*Rome's the capital of Italy*). The first student to give the correct tag (*isn't it?*) gets a point, and the first student to respond correctly (*Yes, it is*) wins a point.
6. The student with the most points in the end is the winner.

Transport : TOPIC

1. Getting to class

No preparation

1. Elicit different ways people get to class (e.g. *bus, bike* etc).
2. Divide students into groups.
3. Instruct one person in each group to write. Ask them to copy the following grid:

type of transport	price	comfort	time	other

4. Ask students to list the types of transport in the left hand column, and make notes on the other columns.
5. Ask them to put the grids away.
6. Reconfigure the groups so students are sitting with new people.
7. Ask students to share how they get to class now. The other students should try to convince them to try a different way.
8. With the whole class, ask who was convinced to try a different form of transport.

2. Transport solutions

Before class: Bring one sheet of A3 for each group.

1. Use a real experience to elicit some examples of transport problems (e.g. *traffic jams*). Speak generally rather than criticising your students' countries.
2. Divide students into groups. Give one sheet of A3 to each group.
3. Draw a grid on the board for students to copy:

Problem	Solution

4. Students work together to make notes in the table.
5. Ask a representative from each group to fill in the grid on the board.
6. Students work together to summarise the ideas (e.g. as a report with recommendations).

TV : TOPIC

1. Favourite shows

No preparation

1. Introduce the topic of TV. Ask the class to tell you five popular local programs, and five popular imported programs.
2. Ask each student to write down two programs they love, and one they can't stand.
3. Get students to stand. They have to find someone with exactly the same programs they love and can't stand.
4. Have students share with the whole class what they found about other students.

2. What will we watch?

Before class: Photocopy a TV guide to the evening's TV. You'll need one for each group.

This works well if you know your students have very different tastes.

1. Ask students to note down what sort of TV they like, and what they don't like.
2. Divide students into groups.
3. Tell them they live together and have to decide what to watch on TV tonight. There's only one TV in the house/apartment.
4. Hand out one TV guide to each group.
5. Students discuss what to watch.

Variation:

- Depending on what's in the guide, you can make role play cards to ensure there is disagreement: e.g. 'You love sport and hate movies', 'You love movies and hate sport' etc.

W Weather : TOPIC

1. Weather report

Before class: Record a weather forecast from the radio. You'll need a CD player and the CD (or, if possible, two or three CD players and copies of the CD).

1. Tell students they're going to listen to a weather forecast.
2. Pre-teach any difficult vocabulary in the recording.
3. Tell students in fifteen minutes you want to see the forecast written perfectly on the board.
4. Give students control of the CD player. Tell them they can listen as many times as they like.
5. Observe from the back of the classroom; only intervene if they are having extreme difficulty.
6. If you have room and equipment, let several groups listen separately and then pool their ideas at the end.
7. Finally praise the class for their achievement. Correct sensitively any errors in the transcript on the board.

Variation:

- This activity can be used for any short listening text. However, make sure students understand the main idea before they listen for each word.

2. Weather around the world

Before class: Find climate information about five very different cities. Look for text, tables and graphs describing temperature, humidity, rainfall etc. Remove any well-known placenames. Photocopy and then cut up the information into a few pieces. You'll need one set per group.

1. Tell students we're planning to visit interesting places around the world – but before we book our trip we need to think about the weather.
2. Write up the names of the five cities.
3. Divide the students into groups. Give one set of weather information to each group.
4. Students try to match the information to each city.
5. Ask groups to discuss what time of year we should visit each destination.
6. Ask groups to share their ideas with the whole class.

Work : TOPIC

1. What's the job like?

Before class: Make two lists of six adjectives that can describe jobs. Number the words in each list 1-6. Make one list quite literal, and the other more abstract, for example:

1 nine-to-five	1 challenging
2 well-paid	2 menial
...	...
6 IT-related	6 fulfilling

You also need one die for each group.

1. Write the lists on the board. Elicit as you go: *What do you call a job that has regular hours?* etc.
2. Divide students into groups.
3. In turn, one student throws the die twice, which gives them two adjectives (e.g. 2 & 1 give 'well-paid' and 'challenging'). The student has to think of a job that both adjectives describe (e.g. 'surgeon').
4. Then the group has to talk about the job: lower levels might discuss what the person does, and higher levels might discuss whether they would like this job, and why/why not.
5. Finish with a whole-class discussion arising from the group conversations.

2. Job dilemmas

Before class: Photocopy and cut up the series of job dilemma cards on page 109. You'll need one set per group.

1. Decide the students into groups.
2. Give one set of dilemma cards to each group.
3. Students have to sort the cards into two groups, whether they agree or disagree.
4. If the group can't come to a consensus they should first try to convince each other. If there is still disagreement, the group should vote.
5. As a class, share ideas from the discussion.

Using writing : VARIOUS

Your students – especially speakers of languages not written in roman script – face challenges when they begin to write. They need to develop skills ranging from basic letter formation, to sentence-level capitalisation and punctuation, and text-level conventions of organisation and paragraphing.

You'll need to assess your students' needs for writing, and their current level, and plan writing activities accordingly. It's a good idea to set a writing task in the first few lessons to determine this.

Conventions are very important in writing: for example, if a student writes an essay as disconnected sentences rather than paragraphs, a reader immediately feels something is 'wrong'. Therefore it's helpful for students to have a model of a genre (whether an essay, a story or a business letter) which they first analyse and then base their writing on. As they're exposed to more examples of the genre, and become more confident, they can move further away from the template.

Writing is, in many ways, the undervalued skill in TEFL. There's a much narrower range of resources and activities for writing than, for example, speaking. A lot of writing materials still used in schools are pretty dry. You may find the best thing is to write your own. You can use your own sources for inspiration and tailor activities to the level and interests of your students. A key benefit of doing this is that it will help you to anticipate what your students will struggle with.

1. Consequences

Before class: Copy and cut up sentences, appropriate for the level of your class. For example:

A cat's watching television.

or

The CEO heads the organisation, which comprises two business units.

Make a set for each group. You'll also need some cut-up scraps of paper.

1. Put students into groups of about four.
2. Give each group one set of sentences face-down, and some paper.
3. Tell students they can't show anyone else a sentence or a picture someone gives them until the end of the activity.
4. Demonstrate the procedure with one group:
   - Student A takes a sentence and draws it.
   - Student A gives their drawing to Student B, who then writes a sentence to describe the picture.
   - Student B gives their sentence to Student C, who draws it.
   - Student C gives their drawing to Student D, who then writes a sentence to describe the picture.
   - All students then reveal their drawings and sentences.
5. At this stage students compare the writing with the original sentence. Write the prompts
   - word choice
   - spelling
   - punctuation
   - grammar

on the board as areas for discussion.

2. Hidden story

Before class: Bring pens and paper.

1. Tell students they're going to write a story.
2. Present the structure on the board:

A boy (name 1) met a girl (name 2)
in/at (place) because (reason).

He said to her (quotation 1)
so she said to him (quotation 2).

So he (action 1)
and she (action 2).

As a consequence (final event).

3. Divide students into pairs (students can be on their own in a small class).
4. Demonstrate the activity:

- The first pair writes the first line at the top of the paper, following the structure (e.g. *A boy, Simon, met a girl, Helen*). They fold it over so no-one else can see it, and pass it to the pair on their right.
- The second pair writes the second line (e.g. *in Mongolia because they were Mongolian*). Again they fold it over so no-one else can see it, and pass it to the pair on their right.
- This continues around the class until the end of the story.

5. On completing the game, students read out the unfolded papers in turn. Resulting stories are invariably funny, and students enjoy this activity.

Extension:
- Once students have written their story, ask them to switch it from the past to the future tense. Silly stories get even sillier when you're predicting what will happen!

3. Creative writing

No preparation

This is a technique to draw further vocabulary and writing practice from a text students have read.

1. After students have read a text in class, ask each student to make a list of ten words in the text they think are useful.
2. Students then swap their list with the person next to them (indicate who the pairs are).
3. Students create a story. They must use all the words on the list.
4. Students compare stories with their partner.

4. Letter of complaint

Before class: Copy the two versions of the same letter of complaint on page 110. One has no spaces between paragraphs; the other has correct paragraphing, but with key signal words and formulas gapped. Make one copy of both letters for each student.

This activity uses a model of a genre as a basis for writing.

1. Divide students into groups.
2. Ask them to discuss when they would complain about something, and how they would do it. When would they write a letter, telephone, or visit in person? Elicit some whole-class feedback.
3. Give students the first version of the letter. Ask them to identify where the paragraphs should begin and end. Have students compare, and then confirm the answers (showing the letter on an OHT is useful for this).
4. Ask students to read each paragraph in detail and summarise what sort of information goes in each.
5. Tell students to put the first version away. Give them the second version. Ask them to fill in the missing words. Have students compare, and then confirm the answers.
6. Have students write their own letter of complaint based on the model.

Variation:

- This activity can be used to teach any genre of written English.

5. Editors

No preparation

This is a variation on process writing where, like in real life, writing goes through several stages.

1. Divide the class into pairs or small groups. The number of groups needs to be a multiple of three.
2. Ask each group to make a rough outline for a piece of writing (depending on what the course requires them to do).
3. Groups pass their notes clockwise to the next group, who then write the notes out as a complete draft one of the text.
4. Groups now pass what they've written to the next group, who edits the text. To guide students, write an editing checklist on the board, for example:
   - paragraphs
   - word choice
   - spelling
   - punctuation
   - grammar
5. Groups then pass the edited version back to the original writers. They read it carefully and make any changes.

Variation:

- This can be used for any genre of writing.

6. Culture story

No preparation

1. Tell students you want to know more about their traditions.
2. Divide student into pairs.
3. Students write a story or song in English that they know from their first language (from memory – not translating a text).

Variation:

- Students could present these as illustrated posters for the walls. Monitor while students are first writing to address errors before the final version.

2.4 Activities: A-Z Photocopiable Materials

These are the photocopiable activities and other materials we refer to in the Activities A-Z. They're cross referenced to each activity.

Some are for you to hold up when presenting new language, or encouraging discussion – these just need enlarging on a photocopier. You can, of course, just use these for ideas and create your own to match your students' needs and interests.

Activities for students generally need copying and cutting up. To avoid doing this every couple of months it's worth getting access to a laminator and make copies of games and activities you'll use frequently.

Adverbs in -ly

Adverb mime

A	B
paint a ceiling	quickly
cook pasta	slowly
check a car engine	nervously
fill in a form	dangerously
do the ironing	cheerfully
play pool	excitedly
ride a motorbike	badly
wash the dishes	angrily
play saxophone	carefully
try on clothes	carelessly

Giving advice

Study advice

I can never remember new words.
People don't understand my pronunciation.
I want to speak more fluently.
English verbs confuse me.
I can understand the teacher, but not people on the street.
I'm too shy to speak up in class.
My spelling's not very good.
My reading's really slow.
Sometimes I don't understand what to do in class.
I always make mistakes when I write.

Agreeing and disagreeing

Heads I agree, tails I disagree

sport	politics
fashion	food
travel	shopping
the Internet	men and women
young people	old people

Apologising

One-minute apologies

You broke a window.
You arrived late for work.
You arrived late for a date.
You stepped on someone's foot.
You drove into someone's car.
You played music too loudly last night.
You forgot someone's birthday.
You took someone else's bag of shopping by mistake.
You're at a restaurant and you left your money at home.
You didn't do your homework.
Your dog dug up someone's garden.
You got angry and called someone stupid.

Business

Business meetings

Director

Your members of staff aren't happy at the moment. They work long hours and they're suffering from stress. You can't afford to pay them more, so you need to think of a way to keep your staff motivated and happy at work.

In the meeting, you will listen to your employees' ideas on how to do this. Think about what will:

- keep staff happy
- help motivate staff
- be cost-effective

You need to choose one idea.

Employee 1

You want to motivate the staff.

You've noticed a lot of staff have back problems as they sit in front of a computer all day. You want to employ a masseuse, to give massages to staff, in order to:

- reduce stress
- increase people's physical wellbeing
- clear people's minds

Write other reasons here:

-
-
-

Employee 2

You want to motivate the staff.

A lot of your team don't like sitting in peak hour traffic. You want to employ chauffeurs to pick staff up and drive them to work. This would:

- reduce stress getting to work
- help staff save money on petrol
- enable staff to give up their cars – this will save them a lot of money!

Write other reasons here:

-
-
-

Employee 3

You want to motivate the staff.

There are a lot of pregnant women and new mothers working in the office. You want to have a childcare facility in the building so that:

- staff can see their children
- maternity leave will be reduced
- staff will be less stressed, as they don't have to worry about their children.

Write other reasons here:

-
-
-

Maintaining a conversation

Start a sentence

I think ...	Yes.
I'm not sure.	No.
Why ...?	Let me think.
Um ...	Ha ha ha.
Wow!	If you ask me, ...
By the way, ...	Do you know something?
In fact, ...	What ...?
Do you ...?	How ...?

The countryside

New life

I might move to the country to have a quiet life.
I might move from the country to work in a factory.
I want to go back to nature and grow my own food.
I want my children to move from the country to the city to study.
I'd like to have my own cows and slaughter them for food.
When I retire I'm going to live in the country.
I'm going to buy a farm and have lots of kids to help out.
I'm going to live the simple life with no phone or Internet.
I want to marry someone from the country. They're more honest.
I insist my children leave the farm so they have more opportunities.

Crime

The punishment fits the crime

murder
stealing a car
stealing intellectual property (e.g. music from the Internet)
workplace bullying
lying to the Taxation Department
shoplifting
assault
manslaughter
verbal assault
drink driving

Education

Successful student

organise my study schedule and workspace
do lots of written grammar practice
just speak – errors don't matter
look up the dictionary and grammar books
use English whenever I can
keep a vocabulary book
organise my notes
be sure I know the rules before I speak
do fun things with English
ask for feedback
translate

Practising fluency

Don't say it

yes	no
today	yesterday
want	English
teacher	classroom
every	big
now	am
is	have
do	say

Invitations and suggestions

The perfect invitation

cinema	restaurant
cafe	bar
library	someone's home
theatre	park
shops	swimming pool
gym	countryside
beach	downtown

Men and women

Ideal spouse

Ideal husband

Number the characteristics from 1 (= most important) to 10 (= least important).

___ wealthy
___ good-looking
___ hard-working
___ loyal
___ good with children
___ intelligent
___ fun
___ romantic
___ good at cooking
___ fashionable

Ideal wife

Number the characteristics from 1 (= most important) to 10 (= least important).

___ wealthy
___ good-looking
___ hard-working
___ loyal
___ good with children
___ intelligent
___ fun
___ romantic
___ good at cooking
___ fashionable

Passives

Passives board game

Use these verbs (in the correct passive form) to complete the sentences.

build invent take add wear write cut break bring deliver spend shoot sell
give play steal fly leave grow make increase release buy win pay

START 1. Apples ____________ on this farm every year.	2. Tennis ____________ in Wimbledon.	3. This table ____________ by my grandfather.	4. Many things ____________ in this shopping centre	5. The first telephone ____________ in 1876.
6. The police think the painting ____________ around 8pm last night.	7. Millions of letters ____________ every day.	8. Her leg ____________ in five places.	9. The sandwiches ____________ at around 12pm every day.	10. The Pyramids ____________ by The Pharaohs.
11. Taxes ________ usually ________ every year.	12. This book ____________ by Shakespeare!	13. Milk ____________ to tea in Britain!	14. "I'm sorry, this car ____________ by someone about an hour ago."	15. Those trees ____________ down a few years ago.
16. The money ____________ by a very lucky woman.	17. "I ____________ late this month, so I had no money to pay for my rent!"	18. The plane ____________ by a new pilot!	19. Her bag ____________ from her chair when she wasn't looking.	20. A lot of money ____________ to the hospital every year by the public.
21. The man ____________ by his wife and died later in hospital.	22. All of our money ____________ on a car?!!	23. The dogs ____________ in the house by itself for 1 week.	24. Many songs ____________ every year. **FINISH**	25. This dress ____________ by Madonna!

Past simple

Table competition

am/are/is	take	was/were	took
have/has	see	had	saw
do	like	did	liked
get	love	got	loved
go	want	went	wanted
buy	live	bought	lived
bring	leave	brought	left
make	lie	made	lay

Personal information

Hidden personality

Does my partner ...

- excel at sport?
- tell jokes?
- dance?
- love animals?
- play practical jokes?
- love books?
- cook?
- travel?
- love languages?

Is my partner ...

- quiet?
- generous?
- sensitive?
- creative?
- romantic?
- rebellious?
- kind?
- untidy?

Using stories

Draw the story

Once upon a time there was a farmer, who lived on a big farm in the mountains with his wife. On his farm he grew carrots, cauliflowers, onions, potatoes, wheat, beans and peas.

Every morning he and his wife drank coffee and looked at the beautiful mountains. One day, while he was drinking his coffee, he saw a goat eating his vegetables. So the farmer went outside and said to the goat, "Please stop eating my vegetables!". But the goat just looked at him and carried on. So the farmer went back into the house to tell his wife what had happened.

While he was telling his wife, they heard a little voice. They looked down and saw an ant. "I will help you if you give me a sack of flour and a sack of sugar," said the ant. The farmer and his wife didn't think that something so small could help, but they agreed.

The farmer and his wife watched from the window as the ant went to the vegetable patch. The ant crawled up the goat's leg and onto its neck and it stung the goat! It then jumped off the goat. The goat thought , "Oh no, I must be standing on an anthill' and so it started rolling down the hill ... and it continued rolling and rolling and rolling! Maybe it is still rolling!"

Using stories

Story gestures

Once upon a time, there was a beautiful girl called Helen. She lived in a small house in a small town with her grandmother. Her parents had died when she was a little girl. Helen didn't like her grandmother as she was very mean to her. One day, Helen decided to run away. So she packed her clothes and left home.

It was a nice day, the birds were chirping and the sun was shining. At first Helen was very happy. But she started to feel lonely so she started to cry! Suddenly, she noticed a cow in front of her ... but it wasn't a normal cow. The cow was singing! Helen walked to the cow and said "Excuse me, but why are you singing and how are you singing? Cow's can't sing!". The cow looked at her and said ...

Work

Job dilemmas

Job satisfaction is more important than money.
I'd let my child have any job they wanted.
It's my duty to earn as much as possible to support my family.
Men and women are clearly suited to different jobs.
Changing jobs frequently gives you rich experience.
There are some jobs I would never do.
Family should come before work.
The best way to get people to work hard is to encourage them.
It's not up to the government to look after the unemployed.
The military is an honourable career.
It's unfair that some people are paid much more than others.
Teaching is an easy job.

Using writing

Letter of complaint

A

Felicity Kent 11 Birch Way London L11 7JW 30/10/2010 Dear Sir/Madam I am writing to complain about our recent stay at your hotel on 11th October 2010. Firstly, when we arrived we had to wait for 10 minutes as there was no one in reception. Secondly, when we got to our room, we realised we were given the wrong key. When we finally got into our room, we were extremely disappointed to find that it was small and instead of a double bed, there were two single beds. In addition to this we paid for a sea view room, but instead we had a view of the car park. When we told the reception staff, they were very rude and told us we couldn't change the room. What's more, our room was above the hotel bar and it was extremely noisy. Your website states that you offer 'excellent' service and your hotel is 'a four star hotel, ideal for a quiet, relaxing holiday'. However we are extremely disappointed with the service we received. As a result, we would like a refund for the extra we paid for the sea view room and also compensation of one night's stay (£178 in total). I look forward to hearing from you and receiving the compensation I have requested above. Yours faithfully Felicity Kent

B

Felicity Kent
11 Birch Way
London
L11 7JW

30/10/2010

Dear Sir/Madam

I am writing to ___________ about our recent stay at your hotel on 11th October 2010.

___________, when we arrived we had to ___________ for 10 minutes as there was no one in reception. ___________, when we got to our room, we ___________ we were given the wrong key. When we ___________ got into our room, we were extremely ___________ to find that it was small and instead of a double bed, there were two single beds. In ___________ to this we paid for a sea view room, but ___________ we had a view of the car park. When we told the reception staff, they were very ___________ and told us we couldn't change the room. What's ___________, our room was above the hotel bar and it was extremely noisy.

Your website ___________ that you offer 'excellent' service and your hotel is 'a four star hotel, ideal for a quiet, relaxing holiday'. ___________ we are extremely disappointed with the ___________ we received.

As a ___________, we would like a ___________ for the extra we paid for the sea view room and also compensation of one night's stay (£178 in total).

I look ___________ to hearing from you and receiving the compensation I have requested above.

Yours faithfully

Felicity Kent

Lesson Plans

Introduction

A major challenge for teachers is how to bring ideas and activities together quickly to plan a cohesive lesson. This section proposes two essential models of lesson plans which you can use as the basis for any sort of lesson. In addition, the section includes a range of complete lesson plans (Lesson Plans A-Z), together with photocopiable materials ready to use.

1. Two Lesson Plans Structures You Must Know

These are two lesson types and stages you can use to teach just about anything.

2. Lesson Plans A-Z

Here we've put the theory into practice, with a range of complete lesson plans, including photocopiable materials.

3. Lesson Plans A-Z: Photocopiable Materials

These are the photocopiable materials referred to and cross-referenced in Lesson Plans A-Z.

3.1 Two Lesson Plan Structures You Must Know

These are templates for you to use as a starting point for your own lessons. Feel free to mould them to your own teaching style and your students' needs.

The first looks at receptive skills (for teaching listening and reading) and aims to engage students with a text, before using activities to lead them to a more detailed level of understanding.

The second is for teaching new language, such as grammar, functions and vocabulary, and is based on the widespread PPP model (see TEFL A-Z).

These are suggested structures for two types of lessons (many English coursebooks use similar frameworks for sequencing activities):

1. **receptive skills** (reading or listening)
2. **new language** (vocabulary, function or grammar)

These lesson plans aim to make sure students have lots of practice. In any lesson, students should spend at least half of the lesson practising (ideally more).

It's always a good idea to start any lesson with a warmer (see page 36), where students talk to other students in a simple activity. This establishes a relaxed and communicative atmosphere for the whole class.

1. Teaching Receptive Skills (Reading or Listening)

This structure works for any reading or listening text. It aims to engage students, and to ensure they understand the text in detail. To make sure this happens they need to do quite a lot before they ever see or listen to the text.

Aim: Students listen/read for main ideas and detailed understanding	
Level: All	Time: 60 mins
Assumptions: (This is what students can do already e.g. they know most of the vocabulary on this topic)	Anticipated problems and solutions: (This is what you think students will have *trouble* with, and what you will do to help them – e.g. they may not know a key word in the text, so you will teach it before they read)

Main stages	Microstages	Sample activities (Just choose one or two at each stage)
Pre-text Aim: students are interested and ready to listen/read **Time: 10 minutes**	**Introduce the topic (lead-in)** Aim: students are interested in the topic	• Students look at pictures related to the topic, and discuss/brainstorm what they know • Students discuss questions related to the topic • Students create a mind map related to the topic • Students see a headline and visuals from the text, and predict what it will say
	Introduce the text Aim: students know where the text comes from	• Show students the source of the text (e.g. hold up the newspaper where the article is from)
	Pre-teach vocabulary Aim: students know the vocab they will need to complete the listening/reading activities	• Teach students words they need • Students look up the words and teach each other
Text Aim: students achieve a detailed understanding of the text **Time: 40 minutes**	**Listening/reading for main idea** Aim: students understand the main ideas(s) in the text	• Give a simple question – 'Is it about X or Y' • Tick the topics you hear • Sequence the topics you hear • Match a topic to each paragraph Give students a time limit (for a reading text)
	Listening/reading for detail Aim: students understand the text in detail	• Short answer questions • True/false questions • Multiple choice questions • Matching halves of statements • Completing a table • Students write questions for each other • Underlining a feature of the text (e.g. numbers) and working out what they mean • Correcting a summary (There could be several activities as long as they become more detailed)
	Analysing language Aim: students learn and practice language naturally occurring in the text	• Have students underline certain language in a reading text • With a listening text, ask students, 'What do they say exactly about X' and play one section of the recording several times • Analyse the meaning, structure and pronunciation • Give students a practice activity
Post-text Aim: students respond to ideas in the text **Time: 10 minutes**		• Students discuss their reactions to the text (e.g. what they liked, whether they agreed) • Students role play characters from the text • Students write a response to the text

2. Teaching new language (vocabulary, function or grammar)

This structure works for any new language: students learn a language feature, analyse its form, meaning and pronunciation, and then practise it. Make sure practice takes up at least half of the class.

The model plan below uses a context – a real-life situation that you create, with a location and characters – to elicit the target language. However, you can also elicit the target language from a reading or listening text.

Aim: Students learn and practise X (for example 'telling the time' or 'present perfect for life experience')	
Target language: (Describe exactly what details of X you will teach: e.g. *hours, a quarter past, half past; no minutes*)	
Level: All	Time: 60 mins
Assumptions: (This is what students *can* do already – e.g. they already know numbers up to 12)	Anticipated problems and solutions: (This is what you think students will have *trouble* with, and what you will do to help them. Usually this will be what they confuse the target language with i.e. something similar in English or in their first language. Think of questions that will clear up this confusion – e.g. write 3.15 on the board and ask 'Is this a quarter past three, or a quarter to three?')

Main Stages	**Microstages**	**Sample Activities** (Just choose one or two at each stage)
Presentation Aim: students learn form, meaning and pronunciation of the target language **Time: 20 minutes**	**Set context** Aim: students understand what the target language means and where it is used	• Set a visual context (e.g. pictures, video, drawing on whiteboard, realia) • Establish where it is, and who the speakers are • Students discuss their experience of similar situations, what they predict will happen etc
	Elicit the target language Aim: students are exposed to meaningful examples of the target language	• Use a visual prompt to elicit an example of the target language
	Analyse the target language Aim: students understand the form, meaning and pronunciation of the target language	• Ask concept questions to check students understand the target language • Analyse relevant aspects of the structure (e.g. words, endings, parts of speech) • Model the pronunciation (at natural speed)
Controlled practice Aim: students are confident and correct producing the target language **Time: 20 minutes**	**Whole class** Aim: students develop confidence producing the target language	• Model the pronunciation • The whole class repeats • Choose individual students at random to repeat
	Pairs Aim: students further develop confidence producing the target language	• Students practise sentences in pairs, with substitutions (e.g. on cards or on the whiteboard)
Free practice Aim: students are able to use the target language fluently and meaningfully in a life-like situation **Time: 20 minutes**		• Set up a life-like situation in the class • Assign roles • Give students a real-life motivation to participate (There could be several activities as long as they are free practice activities)

.2 Lesson Plans A-Z

Take the pain out of planning with this range of complete lesson plans, along with photocopiable materials (in section 3.3). Either use them straight off the shelf – photocopy and cut up the activities, and off you go. Or, use them as a base to adapt to different target language or topics.

Some follow a lesson plan framework from section 3.1. Others are noticeably different: creating an animation in a computer class; solving a real-life problem (as the aim of the lesson); and writing a real job application (based on a model letter).

Lesson Plans A-Z: contents

Animation: using the Web

Aim: Students practise creative writing using a Web animation application	
Assumptions: Students are familiar with basic navigation on the Web; students will enjoy the chance to be creative and to use text-to-speech	
Materials: Computers connected to the Internet (one between two students, with a recent operating system)	
Level: All	Time: One hour
Anticipated problems: • Students do not understand how to use the application	Solutions: • Ask IT-savvy students to monitor and help less confident students

Activities	**Procedure**
State lesson aim Aim: students understand objective of using Internet in the lesson **Time: 5 minutes**	• Have students work in pairs on one computer. • Tell students they're going to make a movie!
Familiarisation with application Aim: students become familiar with what the application can do. **Time: 15 minutes**	• Direct students to a free Web animation application, that includes text-to-speech, such as www.xtranormal.com. • Students look at some of the movies posted by other users. • Students experiment with a two-line dialogue (including text-to-speech) to become familiar with using the application.
Create animation Aim: students prepare and create their animation **Time: 30 minutes**	• Set a very general topic (possibly related to what they've been doing in class). Pairs work on a script in Word. Monitor and correct errors. • Students create and upload their finalised movies.
Follow-up Aim: students see result of their work; students receive feedback **Time: 10 minutes**	• Students watch (and vote on) each others' animations. • Give individual and whole-class feedback. • Students visit later to see feedback posted by other viewers.

Variations: Students can produce new episodes of a story each week, like a TV series.

Blog: using the Web

Aim: Students practise reading for main idea and detail on the Web; students practise writing opinions.	
Assumptions: Students are familiar with basic searches on the Internet; students enjoy expressing opinions.	
Materials: Computers connected to the Internet (one between two students).	
Level: Pre-intermediate and higher	Time: One hour
Anticipated problems: • Students will not understand ungraded material (language could be above their level/ability). • Students will access inappropriate material. • Students will post inaccurate language.	Solutions: • Suggest concrete topics appropriate to their level (e.g. language learning, food, pets rather than politics for an elementary class etc). • Research and guide students to trusted sites with links to blogs. • Help students with accuracy before posting.

Activities	Procedure
State lesson aims Aim: students understand objective of using Internet in the lesson **Time: 5 minutes**	• Have pairs sit at one computer. • Tell students they're going to: - look at different blogs - find a blog they like - post a comment
Survey blogs Aim: students become familiar with the range of blogs on the Web **Time: 15 minutes**	• Get students to look at a blogs links page, for example: bloggerschoiceawards.com/ (students choose a blog) technorati.com/blogs/top100/ (students choose a topic, then a blog from a list) • Alternatively they can google 'language learning blogs'. • Let students play around for some time. • Ask them to share anything funny or interesting they find.
Choose blog Aim: students choose a blog they will read and post a comment on. **Time: 15 minutes**	• Tell pairs to find a blog they both like. They can use an online dictionary to help them, for example: dictionary.cambridge.org www.macmillandictionary.com • Students find out whether the site allows comments, and if so, how to post one (if the blog is moderated students may not see their comment for some time).
Follow-up Aim: students see result of their work; students receive feedback **Time: 10 minutes**	• Give individual and whole-class feedback. • Students visit later to see feedback posted by other viewers.
Post comment Aim: students write and post a comment **Time: 15 minutes**	• Students find a post they think is interesting. Using an online dictionary they draft a comment in Word. • Monitor and help students correct errors. • Students post their comments.
Follow-up Aim: students see result of their work; students receive feedback **Time: 10 minutes**	• Give students oral feedback. • Students visit later to see their comment on the blog.

Variations: Students can start and write their own blog as an ongoing project.

Car Parts (Vocabulary): from visuals

Aim: Students learn and practise using vocabulary to describe parts of a car	
Assumptions: Students are familiar with basic car parts	
Materials: Picture (enlarged) for lead-in Pictures (enlarged) for presentation One picture for each student (A or B) for controlled practice A3	
Target language: See answers on next page	
Level: Pre-intermediate to intermediate	Time: 40 minutes
Anticipated problems: • Students are not interested in cars. • Adult students think drawing is childish.	Solutions: • Activities are intrinsically interesting (e.g. info gap, designing a car); car vocabulary is useful in every day life. • Avoid coloured pencils; stress the fact the students are 'engineers'.

Main Stages	Microstages	Activities
Presentation Aim: students learn form, meaning and pronunciation of the target language **Time: 15 minutes**	**Lead-in** Aim: students are engaged with the topic	• Show the picture of the car and ask, 'What are we talking about today? (cars) • Ask, 'When you buy a car, what do you look for? Price? How it looks?'. Elicit several responses. • Students work in groups to rank what's important in a car. Write the criteria on the board: STYLE COMFORT ECONOMY PRICE RELIABILITY
	Elicit & analyse the target language Aim: students are exposed to meaningful examples of the target language; students understand the form, meaning and pronunciation of the target language	• Tell students, 'Today you're engineers at (Ford/Mercedes) and you're going to design the perfect car that has all of these (point to criteria). But you need to learn some vocabulary'. • Elicit the vocabulary in short sentences (*It's the…/They're the…*) by holding up the pictures (e.g. *What's this? It's the body*). • Model and drill the sentences. • Write the vocabulary on the board.
Controlled practice Aim: students are confident and correct producing the target language **Time: 5 minutes**	**Testing in pairs**	• Students work in pairs. One student draws a picture of a car with the six parts. They test each other (*What's this?*).
Less-controlled practice Aim: students are able to use the target language fluently **Time: 10 minutes**	**Small-group drawing info-gap**	• Put students in groups of four. Two students look at picture 1 and describe it. The other students draw it and see whose drawing is closest to the photo.
Free practice Aim: students are able to use the target language meaningfully in a way that is relevant to their lives **Time: 10 minutes**	**Role play**	• Tell students again, 'Today you're engineers at (Ford/Mercedes) and you're going to design the perfect car'. Divide students into design teams. Tell students they have to meet all five criteria: STYLE COMFORT ECONOMY PRICE RELIABILITY • Give each group A3. They design and present their car to the class.

Answers: **a** body, **b** bonnet (UK)/hood (US), **c** bumper, **d** tyres, **e** mirror, **f** headlights

Celebrities (Information Search): using the Web

Aim: Students practise reading for main idea and detail on the Web (using the topic of celebrities); students extract and present information

Assumptions: Students are interested in celebrities; students can do basic searches on the Internet; students can perform basic functions (typing, pasting images and printing) in Word.

Materials: Computers connected to Internet (one between two students)
Printer with paper, Tape/Blu-tack

Level: All	Time: One hour
Anticipated problems: • Students find it difficult to evaluate websites. • Students have difficulty with names of celebrities. • Students will plagiarise material.	Solutions: • Provide addresses of recommended sites. • Model and drill names if different in students' L1. • Give very basic guidelines for sourcing information.

Activities	Procedure
State lesson aim Aim: students understand objective of using Internet in the lesson **Time: 5 minutes**	• Have two students work on one computer. • Ask them to look at a page (that you have researched and bookmarked) showing a picture of a celebrity they know. Elicit what sort of person it is (e.g. a famous person, a celebrity). • Tell students the aim of the session is to find out about a celebrity they like.
Choose celebrities Aim: students choose three celebrities to research **Time: 5 minutes**	• Ask the class to go to a celebrity site (like www.people.com where they can click on 'photos') and choose the celebrities they're most interested in. • Tell the class they have to choose three only (they can vote).
Analyse Wikipedia entry Aim: students understand basic categories for presenting their research **Time: 5 minutes**	• Tell students we now need to decide what sort of information we should look for, and how to organise it. • Have the class race for a specific celebrity page on Wikipedia. • Elicit headings and write them on the board such as: - family and early life - early work - breakthrough - international success - personal life • Tell students that is how their research will be presented.
Locate websites Aim: students locate sources of information **Time: 5 minutes**	• Ask students where they think they can find information. • Recommend a small number for information and images, and write them on the board, for example: www.wikipedia.org www.okmagazine.com images.google.com
Research Aim: students research and prepare their presentation **Time: 30 minutes**	• Tell students they have to choose one of the three celebrities. • Students need to present their research in Word • They must reference sources of information and images. • Give students a time limit
Follow-up Aim: students see result of their work; students receive feedback **Time: 10 minutes**	• Have students post their work on the wall. They can compare their finished product with other pairs who researched the same person and find out about the other two celebrities. • Give students oral feedback. • Students can vote on the best presentation.

Variations: Students could contribute to an interactive celebrity website.

Cooking Verbs (Vocabulary): using realia and mime

Aim: Students learn and practise verbs for preparing food	
Assumptions: Students are familiar with basic cooking terms; they see cooking as relevant to their lives	
Materials: Potatoes, chopping board, knife, peeler & grater (ensure knife is used safely!)	
Target language: Peel, grate, cut (e.g. in half), chop, slice, dice	
Level: Pre-intermediate to intermediate	Time: 40 minutes
Anticipated problems: • Some students aren't interested in cooking • Students have difficulty remembering a set of similar items	Solutions: • Use activities with intrinsic interest (e.g. showing off food knowledge); keep lesson short • Have written practice where students can refer to notes

Main Stages	Microstages	Activities
Presentation Aim: students learn form, meaning and pronunciation of the target language **Time: 10 minutes**	**Set context** Aim: students understand what the target language means and where it is used	• Tell students, 'Today we have a cooking lesson'. • Reveal the potatoes. Ask, 'What are they?'. In three minutes groups brainstorm as many things as possible you can make with potatoes. • Elicit some ideas to the board (e.g. *fries, mash etc*).
	Elicit & analyse the target language Aim: students are exposed to meaningful examples of the target language; students understand the form, meaning and pronunciation of the target language	• Tell students, 'You're going to learn some cooking words'. • Teach the five verbs by doing the real actions with the potatoes. Elicit short instructions (e.g. *Chop the potato*). • Concept check with actions and questions: - What's this? (Cutting it in half.) *Cut the potato in half.* - *Which means large pieces? Chop the potato. Which means small pieces? Dice the potato. Which means thin pieces? Slice the potatoes.* • Model and drill as you go. • Write the words on the board.
Controlled practice Aim: students are confident and correct producing the target language **Time: 10 minutes**	**Mime activity**	• Have a volunteer out the front. Students call out instructions and the student does it. • In pairs, one student instructs and the other mimes the action.
Free practice Aim: students are able to use the target language meaningfully in a way that is relevant to their lives **Time: 20 minutes**	**Writing & mingling activity**	• Ask each student to write down a recipe they know. They need to list the ingredients, and the method. • Students mingle and ask other students if they know how to prepare their dish. • Students post their recipes on the wall. Students can copy down any recipes they want to keep.

Cultural Dos And Don'ts (Speaking): using a text to stimulate discussion

<table>
<tr><td colspan="2">Aim: Students practise speaking for fluency (in response to a text on cultural Dos and Don'ts)</td></tr>
<tr><td colspan="2">Assumptions: Students are interested in cultural differences.</td></tr>
<tr><td colspan="2">Materials: Pictures (enlarged) for presentation
One copy of text for each pair
A3 and markers</td></tr>
<tr><td>Level: Pre-intermediate and higher</td><td>Time: One hour</td></tr>
<tr><td>Anticipated problems:
• Lower-level students may find text difficult
• Students may find topics sensitive, especially in a text written by an 'outsider'</td><td>Solutions:
• Pre-teach any key words to be able to achieve task
• Keep lesson completely non-judgemental; ask for students' opinions after they read if their country is mentioned in text</td></tr>
</table>

<table>
<tr><th>Main Stages</th><th>Microstages</th><th>Activities</th></tr>
<tr><td rowspan="4">Reading & information exchange
Aim: students engage with a text and exchange information
Time: 20 minutes</td><td>Lead-in
Aim: students are engaged with the topic</td><td>• Show the map and pictures. Elicit the two countries (Thailand and Indonesia).
• In groups, students discuss what they know about the two countries. Write prompts on the board:
- dress
- greetings
- religion
- other behaviour</td></tr>
<tr><td>Introduce the text
Aim: students know the main idea of what they're going to read</td><td>• Tell students they're going to read a short text about Thailand and Indonesia, and see if they were right (the text is typical of information for business people on the Internet describing 'Dos and Don'ts' in different countries).</td></tr>
<tr><td>Pre-teach vocabulary
Aim: students are ready to understand the text</td><td>• Pre-teach any key words your students probably won't know.</td></tr>
<tr><td>Reading for detail
Aims: students understand the text in detail; students exchange information</td><td>• Ask each pair to copy the following grid:
<table><tr><th></th><th>Dos</th><th>Don'ts</th></tr><tr><td>Indonesia</td><td></td><td></td></tr><tr><td>Thailand</td><td></td><td></td></tr></table>• Assign each pair to read about one country only.
• Tell students to take notes; elicit an example (e.g. religion – Buddhism).
• When they've completed their notes, they exchange information: pairs mingle and ask another pair, and take notes.
• Students look at the text again to see if their notes are correct.</td></tr>
</table>

Main Stages	Microstages	Activities
Discussion Aim: students practise speaking for fluency in staged activities **Time: 20 minutes**	**Whole-group discussion** Aims: students respond to concrete details in the text	• Ask students whether anything surprised them in the texts.
	Small-group personalised discussion Aims: students discuss the ideas in the text in depth	• In small groups students discuss (write questions on board): - How can you find out what to do when you visit another country? - What topics are often sensitive in a culture? - Should a visitor always follow local customs?
Follow-up Aim: students see result of their work; students receive feedback **Time: 20 minutes**	**Writing activity** Aims: students apply ideas they've discussed	• Students work in groups and create a poster giving cultural advice for travellers (in general or to their country). • Students present their posters to the class.

Variations: Jigsaw reading greatly increases opportunities for oral practice; the technique can be used for any text.

D Demotivation (Problem Solving): task-based approach

Aim: Students give a presentation (proposing solutions to demotivation in language learning); students practise speaking for fluency; students practise expressing opinions and reaching consensus	
Assumptions: Students can relate to the topic of demotivation in language learning.	
Materials: One copy of sorting activity strips for each group A3 and markers	
Level: Intermediate and above	Time: One hour
Anticipated problems: • Students are confused by high-level lexis such as demotivation • Demotivation is a depressing topic • Many different language issues come from the task	Solutions: • Start with concrete examples of students' experience • Aim of lesson is for students to propose solutions to the problem • Focus on one or two issues central to (a) completing the task and (b) intelligibility

Main Stages	Microstages	Activities
Pre-task Aims: students are engaged with the topic; students are ready to perform the task **Time: 20 minutes**	**Lead-in** Aim: students are engaged with the topic	• Ask students how it feels to learn a language. 'When do you feel happy? Sad? Angry?' Students discuss in groups. Elicit some responses. • Ask students, 'What do you call it when you **want** to do something; for example, I want to learn English because I love English, or I'll get a good job?'. Elicit *motivated*. • 'What's the opposite? For example, I don't like English because I can't improve, or I don't like the teacher?' Elicit *demotivated*. • Tell students our task is to: - decide what makes us demotivated - present ideas to help us avoid demotivation

<table>
<tr><th>Main Stages</th><th>Microstages</th><th>Activities</th></tr>
<tr><td rowspan="1">Pre-task continued</td><td>Small-group brainstorm
Aim: students start to think about the topic in depth</td><td>• Get students to copy the grid from the board:
<table><tr><th>Good</th><th>Bad</th></tr><tr><td>(e.g. a teacher praised my pronunciation)</td><td>(e.g. I couldn't understand someone)</td></tr></table>• Groups compare their notes.
• Ask students to discuss briefly the most important influences on how they feel about learning English.</td></tr>
<tr><td rowspan="3">Task
Aim: students deliver a presentation that proposes solutions to a problem
Time: 30 minutes</td><td>Task scoping
Aim: students identify scope of the task</td><td>• Give the strips to each group. Students sort them into two groups: important and not important.
• Students sort the 'important' factors into what can influence and what we cannot.
• For the factors we can influence, students discuss what a language learner can do.</td></tr>
<tr><td>Task planning
Aim: students write their presentation</td><td>• Groups now prepare a summary of their ideas in note-form to report to the whole class.
• As students prepare, monitor and make notes of language difficulties students are having (e.g. vocabulary, grammar and opinion language). Don't correct at this stage.</td></tr>
<tr><td>Presentation
Aim: students deliver their presentation</td><td>• Students present their ideas in a three-minute presentation.
• Each group tells the class the most important idea they heard.
• For homework, students write down two things they will do to avoid demotivation as an 'action plan'.</td></tr>
<tr><td rowspan="2">Post-task
Aim: students learn and apply knowledge from the experience
Time: 10 minutes</td><td>Language analysis
Aim: raise students' awareness of language generated by the task</td><td>• Use your notes to discuss any language issues students had.
• If possible, give students a short practice activity.</td></tr>
<tr><td>Follow-up
Aim: students can reflect on & apply the task</td><td>• The following week, form groups different from those on the day of the task.
• Students share their action plans and discuss how successful they've been.</td></tr>
</table>

Variations: This could be adapted for any problem-solving activity.

Dream Holiday (Writing): from pictures

Aim: Students write a story about a holiday	
Assumptions: Students are familiar with basic vocabulary in the pictures; students can form past simple	
Materials: Pictures for lead-in (enlarged) One set of story pictures for each pair	
Level: All	Time: 45 mins
Anticipated problems: • Students using tenses inaccurately	Solutions: • Check what tense students should use for a narrative (past); briefly revise past tense forms if necessary

Microstages	Activities
Lead-in Aim: students understand what the target language means and where it is used **Time: 10 minutes**	• Show first two pictures. Ask, 'What's she doing?' (Going on holiday) Say, 'It was her dream holiday. Last year. You're going to find out what happened in a moment'. • Divide students into pairs. Ask, 'What was your last holiday? Tell each other.' • Elicit some responses.
Preparation Aim: students are exposed to meaningful examples of the target language **Time: 15 minutes**	• Tell the class, 'Here are some pictures from her holiday'. Give one set to each pair. • Ask pairs to describe what's happening in the photos. Elicit some key vocabulary and write it on the board. Remind students it was last year, so elicit the correct tense (past, e.g. *She went to the airport*).
Writing Aim: students understand the form, meaning and pronunciation of the target language **Time: 15 minutes**	• Tell pairs to put the pictures in the order they think the events happened. • Join two pairs together to discuss and see if they agree. • Separate pairs again. Each pair writes the story. Tell them to give it a title. • Pairs swap their stories and make any corrections, then give them back.
Follow-up Aim: students receive feedback **Time: 5 minutes**	• Join two different pairs together. One pair reads their story and the other puts their pictures in the same order. Pairs swap roles. • Give individual and group feedback.

Variations:
- This could be followed by students writing about their last holiday.
- This approach can be used for any series of pictures.

First Lesson: interactive activities

Aim: Students feel relaxed; students practise speaking for fluency; students identify their language needs	
Assumptions: Students are not true beginners	
Materials: *Find someone who...* for each student (can be modified for different levels) Needs analysis for each student (can be modified for different levels) Ball or stuffed toy	
Level: All	Time: 2 hours (could be split over 2 days)
Anticipated problems: • Students are anxious • Students are not used to interactive activities and find instructions difficult to follow	Solutions: • Play music when students are entering/welcome students warmly/students speak to each other in small groups before the whole group • Use succinct instructions and demonstrate

Main Stages	Activities	Procedure
Getting to know each other Aim: students feel relaxed speaking English in class; students practise speaking for fluency **Time: 60 minutes**	**Pre-class**	• Have music playing as students come into class. • Move around and warmly greet students individually; don't make them talk to a large group of strangers.
	Introductions Aim: students meet other students; students understand why speaking in class is important **Time: 15 minutes**	• Welcome the whole class. • Put students in pairs. Ask students to introduce themselves, and say why they're learning English. • Join pairs together. Students introduce their partner (not themselves). • As a whole class, ask for volunteers to tell the whole group about someone they just met. • Ask students why they'll often have speaking activities like this. Elicit some ideas: to feel relaxed, because they need to practise etc.
	Ball game Aim: students become confident with each other's names **Time: 20 minutes**	• Students stand in a circle with a ball. • Students say their name and throw the ball to someone else. • Later, students try to remember the name of the person they're throwing to (*I think you're ...*). • Students now give one piece of information about themselves (e.g. *I can play guitar)*. • Later, students try to remember the information about the person they're throwing to. • Pair students off. Ask them to share more about themselves. • As a whole class, ask for volunteers to tell the whole group something interesting they just heard.
	Find someone who ... Aim: students become familiar with each other **Time: 25 minutes**	• Instruct 'Find someone who'. Elicit an example question (e.g. Have you ever eaten something strange?). • Students mingle and complete the sheet. • Ask the whole group to share something funny someone told them.

Main Stages	Microstages	Activities
Needs analysis Aim: students identify their language needs **Time: 60 minutes**	**Pair interview** Aim: students describe their current and future needs for English, and their current strengths and weaknesses **Time: 30 minutes**	• Tell students you want to know about their English so you can help them. • Put students in pairs. Give each a needs analysis. Ask them to interview each other (elicit some examples of what they should write). • Collect the completed forms.
	Small-group discussion Aim: students share opinions and agree on priorities for the class. **Time: 30 minutes**	• Join pairs together. Ask them to discuss what they want most from the course based on the questions in the needs analysis (students can write and present this; alternatively use pyramid discussion where entire class agrees on priorities). • Ask for some whole-class feedback.

Variations:
- This could be adapted for any level by modifying the handouts.
- Students could write up their own needs analysis for homework.

Going To for Plans (Grammar): test-teach-test

Aim: Students learn and practise *going to* for future plans	
Assumptions: Students have encountered the structure but may be unconfident with its form and meaning.	
Materials: One pair of controlled practice cards for each student One board game, die and counters for each group	
Target language: *Going to* (positive & negative statements; questions; short answers)	
Level: Pre-intermediate to intermediate	Time: 40 minutes
Anticipated problems: • Students use *will* for future plans • Students confuse *going to* with movement somewhere	Solutions: • Contrast the meaning of *will* and *going to* • Avoid examples where meaning could be ambiguous

Main Stages	Microstages	Activities
'Test'/Presentation Aims: evaluate students' knowledge of the target language; if necessary, students learn form, meaning and pronunciation of the target language **Time: 15 minutes**	**'Test'** Aim: evaluate whether students can use the target language appropriately and accurately	me Plans (next ten years) ● new car ● travel • Draw this on the board. Tell the class this is you. Ask, 'What are my plans for the next ten years?'. Elicit language such as 'You need to buy a new car', 'You want to travel' (students do not need to produce *going to* at this stage). • Rub out me and write *you*. Ask the students: 'How about you? What are your plans for the future?' • Students discuss in groups. • Monitor and listen to what language students are using to talk about future plans.

<table>
<tr><th>Main Stages</th><th>Microstages</th><th>Activities</th></tr>
<tr><td>'Test'/Presentation continued</td><td>'Test' continued</td><td>Assess and decide:
1. Are students using going to appropriately and accurately? If yes, then jump to the 'free practice' stage.
2. Are students confusing will and going to? If yes, continue to the next stage: 'analyse the meaning' (of will and going to).
3. Are students are using 'going to' in the correct context but with inaccurate grammar (e.g. 'I going to')? If yes, jump to the 'analyse the form' stage.</td></tr>
<tr><td></td><td>Analyse the meaning
Aim: students understand the meaning of the target language</td><td>• Check meaning:
Going to
Draw a picture of a man holding a book that says 'French'. Ask 'what is he holding?' Elicit the sentence: 'I'm going to learn French'

• Check meaning:
- Are we talking about now or the future? (Future)
- When did we decide? Now or before now? (Before now)

Will
Now mime having a quick thought and walking to close the window. Elicit
'I'll close the window'

• Check meaning:
- Are we talking about now or the future? (now)
- When did we decide? Now or before now? (now)

X decide — X do it
past — now — future going to

X decide — X do it
past — now — future will

Check with examples:
- To talk about important plans, do I say going to or will? (going to e.g. I'm going to buy a new car.)
- If I'm in a café, do I say going to or will? (will e.g. I'll have a coffee.)</td></tr>
<tr><td></td><td>Analyse the form
Aim: students understand the form of the target language</td><td>• Quickly analyse the form on the board:
<table>
<tr><td>Subject</td><td>aux be (pres)</td><td>going to</td><td>infinitive</td><td></td></tr>
<tr><td>I</td><td>'m</td><td>going to</td><td>buy</td><td>a car.</td></tr>
</table>
<table>
<tr><td>Subject</td><td>aux be (pres)</td><td>not</td><td>going to</td><td>infinitive</td><td></td></tr>
<tr><td>I</td><td>'m</td><td>not</td><td>going to</td><td>live</td><td>abroad.</td></tr>
</table>
<table>
<tr><td>aux be (pres)</td><td>subject</td><td>going to</td><td>infinitive</td><td></td></tr>
<tr><td>Are</td><td>you</td><td>going to</td><td>work</td><td>in a bank.</td></tr>
</table></td></tr>
</table>

<table>
<tr><th>Main Stages</th><th>Microstages</th><th>Activities</th></tr>
<tr><td>'Test'/Presentation continued</td><td>Analyse the form continued</td><td>
<table>
<tr><td>yes</td><td>subject</td><td>aux be (pres)</td></tr>
<tr><td>Yes,</td><td>I</td><td>am.</td></tr>
</table>
<table>
<tr><td>no</td><td>subject</td><td>aux be (pres)</td><td>not</td></tr>
<tr><td>No,</td><td>I</td><td>'m</td><td>not.</td></tr>
</table>
</td></tr>
<tr><td>Controlled practice
Aim: students are confident and correct producing the target language
Time: 10 minutes</td><td>Pair info-gap activity</td><td>• In pairs students ask and answer questions according to the prompts, for example:
A: Are you going to buy an apartment?
B: No, I'm not. I'm going to buy a house in the country.
• Students swap roles with new cards.</td></tr>
<tr><td>Less-controlled practice
Aim: students are able to use the target language fluently
Time: 5 minutes</td><td>Pair practice using question prompts from CP</td><td>• Students use the same two sets of question prompts but can give real answers.</td></tr>
<tr><td>Free practice
Aim: students are able to use the target language meaningfully in a way that is relevant to their lives
Time: 10 minutes</td><td>Small-group discussion board game</td><td>• Students play the board game in groups. They throw a die and move their counter accordingly. When they land on a topic they have to talk about it for a minute (I'm going to ...) OR everyone in their group has to guess the player's plans (I think you're going to ...), and the most accurate guess has the next throw.</td></tr>
</table>

Variations: This could be followed by a writing activity; students describe plans and how they're going to achieve them.

Invitations (Function): using a dialogue build

<table>
<tr><td colspan="2">Aim: Students learn and practise inviting people to do something, and responding to invitations</td></tr>
<tr><td colspan="2">Assumptions: Students are familiar with infinitive and V+ing; students can relate to the school context.</td></tr>
<tr><td colspan="2">Target language: Feel like (V+ing)?
Sorry, I've got to (inf).
How about (expression of time)?
Sure, that'd be great.</td></tr>
<tr><td>Level: Pre-intermediate to intermediate</td><td>Time: 40 minutes</td></tr>
<tr><td>Anticipated problems:
• Students unclear of functional meaning of 'Sorry'
• Students ask why there is no subject + auxiliary (e.g. 'Do you feel like...?') in 'Feel like ...?'
• Students have trouble pronouncing that'd</td><td>Solutions:
• Check 'Is she saying 'yes' or 'no'?' (no)
• Check 'Is this formal or informal language?' (informal)
• Backchain when modelling and drilling: /təd/ → /ˈðæ·təd/</td></tr>
</table>

Main Stages	Microstages	Activities
Presentation Aim: students learn form, meaning and pronunciation of the target language **Time: 15 minutes**	**Set context** Aim: students understand what the target language means and where it is used	School • Draw pictures to elicit we're at school and it's the end of the day. • Draw pictures to elicit these are two students (David and Susan). • Ask, 'What are they talking about?'. Students discuss briefly in pairs. Elicit some ideas (e.g. homework, going out).
	Elicit the target language Aim: students are exposed to meaningful examples of the target language	• Use these prompts to elicit the target language: movie? ✗ study tomorrow? ✔ • Check meaning, highlight, model and drill as you go. • **Dialogue:** David: *Feel like seeing a movie?* Susan: *Sorry, I've got to study.* David: *How about tomorrow?* Susan: *Sure, that'd be great.* • **Concept questions:** - Does he want her to do something with him? (Yes) - Is she saying 'yes' or 'no'? (No) Is it rude or polite? (Polite) - Does he suggest another time? (Yes) - Is she saying 'yes' or 'no'? (Yes)
	Analyse the target language Aim: students understand the form, meaning and pronunciation of the target language	• Check: is the language formal or informal? (informal) • Write the dialogue on the board. Highlight 'Feel like' needs verb +ing; 'I've got to' needs an infinitive.
Controlled practice Aim: students are confident and correct producing the target language **Time: 5 minutes**	**Pair practice of original dialogue**	• Rub the dialogue off the board. • Students practise the original dialogue. Monitor and correct errors.

Main Stages	Microstages	Activities
Less-controlled practice Aim: students are able to use the target language fluently **Time: 5 minutes**	**Pair substitution practice using prompts on board**	• Elicit substitutions and write them on the board: *(have) something to eat / (go) for a walk / (watch) TV etc* *(see) my family / (do) something / (finish) some work etc* *Friday / next week / some other time etc* • Students practise dialogue with substitutions in pairs. Monitor and correct errors.
Free practice Aim: students are able to use the target language meaningfully in a way that is relevant to their lives **Time: 10 minutes**	**Real-life simulation**	• In groups, students brainstorm real things they can do in the local area after class. • Students mingle and arrange to do something after class.

Variations: A dialogue build can be used for any functional language where there is logically a statement/question and response.

I Irregular Plurals (Grammar): using dictionaries

<table>
<tr><td colspan="2">Aim: Students learn and practise irregular plurals</td></tr>
<tr><td colspan="2">Assumptions: Students are familiar with regular plurals and have encountered some of these forms before</td></tr>
<tr><td colspan="2">Materials: Dictionaries (that give irregular plurals)
One copy of plural grid (either A or B) for each student
One set of controlled practice cards for each group</td></tr>
<tr><td colspan="2">Target language: 10 irregular plurals (see answers next page)</td></tr>
<tr><td>Level: Elementary</td><td>Time: 40 minutes</td></tr>
<tr><td>Anticipated problems:
• Students have difficulty remembering ten
• Pronunciation of women</td><td>Solutions:
• Have written and spoken controlled practice
• Model and drill</td></tr>
</table>

Main Stages	Microstages	Activities
Presentation Aim: students learn form, meaning and pronunciation of the target language **Time: 15 minutes**	**Warmer/revision** Aim: students practise regular plurals; warm up	• Tell the class: *I've got one* ... (elicit head by pointing). *I've got two ... ears*. Highlight the plural. • Students play in small groups.
	Students learn irregular plural forms Aim: students look up and record the target language	• Elicit irregular plural feet: *I've got two ... feet.* • Tell the class some nouns have a special plural form. • Give handout A to some students, and B to others. • Students use a dictionary to find the plural forms. • Students teach each other.
Controlled practice Aim: students are confident and correct producing the target language **Time: 15 minutes**	**Group card activity**	• Put students in groups. Give each group a set of the picture cards only. As written controlled practice, each student writes *one* ... and *two* ... for each noun in their book. • Give groups the number cards. Students spread out the number cards and picture cards face down in two groups. One student turns one card up in each group, and the first student to say the number and noun correctly keeps the cards.

Main Stages	Microstages	Activities
Free practice Aim: students are able to use the target language meaningfully in a way that is relevant to their lives **Time: 10 minutes**	**Mingling activity**	• Ask each student to write down five sentences that are true about them, using the plurals they've learned (e.g. *I have two missing teeth; I am sitting next to three women;*). • Students mingle and say their sentences, stopping before the noun (e.g. *I have two missing …*). The other person has to guess the missing word.

Answers for presentation

singular	plural	singular	plural
aircraft	**aircraft**	mouse	**mice**
child	**children**	person	**people**
fish	**fish**	sheep	**sheep**
foot	**feet**	tooth	**teeth**
man	**men**	woman	**women**

Variations: Deductive teaching is efficient where the choice of form is not meaningful (e.g. irregular past tenses; use of V+ing or infinitive after verbs). However, there can still be an element of 'discovering' the rules (e.g. students look up a dictionary or reference grammar).

Job Application (Writing): from a written model

Aim: Students learn and practise writing a job application	
Assumptions: Students are interested in work; students are familiar with job advertisements and processes for applying for work in their own country	
Materials: One set of job advertisements for each group One model letter for each group (application 1) One copy of 'analysing the model' for each group One controlled practice handout for each group (application 2)	
Level: Intermediate and higher	Time: One hour
Anticipated problems: • The students' countries have different processes for applying for work • Application letters in the students' L1 have very different conventions from the English-speaking world	Solutions: • Brainstorm the process at the start of the lesson • Students analyse a model and complete a gapfill to familiarise themselves with the text type

Stages		Activities
Elicit lesson aim Aim: students understand writing objective **Time: 10 minutes**	**Lead-in** Aim: students are engaged with topic	• Divide students into groups of three. Ask students, 'What are these?', and hand out one set of job ads to each group. Elicit 'job advertisements'. • Ask, ' What's the process in your country?' Start a flowchart on the board and ask if it's similar. In groups students create a flowchart (e.g. job ad → write letter → receive phone call).
	Introduce text Aim: students understand what the text type is	• Ask students, 'What's this?'. Hand out application 1 to each group. Elicit 'application letter'. • Ask students to match it to the advertisement.

Stages		Activities
Elicit lesson aim continued	**State aim** Aim: students understand what they will produce	• Tell students they're going to learn how to write a job application.
Analyse model Aim: students understand the structure of the model **Time: 15 minutes**		• Pre-teach 'resume', 'qualified', 'experienced'. • Students match the paragraph with its purpose.
Controlled practice Aim: students have guided practice writing an application **Time: 10 minutes**		• Hand out application 2 to each group. Ask students to match it to the advertisement. • Students fill in the missing words and check with application 1. Elicit the correct answers.
Free practice Aim: students write an application **Time: 10 minutes**		• Each group now writes an application for the remaining job advertisement.
Follow-up Aim: students receive feedback **Time: 5 minutes**		• Give individual and group feedback.

Answers (analysing the model)
Write A-D to match each paragraph with its purpose.

B showing she meets the first selection criterion
A explaining what the letter is
D requesting the next step
C showing she meets the second selection criterion

Answers (controlled practice)

Dear Mr Davis

This is an <u>application</u> for the position of Web Designer/Developer, as <u>advertised</u> on www.seek.com.

I have over three years' <u>experience</u> as a Web Developer for Leader Newspaper Group. This was a challenging role in a very competitive, deadline-driven environment.

I have outstanding <u>design</u> skills. I was chosen to lead the design team responsible for the redevelopment of the Leader website, which later won the prestigious Walker Award for best Website Design.

I would be very <u>grateful</u> for the <u>opportunity</u> to discuss my application further. Please find my <u>resume</u> attached.

Yours sincerely

Variations:
- Students may prefer to work in pairs or individually for reading and writing
- As a follow-up, students could find a job advertisement they are really interested in, and write an application.
- This could be linked to resume writing and interview skills.

Permission (Function): from a recording

Aim: Students learn and practice asking for and giving permission

Assumptions: Students are familiar with present and past simple; they relate to the school context.

Materials: Create a recording of the dialogue given at the end of this section
Set of controlled practice cards for each pair or group
One teacher or student role play card for each person
One student timetable per pair

Target language: *Is it all right if I (present simple)?*
I'm afraid not.
Would you mind if I (past simple)?*
Sure, that'll be fine.

Level: Pre-intermediate to intermediate	Time: 40 minutes
Anticipated problems: • Students are confused with use of past tense after *Would you mind if I …?* • Students have trouble pronouncing *that'll*	Solutions: • Check 'Is she talking about the past?' (no) • Backchain when modelling and drilling: /təl/ → /ˈðæ·təl/

* *Present simple is acceptable in informal English.*

<table>
<tr><th>Main Stages</th><th>Micro Stages</th><th>Activities</th></tr>
<tr>
<td>Presentation
Aim: students learn form, meaning and pronunciation of the target language
Time: 15 minutes</td>
<td>Students understand conversation in context
Aim: students understand what the target language means and where it is used</td>
<td>
• Draw a picture and ask, 'Where are we?', to elicit we're at school.
<table>
<tr><td>can do</td><td>can't do</td><td>need to ask</td></tr>
<tr><td>eat in the canteen</td><td>smoke</td><td>leave early</td></tr>
</table>
• Students brainstorm in groups what they can do, can't do and need to ask to do at school (they can copy the grid from the board).
• Draw pictures to elicit this is a student and a teacher talking.
• Tell students, 'Listen to their conversation. Are they talking about what students can do, can't do, or is the student asking to do something?'.
• Elicit the answer (asking to do something).
• 'Listen again. What two things does he ask to do?' Write these prompts on the board:
- bring his bike to school
- miss a test
- go home early
- go home late
- arrive at school late
(Answers: go home early/arrive at school late)
</td>
</tr>
</table>

Main Stages	Microstages	Activities
Presentation continued	**Elicit the target language** Aim: students are exposed to meaningful examples of the target language	• Play sections of the tape again starting just before each piece of target language. Ask: - What does he say exactly about leaving early? (**Is it all right if I leave early?**) - How does she say no? (**I'm afraid not.**) - What does he say exactly about coming late? (**Would you mind if I came late tomorrow?**) - How does she say yes? (**Sure, that'll be fine.**) • Check meaning, highlight, model and drill as you go. • **Concept questions:** - Does he want to do something? (Yes) - Who decides if it is OK – the student or the teacher? (Teacher)
	Analyse the target language Aim: students understand the form, meaning and pronunciation of the target language	• Write the four sentences on the board. Highlight 'Is it all right if I ...' needs present simple; 'Would you mind if I ...' needs past simple. • Check: is he talking about the past? (No)
Controlled practice Aim: students are confident and correct producing the target language **Time: 10 minutes**	**Pair substitution practice with cards**	• Rub the dialogue off the board. • Give one set of cards to each pair or group. Students turn over the prompts to make questions and answers. • Monitor and correct errors.
Less-controlled practice Aim: students are able to use the target language fluently **Time: 5 minutes**	**Pair practice using question cards from CP**	• Students take away the answer prompts and can answer as they like.
Free practice Aim: students are able to use the target language meaningfully in a way that is relevant to their lives **Time: 10 minutes**	**Role play**	• Assign 'teachers' and 'students'. Hand out the role play cards to each person, and the timetables to the 'students' only. • Give them several minutes to read their role play cards before starting. • 'Students' tick any request the 'teacher' says yes to. • 'Students' aim to receive permission for as many items on their list as possible.

Transcript for presentation

Male student:	Excuse me, Ms Ryan, do you have a moment?
Female teacher:	Sure, Van, what is it?
Male student:	Is it all right if I leave early today?
Female teacher:	I'm afraid not. We have a test this afternoon. Is there a problem?
Male student:	I need to get my bike fixed.
Female teacher:	Oh I see.
Male student:	Would you mind if I came late tomorrow?
Female teacher:	Sure, Van, that'll be fine. Don't be too late though.
Male student:	Thank you Ms Ryan.

Variations: Eliciting from a recording can be used for any spoken language.

Present Perfect (Grammar): from a situation

Aim: Students learn and practise present perfect for describing a past event that is still important	
Assumptions: Students are familiar with forming present tense of have and past participles.	
Materials: One set of pictures/prompts (enlarged) for presentation One set of controlled practice cards for each pair/group One set of free practice cards for each group	
Target language: Present perfect simple (statements, questions and short answers)	
Level: Pre-intermediate to intermediate	Time: 40 minutes
Anticipated problems: • Students are confused a verb structure is called 'present' but describes something that happened in the past. • Students use a past time expression with present perfect (e.g. I have lost my bag yesterday) • Students may not know the past participle form of the verbs need to complete the controlled practice exercise.	Solutions: • Explicitly separate what it's called and what it means. • Teach (and check) rule that if you mention the time it happened you must use past simple. • Revise common past participles before the game.

Main Stages	Micro Stages	Activities
Presentation 1: statements Aim: students learn form, meaning and pronunciation of the target language **Time: 10 minutes**	**Set context** Aim: students understand what the target language means and where it is used	• Use pictures to elicit location (at a train station). • Use picture of the man dropping his wallet to elicit problem (He's lost something). • Use picture to elicit where you go when you lose something (lost property office). • Have students discuss when they've lost something. How did they feel? What did they do?
	Elicit the target language Aim: students are exposed to meaningful examples of the target language	• Use the prompts to elicit the first sentence **I've lost my bag.** • **Check meaning:** - Did it happen now, or in the past? (The past) - Is it still important? (Yes) • Highlight the form (subject + auxiliary have (in the present) + past participle). • Model at natural speed and drill. • Do the same with the other sentences: **I've left my glasses somewhere.** **Someone's taken my purse.** **Someone's stolen my laptop.**
	Analyse the target language Aim: students understand the form, meaning and pronunciation of the target language	• Write the four sentences on the board. Label the parts of the structure: (see table below)

subject	aux *have* (pres)	past participle	
I	've	lost	my bag.
I	've	left	my glasses.
Someone	's	taken	my purse.
Someone	's	stolen	my laptop.

<table>
<tr><th>Main Stages</th><th>Microstages</th><th>Activities</th></tr>
<tr><td>Controlled practice
Aim: students are confident and correctly producing the target language
Time: 10 minutes</td><td>Small-group substitution practice</td><td>• Give one set of cards to each pair or group. Students turn over the prompts to make statements.
• Monitor and correct errors.</td></tr>
<tr><td rowspan="2">Presentation 2: questions and short answers
Aim: students learn form, meaning and pronunciation of the target language (statements only)
Time: 5 minutes</td><td>Elicit the target language</td><td>• Elicit the question form (by asking 'What does the railway official ask her?') and short answers:

Have you checked the platform?
Yes I have.
Have you looked in the waiting room?
No I haven't.

• Model and drill.</td></tr>
<tr><td>Analyse the target language</td><td>• Write the sentences on the board. Label the parts of the structure:
<table>
<tr><td>aux have (pres)</td><td>subject</td><td>past participle</td><td></td></tr>
<tr><td>Have</td><td>you</td><td>checked</td><td>the platform?</td></tr>
</table>
<table>
<tr><td>yes</td><td>subject</td><td>aux have (pres)</td></tr>
<tr><td>Yes,</td><td>I</td><td>have.</td></tr>
</table>
<table>
<tr><td>no</td><td>subject</td><td>aux have (pres)</td><td>not</td></tr>
<tr><td>No,</td><td>I</td><td>have</td><td>n't.</td></tr>
</table></td></tr>
<tr><td>Less-controlled practice
Aim: students are able to use the target language fluently
Time: 5 minutes</td><td>Pair writing activity</td><td>• Students work in pairs to write their own dialogue between the passenger and the railway official, then perform it.</td></tr>
<tr><td>Free practice
Aim: students are able to use the target language meaningfully in a way that is relevant to their lives
Time: 5 minutes</td><td>Small-group role play</td><td>• Create groups of three.
• Give each group a set of scenario cards.
• They turn one scenario over and act it out, one person informing the other of a problem and asking for help (e.g. in the scenario police station – a robbery one student is a police officer, and the other is a member of the public reporting a robbery).</td></tr>
</table>

the with place names (Grammar): from a text

Aim: Students learn and practise using *the* with placenames	
Assumptions: Students are familiar with some of the rules (e.g. use of the with countries and cities); students will understand the vocabulary in the text	
Materials: Pictures (enlarged) for lead-in One copy of Japan and French text for each pair One copy of table for each pair	
Target language: See answers on next page	
Level: Pre-intermediate/intermediate	Time: One hour
Anticipated problems: • Students will find the rules hard to remember. • Students are not interested in Japan or France.	Solutions: • Students keep a record of the rules and can refer to it during practice. • Students write about own country in free practice.

Main Stages	Micro Stages	Activities
Presentation 1: statements Aim: students learn form, meaning and pronunciation of the target language **Time: 20 minutes**	**Students read text for understanding** Aim: students understand what the target language means and where it is used	• Show the pictures of Japan. Students discuss whether they've been there, and what they know about it. • Show a picture of David. Say, 'He's been to Japan. Where do you think he's from? And why did he go to Japan?' • Say, 'Let's find out.' Hand out the text to each pair. • Students read quickly and call out answers (the U.S., for work). • 'Let's read in more detail.' Students answer the true/false questions. Elicit the answers.
	Elicit and analyse the target language Aim: students are exposed to meaningful examples of the target language; students analyse rules	• Tell students to underline all the place names in the text. • Tell students, 'Sometimes places need *the*, and sometimes they don't'. Elicit several examples from the text. 'We're going to learn the rules.' • Give the table to each pair. Students fill in the examples of different types of placenames, and work out whether they need *the* or not. • Elicit the answers.
Controlled practice Aim: students are confident and correct producing the target language **Time: 15 minutes**	**Pair correction activity**	• Tell students they're going to read a tourist brochure for France. Unfortunately the writer didn't know the rules for *the* and there are eight errors. • Give one handout to each pair. Students correct the errors. • Elicit the answers (an OHT is good for this).
Free practice Aim: students are able to use the target language meaningfully in a way that is relevant to their lives **Time: 25 minutes**	**Pair writing activity**	• Put students in new pairs. Ask them to write a similar brochure for their country (or the country they're studying in), but they need to get the rules correct! • Ask them to draw a map as well to accompany the information, showing where the places they mention are. • Students role play travel agents and customers. They have to convince customers to visit the country in their brochure.

Answers for text questions: **F F T T F**

Answers for presentation table

		the?
continent	North America, Asia	✗
country: one word	China, Japan, Korea	✗
country (a plural noun or with the word *Republic*)	the United States, the Republic of Korea	✓
state or region	Illinois, southern Japan	✗
city	Chicago, Tokyo	✗
river	the Nihombashi River	✓
mountain range	the Japanese Alps	✓
mountain	Mount Fuji	✗
lake	Lake Biwa	✗

Answers for controlled practice
Visit ~~the~~ beautiful France.

France has everything a visitor could hope to see.

~~The~~ Paris is one of the most beautiful cities in ~~the~~ Europe. There are countless things to do during your stay in the capital. Take a boat ride down the River Seine and visit the city's famous landmarks such as the Eiffel Tower and Notre Dame Cathedral.

You must also spend time in the regions of France. ~~The~~ Champagne, located just to the east of Paris, is of course the home of sparkling wine. See the gorgeous towns of ~~the~~ Provence. The Alps, and ~~the~~ Mont Blanc, on the eastern border, are wonderful for skiing in winter and hiking in summer.

3.3 Lessons Plans: Photocopiable Materials

These are the photocopiable activities and other materials we refer to in Lesson Plans A-Z. They're cross referenced to each plan.

Some are for you to hold up when presenting new language, or encouraging discussion – these just need enlarging on a photocopier. Alternatively just use these for ideas and create your own to match your students' needs and interests.

Activities for students generally need copying and cutting up. We suggest you gain access to a laminator and make copies of games and activities you'll use frequently. Otherwise every two months you'll be copying and cutting up the same pictures!

Car Parts (Vocabulary): from visuals

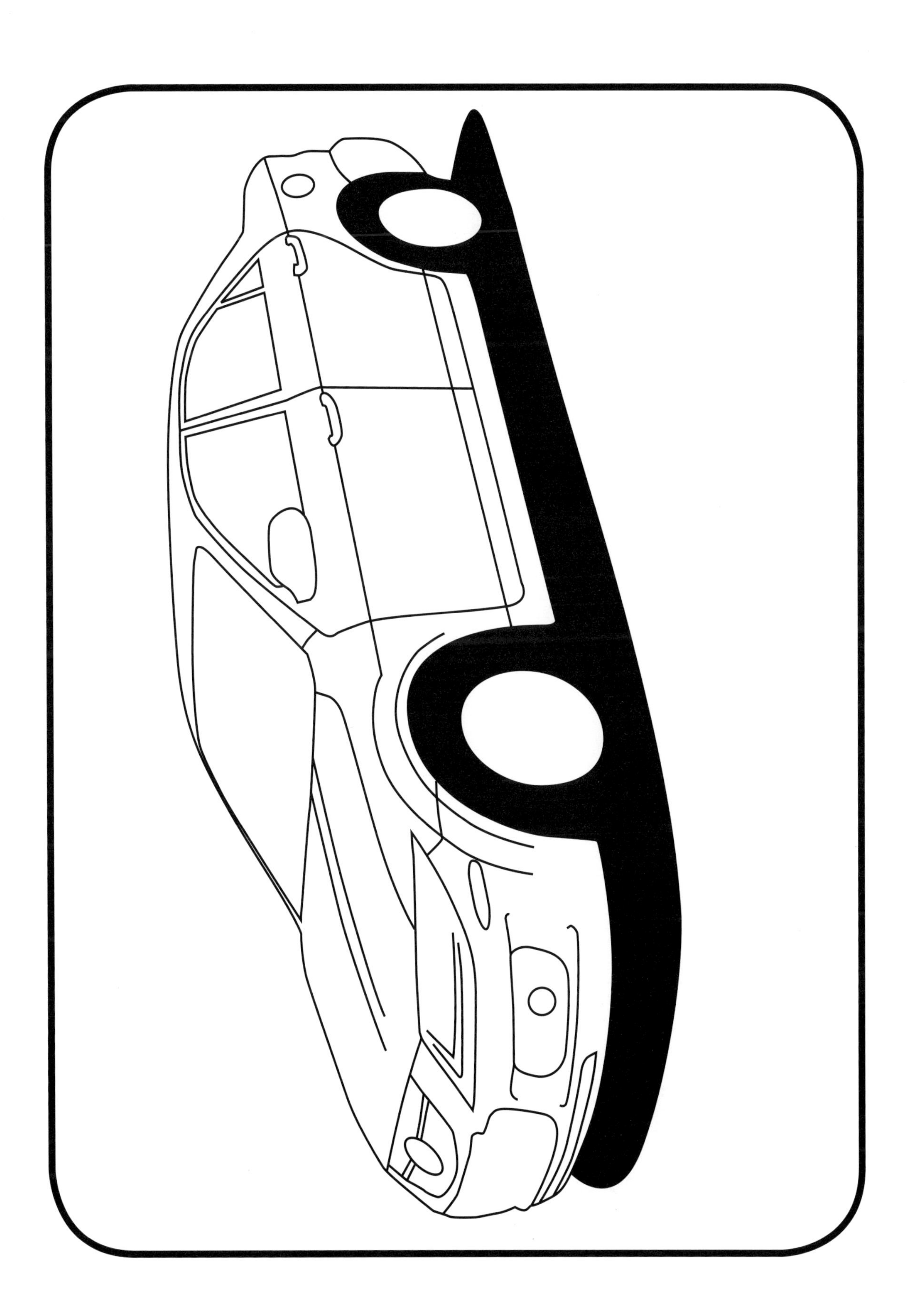

Cultural Dos And Don'ts (Speaking):

1. **Pictures for lead-in**

Cultural Dos And Don'ts (Speaking):

2. Text for discussion

The national religion of Thailand is Buddhism and you should always be respectful towards religious practices, sites and images of Buddha. You should always dress appropriately when you enter a shrine or temple; avoid bare arms, shorts and open-toed sandals.

Thais greet one another with a 'wai', a gesture made by putting your hands together in front of your face. A young person or subordinate should 'wai' to an older person or superior first.

The head is considered the highest part of the body and should be respected. Never pat someone on the head. By contrast, the feet are considered the lowest and most unclean part of the body. Never point the soles of your feet towards another person, or an image of Buddha.

Thais take the concept of 'face' very seriously. Avoid shouting or showing anger; this causes you to lose face. People may react by smiling or laughing as a way of giving you face.

Thais revere the monarchy and disrespect will offend people (and is in fact against the law).

Indonesia is the world's largest Muslim state. You need to be aware of how central faith is to Indonesian life, and to what extent religious beliefs and practices will affect your day-to-day interactions with local people. Many Indonesians are devout Muslims, praying five times a day and fasting for an entire month during Ramadan.

It is a taboo to display affection between members of the opposite sex in public. However, it's very common to touch a member of the same sex.

Initial business introductions in Indonesia are formal. Generally people will shake hands, although with a softer grip than in the West, and accompany this with a slight bow.

Alcohol and pork are strictly forbidden by devout Muslims. When food or drink is served, it is impolite to begin until after the host invites you. Never pass food or drink with your left hand as it is considered 'unclean'.

Indonesians value respect and courtesy. Never lose your temper in public, or treat an older person disrespectfully. The concept of 'face' is very important; anger displays a loss of face.

Demotivation (Problem Solving): task-based approach

1. Sorting activity

I've got no reason to learn English.
Someone criticised my English.
The classroom's uncomfortable.
I feel my progress is slow.
The coursebook's boring.
I don't like the style of teaching.
The technology in the classroom doesn't work.
The best way to get people to work hard is to encourage them.
I never have opportunities to practise.

Dream Holiday (Writing): from pictures

1. **Pictures for lead in**

2. **Picture activity**

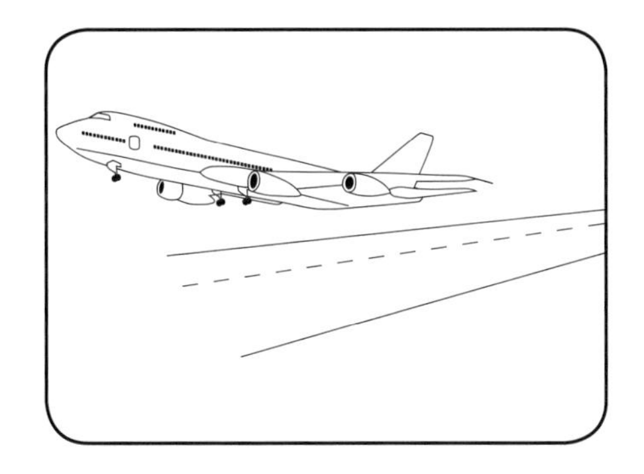

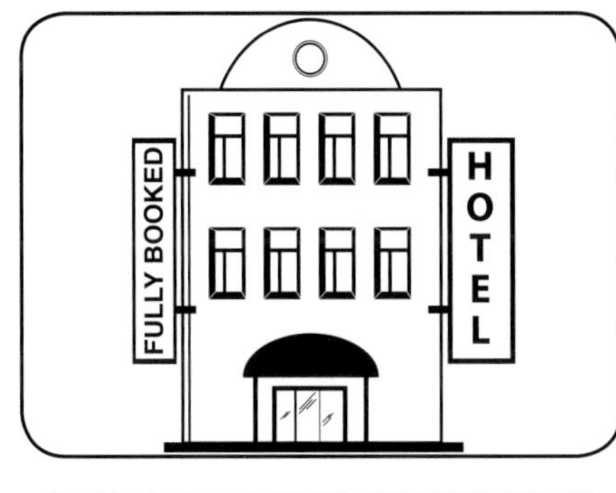

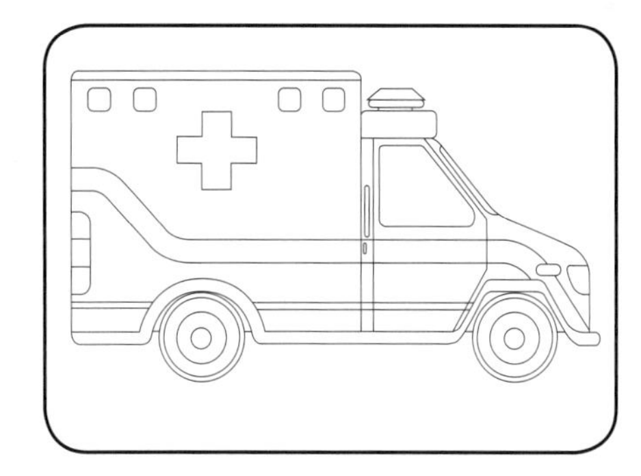

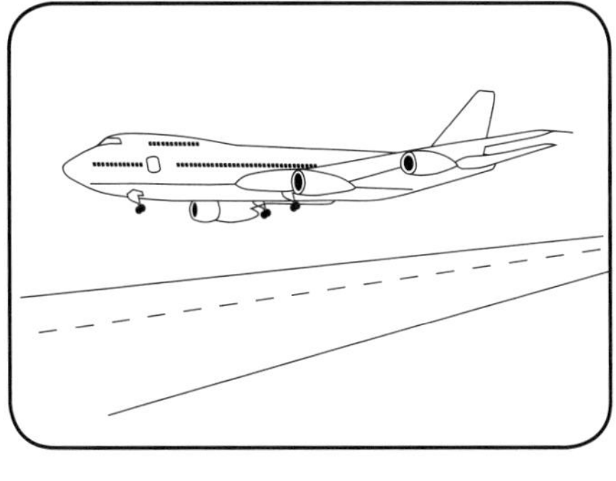

First Lesson: interactive activities

1. Find someone who

Find someone who ...

- has eaten something strange ____________________
- has been to another country ____________________
- has learnt a special talent ____________________
- has flown, but not in an aeroplane ____________________
- has been on TV or the radio ____________________
- has done something dangerous ____________________
- has met someone famous ____________________
- has studied another language ____________________

---

2. Needs analysis

NEEDS ANALYSIS Partner's name: ____________________

What do you need English for now? ____________________

What will you need English for in the future? ____________________

How is your English now? Score out of 5 (1 = poor, 5 = excellent)

- listening ________
- speaking ________
- reading ________
- writing ________

What are you most confident with? ____________________

What do you most need to work on? ____________________

Going To for Plans (Grammar): test-teach-test

1. **Controlled practice**

STUDENT A: Ask your partner about their plans.	**This is what you think**
• (buy) an apartment?	✗ (work abroad) live in the US
• (go on vacation)?	✗ (move house) I'm happy in my current home
• (get) married?	✓ start a business
• (start) your own business?	✗ (study) I don't need to do any more study
• (work) abroad?	✓ have three children
• (get) another qualification?	✗ (learn another language) English is enough for me

STUDENT B: Ask your partner about their plans.	**This is what you think**
• (move) house?	✗ (study abroad) go on vacation
• (live) in an English-speaking country?	✓ a house in the country
• (have) children?	✓ work in Korea
• (work) for a large company?	✗ (get married) I'm already happily married!
• (learn) another language?	✓ do a Master's degree
• (study) abroad?	✓ work for a large organisation

2. Free practice

MY PLANS

	religion		politics		languages	**FINISH**
		friends		home decoration		money
	family		study		interests	go back three spaces
START		career		travel		possessions

Irregular Plurals (Grammar): using dictionaries

1. **Grids for presentation**

singular	Plural
aircraft	
child	
fish	
foot	
man	

singular	Plural
mouse	
person	
sheep	
tooth	
woman	

2. **Controlled practice**

1	1	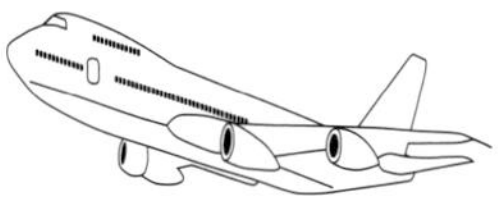	
1	1		
2	3	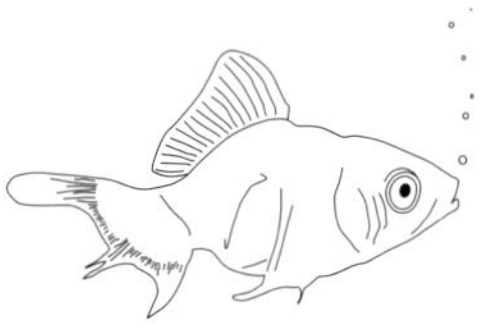	
4	5	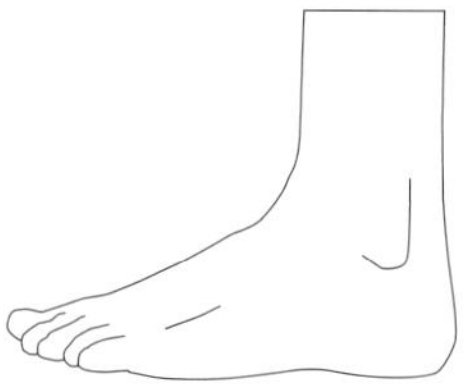	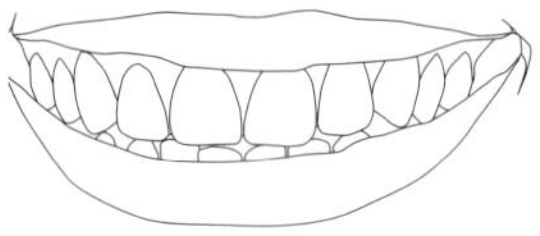
6	7		

Job Application (Writing): from a written model

1. Three job advertisements for lead-in

Web Designer / Developer

Brown Media Group is looking for a full-time Web Developer to join our digital team.

This is a great opportunity to work with one of the country's most well-known media organisations in a friendly and creative environment. This is a permanent full-time role.

The candidate we're looking for has:

- experience as a web developer in a commercial environment
- excellent design skills

For further information visit www.brownmediagroup.com or contact Jim Davis, HR Manager on 020 3564 2919.

STORE MANAGER, ASL ELECTRONICS

Do you enjoy the thrill of a high-paced sales environment?
Are you enthusiastic and enjoy working with people?

If so, then this is the role for you!

We're looking for a full-time Store Manager who will:

- **lead and manage a profitable business**
- **motivate, coach and develop a dynamic sales team**

Please direct applications to:
Mr Peter Wu
Regional Sales Manager
ASL Electronics
404 Edgware Road
London, W2 1ED

Office Manager, Welland Consulting

This market leader in their field currently seek an Office Manager to join their organisation on a 12 month maternity leave position.

Reporting to the CFO you will be responsible for:

- Full function Payroll and end of month reporting
- Supervision of accounts and administration staff

For further information contact **Suzy Gordon, HR Manager on 020 4635 2998.**

2. Model letter (application 1)

Ms Suzy Gordon
Human Resources Manager
Welland Consulting
13 Liverpool Road
Islington
London N1 0RW

Zhang Xin
1/130 Coventry St
Balham
London SW12 9RT

20 December 2010

Dear Ms Gordon

This is an application for the position of Office Manager, as advertised on www.seek.com. **A**

Having worked for over six years as an Office Manager for two large organisations, I have extensive experience in managing financial operations, including payroll and financial reporting. **B**

In these two positions I was also responsible for supervising office staff and ensuring a productive work environment. **C**

I would be very grateful for the opportunity to discuss my application further. Please find my resume attached. **D**

Yours sincerely

Zhang Xin

Zhang Xin

3. Analysing the model

Analysing the application letter

1. Write A-D to match each paragraph with its purpose.

______ showing she meets the first selection criterion

______ explaining what the letter is

______ requesting the next step

______ showing she meets the second selection criterion

4. Controlled practice (application 2)

Dear Mr Davis,

This is an ____________ for the position of Web Designer/Developer, as ____________ on www.seek.com.

I have over three years' ____________ as a Web Developer for Leader Newspaper Group. This was a challenging role in a very competitive, deadline-driven environment.

I also have outstanding ____________ skills. I was chosen to lead the design team responsible for the redevelopment of the Leader website, which later won the prestigious Walker Award for best Website Design.

I would be very ____________ for the ____________ to discuss my application further. Please find my ____________ attached.

Yours sincerely,

Permission (Function): from a recording

1. Controlled practice

✔	✗
✔	✗

(leave) early	(come in) late
(go) to the toilet	(eat) in the classroom
(swap) classes	(go) out for a minute
(give) you my homework tomorrow	(miss) class
(bring) water into class	(use) my phone
(sit) near the window	(get) my bag

2. Role play cards

You're a Teacher

A student is going to ask you for permission to do some things during the week.

You think rules are important, but you will be flexible if the student has a good reason.

You're a student

You want to do these things:

- take half a day off to go shopping for a friend's birthday
- watch a great TV program from 10.00 to 11.00 on Tuesday morning
- go to a football match that starts at 3.00 pm Wednesday
- skip Japanese because you're scared of the teacher
- take a friend from another country sightseeing for half a day
- skip sport because you've hurt your ankle
- be excused from the test because you want to go out Thursday night to a party

Ask the teacher for permission. If they say yes to an item, tick it.

3. Student timetable

	Mon	Tues	Wed	Thurs	Fri
9.00 - 11.00	English	English	English	Test Revision	Writing Test
11.30 - 12.30	Japanese	History	Japanese	History	Speaking Test
1.30 - 3.00	Maths	Maths	Science	Science	Maths Test
3.00-4.00	Sport	Geography	Art	Geography	Sport

Present Perfect (Grammar): from a situation

1. **Pictures/prompts for presentation**

(lose)

(leave ... somewhere)

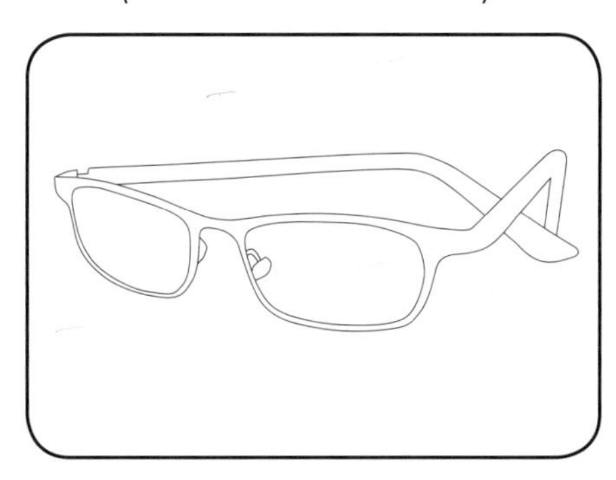

someone (take)

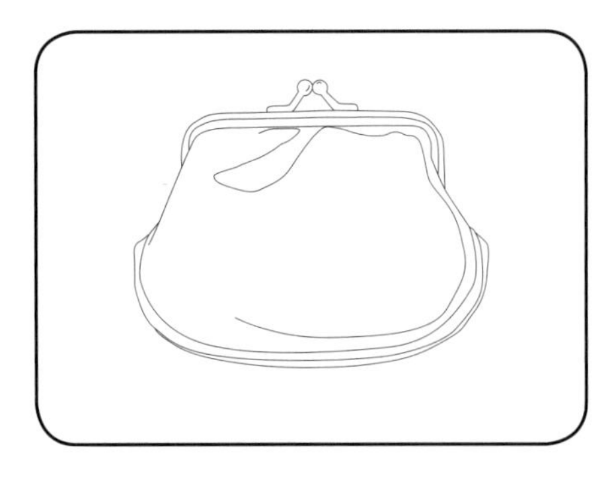

someone (steal)

2. Controlled practice

lose	take		
misplace	steal	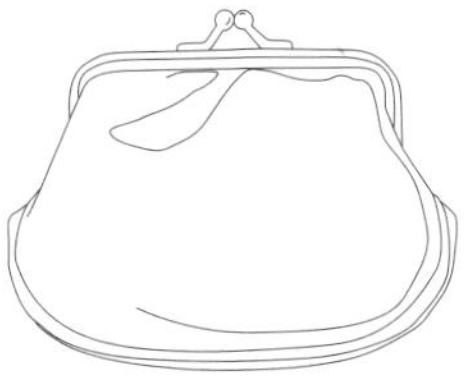	
drop	leave ... somewhere		
lose	take	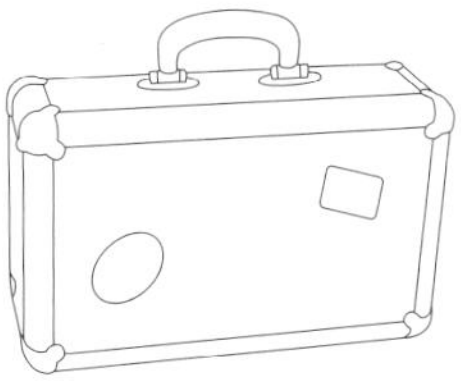	
misplace	steal	PASSPORT	
drop	leave ... somewhere	Class STD Ticket type DAY RETURN Adult ONE Child NIL RTN Start date 02·NOV·05 Number 13303 1226⊕5568S30 From BOGNOR REGIS ✱ Valid until 02·NOV·05 Price £19·50X To NORBITON ✱ Route NOT LONDON 0919	

3. Free practice

police station - a robbery

doctor - an injury

hospital - an accident

dentist - a problem

fire brigade - a fire

teacher - late homework

the with placenames (Grammar): from a text

1. Pictures for lead-in

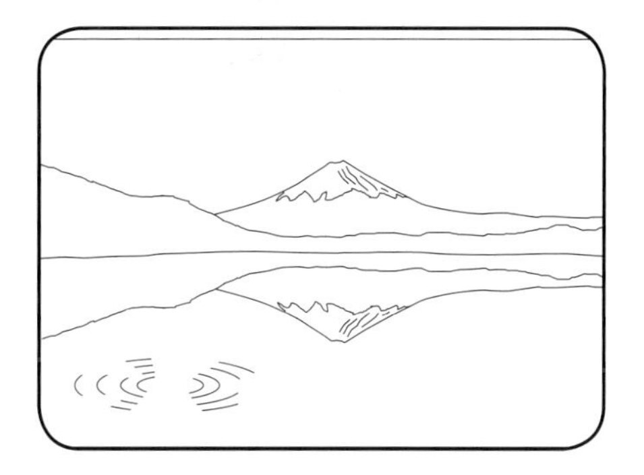

2. Text for presentation

David's story

I was born in the US and while I was growing up I never left North America – in fact I hardly ever left Chicago. I always wanted to go to Asia; we learned a lot about China at school. So last year my employer told me I'd be going to either Japan or Korea (the Republic of Korea, of course) to work in one of our regional offices.

As it turned out I was sent to Tokyo, Japan.

The office was located near the Nihombashi River in the middle of Otemachi, an ancient part of the capital and now a big financial centre.

Nature's my passion and Tokyo is a great base to explore the Japanese countryside. There are so many beautiful places close by in southern Japan such as Lake Biwa, Mount Fuji and the Japanese Alps.

I was sad when my time in Tokyo was up.

True or false?

He traveled a lot when he was young. ______

The company first sent him to Korea. ______

He lived in Tokyo. ______

He likes visiting places outside the city. ______

Lake Biwa and Mount Fuji are in Tokyo. ______

3. Table for presentation

the with placenames

		the?
continent	*North America*	✗
country (one word)		
country (a plural noun or with the word Republic)		
region		
city		
river		
mountain range		
mountain		
lake		

4. Text for controlled practice

Visit the beautiful France.

France has everything a visitor could hope to see.

The Paris is one of the most beautiful cities in the Europe. There are countless things to do during your stay in the capital. Take a boat ride down River Seine and visit the city's famous landmarks such as the Eiffel Tower and Notre Dame Cathedral.

You must also spend time in the regions of France. The Champagne, located just to the east of Paris, is of course the home of sparkling wine. See the gorgeous towns of the Provence. Alps, and the Mont Blanc, on the eastern border, are wonderful for skiing in winter and hiking in summer.

Grammar

Introduction

Grammar can be a sore spot for even the most experienced of teachers. This section demystifies virtually all the grammar you'll be asked to teach up to upper-intermediate level (and includes much that advanced students will be hazy about).

It's designed to be easy for you to find a grammar point, understand it, and know how to teach it.

1. Nouns & Determiners

This section looks at nouns – the names of things (car), people (Catherine), animals (rabbit), places (Los Angeles) and concepts (envy) – and determiners, the small words like articles (a/the) and possessives (my) that come before nouns.

A Countable/uncountable nouns

Level:
Elementary
Pre-Intermediate
Intermediate

What are they?

Countable	Uncountable

Words like hamburger, car, friend that you can count: one hamburger, two hamburgers.	Words like rice, air, petrol, that you can't count: one rice, two rices.

(!) 1. A learner's dictionary tells you whether nouns are countable or not: hamburger [C], rice [U].
2. Most abstract nouns (e.g. poverty, democracy etc) are uncountable.
3. The rules are different for different languages. Information is uncountable in English, but countable in other languages. Grape is countable in English, but not in other languages.

Using countable and uncountable nouns

Countable	Uncountable
Singular countable nouns: • don't use a singular countable noun on its own (it needs a determiner, see X): ~~I have car~~ → I have **a** car.	• don't use **a(n)** or plural **-s** with an uncountable noun: ~~A petrol is expensive~~ → Petrol is expensive. ~~I like rices~~ → I like rice.
Plural countable nouns: • use a plural noun on its own to mean 'all' or 'in general': Cars are expensive to run. (= all cars)	• use an uncountable noun on its own to mean 'all' or 'in general': Rice is good for you. (= all rice)
• use **some/any** (see X) with a plural noun when you don't mean 'all': I saw some friends on the weekend. (= a certain number)	• use **some/any** (see xxx) when you don't mean 'all': I bought some petrol. (= a certain amount)

Countable	Uncountable
	• use **a(n) X of** … to say 'how much': a packet of cigarettes, a bottle of milk (containers) a kilo of rice, a litre of petrol (quantities) a slice of bread, a piece of information (units)

Some nouns can be countable and uncountable.

1. They are uncountable when they describe a material: How many eggs do you need? vs Is there egg in this pastry?
2. They are uncountable when you speak generally: I heard a noise vs I don't like noise.
3. In informal English drinks can be countable. You can say 'a beer' and 'two coffees' to mean 'a glass of beer' and 'two cups of coffee'.
4. Some nouns have different meanings, depending on whether they're countable or uncountable:

	Countable	Uncountable
country	a nation Russia's a big country.	outside the city There are lots of farms in the country.
experience	an interesting event I had lots of experiences on my trip.	doing something specialised I have six years' experience as a nurse.
hair	a single strand I found a hair in my lasagna.	all the hairs on your head I had my hair cut.
iron	an appliance I unplugged the iron.	a type of metal It's made of iron.
light	a device Can you turn on the light?	e.g. from the sun The trees block the light.
paper	newspaper Did you read today's paper?	material you write on Could I borrow some paper to take notes?
room	part of a house I like your living room.	space for something There's no room in my bag.

Teaching ideas

From visuals or realia

Elicit **a** vs **some** from pictures or realia (related to one topic). Teach it in a sentence (rather than disembodied words and phrases): What's on the shopping list?

We need an egg.

We need some flour.

Pronunciation hint

- The final /n/ in an links with the following word, so an egg sounds like /əˈneg/.

B Plural Nouns

Level:
Elementary
Pre-Intermediate
Intermediate

What are they?

We use the plural form of nouns

1. if there is more than one (often after a number, or a quantifier like some or many).
2. to mean 'all' or 'in general': Cars run on petrol (a car runs on petrol is possible, but more formal; the car runs on petrol is possible, but very old-fashioned).

Form

Most nouns add -s (car → cars). Exceptions:

Nouns ending in	Singular	Plural
consonant + **y** → delete the y, add **ies**	country	countr**ies**
-ch, **-s**, **-sh**, **-x**, **-z** → add **-es**	bus	bus**es**
some nouns in **-o** → add **-es**	tomato (also potato, hero, echo, volcano)	tomat**oes**
most nouns in **-f** or **-fe** → **-ves**	loaf, wife	loa**ves**, wi**ves**
-is → delete the -is, add **-es**	crisis	cris**es**
-us →**-i** (delete the **–us**, -i (now commonly **–es** is added))	cactus	cact**i** (or cactu**ses**)

Common irregular plurals:

Singular	Plural	Singular	Plural
aircraft	**aircraft**	mouse	**mice**
child	**children**	person	**people**
deer	**deer**	series	**series**
fish	**fish**	sheep	**sheep**
foot	**feet**	species	**species**
goose	**geese**	tooth	**teeth**
man	**men**	woman	**women**

(!)
1. Only countable nouns can have a plural ending.
2. Police, cattle and staff are plural, with no singular form: The police are looking for him. (to describe an individual, you can say a police officer, a cow, a staff member.)
3. News is always singular: The news is on now.
4. Some other nouns (mainly subjects and sports) ending in **-s** are singular (economics, maths, physics, politics; athletics, billiards, draughts): Maths is hard.

Special cases:

1. In UK English, nouns that describe groups of people (e.g. company, family, team, government) can be used as plural nouns if you want to stress they are made up of individuals: Manchester United is a good team vs Manchester United aren't all playing well.
2. Amounts and quantities with numbers are singular, when treated as a unit: Twenty dollars isn't enough; five years is a long time to wait.

Teaching ideas (see Lesson Plans page 130)

(see Lesson Plans page 130)

From visuals or realia

Flash pictures (or reveal real objects) to elicit the singular and plural forms:

It's a tree.

They're trees.

Pronunciation hint

- Plural **-s** is usually pronounced /z/.
- **-s** is only pronounced /s/ after unvoiced sounds.
- The **-es** fter **ch**, **s**, **sh**, **x** and **z** is pronounced /ɪz/ (or US /əz/).

C Noun + noun/*of*/*'s*

Level:
Pre-Intermediate
Intermediate

What are they?

There are three ways of putting nouns together:

noun + noun	noun + **of** + noun	noun + **'s** + noun
a coffee cup	a cup of coffee	Pete's cup

• the first noun is like an adjective: it tells you *what sort of cup*	• this tells you *how much coffee* • **of** is usually used with things, not people	• this tells you *who owns the cup* • **-'s** is usually used with people, not things

Other uses:

	• to show things belong to things: the door of the car (not this car's door)	• with organisations: China's decision (meaning the government) • with times: **tomorrow's class** (note **Sunday's newspaper** means one newspaper last or this Sunday; a Sunday newspaper means this type of newspaper)

Form

noun + noun	noun + **of** + noun	noun + **'s** + noun
• there are no clear rules whether the words are separate (bread shop), joined (bookshop), or joined with a hyphen (post-office). If in doubt, write them separately, or check a dictionary. • the first noun is always singular (~~cars factory~~ → car factory), except for clothes, sales and sports (e.g. clothes shop). • you can often use a verb **+ing** to show what the second noun is used for (e.g. dining room).	• use **a** for amounts: a kilo of rice • use **the** to say something belongs to something: the colour of the roof	• use **-'s** after a singular noun: Susan's house. • just add an apostrophe after plural **-s**: my parents' house meaning 'two parents'. • use **-'s** after an irregular plural: my children's school.

Teaching ideas

From a visual

Use a family tree to elicit possessive -**'s**:

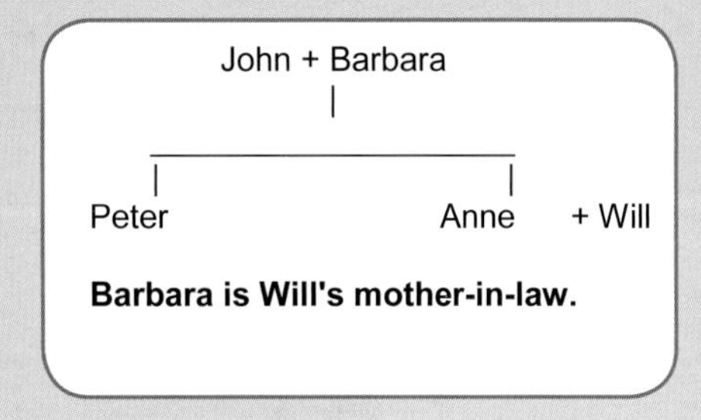

Using real objects

Elicit the names of objects around the classroom or school, contrasting the three forms:
it's a water bottle / it's a bottle of water / it's Susan's bottle

Pronunciation hint

- In an adjective + noun phrase, the noun is stressed: a large **cup**.
- In a noun + noun phrase, the first noun is stressed: a **tea** cup.

D Articles 1: *a* vs *the*

Level:
Pre-Intermediate
Intermediate
Upper Intermediate
Advanced

What are they?

English has two articles. They go before nouns.

a(n) (indefinite article)	**the** (definite article)
• can only go before singular countable nouns (see page 160)	• can go before singular and plural, countable and uncountable nouns (see page 160)

Form

1. Use **a** before a consonant sound: a car, a uniform; use **an** before a vowel sound: an egg, an hour

2. There can be an adjective between an article and a noun: I've got a new car.

Using *a* and *the*

a = new information
the = something the listener knows about

Don't start a sentence with **a** (e.g. ~~A car is outside~~ → There's a car outside) (see page 180)

1. Use **a** the first time you mention something. Use the when you mention it again.
   I bought **a** dictionary and a grammar book yesterday. **The** dictionary cost me $90!
2. Use **the** when there's typically only one in a situation, and the listener can imagine it.
   I forgot to turn off **the** oven before I came. He never cleans **the** bathroom.
3. Use **the** if you give extra information to specify what you're talking about (e.g. with a relative clause).
   The book I got is really good. I like **the** people next door.
4. Use **the** when there's only one in nature
   the country(side), **the** environment, **the** earth, **the** equator, **the** ground, **the** sea, **the** sky, **the** wind

Other uses of *a*

1. to say what something is
   That's **a** computer.

2. to describe someone's job
   She's **a** teacher.

3. to describe an amount in the phrase **a... of...**
   I need **a** kilo of sugar.

4. to say how much a quantity of something costs
   five dollars **a** kilo

5. to describe how often something occurs
   three times **a** month

6. in the phrase **a... of mine**
   She's **a** friend of mine.

7. in the phrase **What a...!**
   What **a** great party!

Other uses of *the*

1. in a phrase containing **of** (except with amounts: a kilo of sugar)
   the end of the month, **the** history of Europe

2. with superlatives (see page 185)
   the best, **the** longest

3. in the phrase **the same**

4. for shops and typical places in a town.
   I need to go to **the** bank (**the** station, **the** post-office, **the** supermarket, **the** doctor, **the** dentist).
   (But use **a** if there's something unusual: I found a new Italian supermarket near me.)

5. in the phrases **the radio, the (news)paper** (but not TV)

6. for most organisations
   the United Nations, **the** BBC

7. for most newspapers
   the New York Times, **the** People's Daily

8. to describe a nationality ending in **-ch, -sh, -ese**
   the French, **the** Chinese
   (others use **-s** without **the**: Americans like hamburgers. It's now common to say French people, Chinese people without **the**: French people like wine.)

No article

1. to mean 'all' or 'in general' (see page 160)
   ~~I like the music~~ → I like music.
   ~~She doesn't eat the meat~~ → She doesn't eat meat.

2. with another determiner
   ~~the my car~~ → the car or my car

3. before people's names
   ~~the Jane~~ → Jane
   ~~the Dr Wang~~ → Dr Wang
   (except before a surname with **-s** to mean a family: the Smiths)

4. for meals
   ~~I had a lunch earlier~~ → I had lunch earlier.
   (unless you use an adjective: I had a fantastic lunch.)

5. before a noun with a number
   ~~Catch the bus 126~~ → Catch bus 126.
   ~~Do the exercise 3~~ → Do exercise 3.

6. for most companies
   BP, i-to-i

7. for most magazines
   Time, Newsweek

8. for musical instruments
   She plays cello (the cello is used for classical music)

Teaching ideas

The rules for articles are complicated so it's a good idea to treat them as they arise in context. You can keep a poster on the wall divided into THE/A/NO ARTICLE, and add rules to it as students come across them.

Pronunciation hint

- **the** before a vowel is pronounced /ðiː/ with a linking /j/, so the egg sounds like /ðiː'jeg/.
- **the** before a consonant is pronounced /ðə/, so the cheese sounds like /ðə'tʃiːz/.
- The final /n/ in **an** links with the following word, so an egg sounds like /ə'neg/.

E Articles 2: *to school* or *to the school*?

Level:
Pre-Intermediate
Intermediate

What are they?

With certain places you sometimes need **the**, and sometimes don't:

school, university, college, hospital, prison, church

go to...	go to the...
She's going to school.	He's going to the school.

• this is the normal purpose for going there (to study)	• this is for some other reason (e.g. to drive children home or to teach)

Using *to school* etc

1. When you **go to...** (without **the**):

Place	It means to...
school	study
university	study
college	study
hospital	be a patient
prison	be a prisoner
church	worship

2. You can also **be in** or **be at** these places (you have to learn which one): ~~He's been in the university for two years~~ → He's been at university for two years.

be at	be in
school	hospital
university	prison
college	church

Similarly, don't use **the** with **bed, home** and **work.**

phrases	example
go to bed/be in bed	I went to bed early last night.
go to work/be at work	He's at work from nine till five.
come home/get home/be (at) home	I got home after dark.

Teaching ideas

From visuals

As in the examples above, use visuals that clearly show the different reasons people are somewhere to elicit the language:

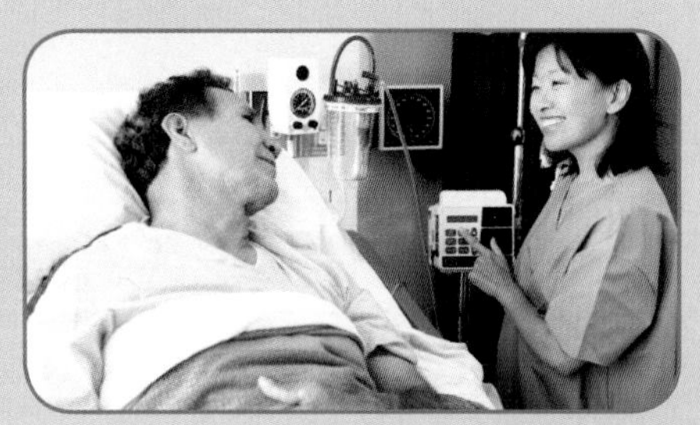
He's in hospital.

He's in the hospital.

F Articles 3: place names

Level:
Pre-Intermediate
Intermediate

What are they?

Some place names need **the** (e.g. the Pacific Ocean), but others don't (e.g. Asia).

Using the with place names

1. Don't use **the** for:

Type of place	Example
Areas	
• continents	Europe
• countries	Japan (but with Republic, Kingdom and plurals use **the**: the United Kingdom, the Netherlands, the USA)
• regions	Northern Africa
• states and provinces	Sichuan
• towns and cities	Jakarta
Geographical features	
• mountains	Mount Everest
• single islands	
• lakes	Lake Victoria
In the city	
• streets, squares, parks	Austin Street, Tiananmen Square, Gorky Park
• buildings (if the first word is a place)	London Bridge

2. Use **the** for:

Type of place	Example
Geographical features	
• mountain ranges	the Urals
• groups of islands	the Bahamas
• oceans	the Atlantic
• seas	the Mediterranean
• rivers	the Amazon
• deserts	the Sahara
In the city	
• buildings	the White House, the Eiffel Tower, the Kremlin

Maps usually don't show **the**, even though you need to use it when you speak or write.

Teaching ideas (see Lesson Plans page 115)

From visuals
Use a map and visuals to elicit a real or imaginary trip.

From a text
Give students a text about a journey that contains examples of placenames with and without **the**. First have them read for understanding, and then ask them to underline the placenames and work out the rules. Alternatively, as they read, students can fill in a table like this:

	Example	the?
river sea mountain		

some vs any?

Level:
Pre-Intermediate
Intermediate

What are they?

You can use **some** and **any** to refer to a small number or amount.

Using some and any

1. You can use **some** and **any** before uncountable or plural countable nouns (e.g. rice or sandwiches). You can't use **some** and **any** before a singular countable noun (e.g. pea). (See page 160).

2. The difference between **some** and **any** is the same as the difference between **somebody** and **anybody** (see page 177).

some	any
• In positive statements: I feel like some junk food. **But** • In offers: Have some tea. • In requests: Can you bring some menus? • In questions when we think the answer is yes: Are there some good places to eat around here? (You've seen lots of restaurants nearby.)	• In negative statements: I don't have any money. • In questions: Do you have any seafood dishes? **But** • After **if**: If you have any requests, just ask. • In statements to mean 'it doesn't matter what': – Order me a curry. – What sort? – Any.

Special cases:

1. Its normal to bring a negative to the start of a sentence: ~~I have no money~~ → I don't have any money.
2. You can use **some of** and **any of** before another determiner:
   I've invited some of my friends to the restaurant.
   Have you been to any of the restaurants in the main street?
3. English has a number of 'pair words':
   glasses, jeans, pants, shorts, scissors, trousers, stockings
   Before these you can use **some** or **a pair of**: I bought some jeans/I bought a pair of jeans.

Teaching ideas

From a dialogue

Teach students a dialogue in a real-life context where questions about availability, offers and requests naturally occur (such as a shop, a restaurant or an office):

water? desserts?

else?

Customer: Could you please bring us some water?
Customer: Do you have any desserts?

Waiter: Certainly. Anything else?
Waiter: Yes madam.

Pronunciation hint

- **Some** and **any** are not usually stressed, and **some** is reduced to /səm/.

H *each/every/whole/all*

Level: Intermediate

What are they?

1. **Each** and **every** both describe all people or things in a group.

each	every
Each student plays a different sport.	Every student plays sport.
Use **each** to show people or things are different.	Use **every** to show people or things are the same.

2. **Whole** and **all** are similar in meaning:

whole	all
I ate a whole loaf of bread.	I finished off all the fries and all the juice.
Whole only describes one; never more.	**All** describes more than one, or an uncountable amount.

Using *each/every/whole/all*

1. Each and every
   a. Use them both with a singular countable noun (see page 160).
   b. **Each** can describe a group of two or more; **every** describes a group of three or more.
   c. The meanings are very similar. If you mix them up you'll still be understood.

2. Whole
   a. Use **whole** with a singular countable noun (see page 160).
   b. You need to use it after a determiner like **a** or **the**: ~~I watched whole match~~ → I watched the whole match.

3. **All**
   a. Use **all** with an uncountable or plural noun (see page 162).
   b. Use it alone to mean 'in general'; use it before a determiner like **the** or **my** to specify which ones: I like all sports (in general); I liked all the sports we played at school (specific). (See page 164.)
   c. You can use all of before the or my etc; the meaning is the same: All my friends play basketball or All of my friends play basketball.

Special uses:

1. Use **each** after a noun to say the price: They cost two euros each.
2. Use **every** with times to show how often something happens: I play football every Tuesday.
3. Use **the whole** or **all** to say the period something lasted: We practised the whole afternoon or We practised all afternoon.
4. **All the time** = 'very often': He goes jogging all the time.

Don't use **all** to mean 'all the people somewhere' – use **everyone/everybody**: ~~All played well yesterday~~ → Everybody played well yesterday (but you can use **all** after a pronoun: They all played well).

Teaching ideas

From a class survey

Have students conduct a class questionnaire: who has what, who does what. Summarise the results on the board. Elicit the language from the results: Every student has a mobile. Each student has a different mobile.

I *much/many/a lot/little/few*

Level:
Elementary
Pre-Intermediate

What are they?

Use **much** with an uncountable noun, and **many** with a plural (see page 160-163):

much	many
much oil	many onions 

You can use **a lot of** for both: a lot of oil/a lot of onions

Using *much/many/a lot*

1. Use **much** and **many** in negative statements and questions:
   We don't have many onions. Do we have much oil?

2. **Much** and **many** are rarely used in positive statements: ~~I bought much rice~~.

3. You can use **a lot of** in positive as well as negative statements, and questions:
   I bought a lot of rice. They don't sell a lot of vegetables. Do we need a lot of potatoes?

4. However, **a lot of** is informal and is not often used in writing. Instead you can use a phrase such as **a large amount of** (with uncountable nouns) and **a large number of** or **a wide range of** (with countable nouns): China exports a large number of food products.

(!) The verb agrees with the noun, not with **a lot** or **a large number:** ~~A lot of cars is expensive~~ → A lot of cars are expensive.

What are they?

Use **little** with an uncountable noun, and **few** with a plural (see page 162):

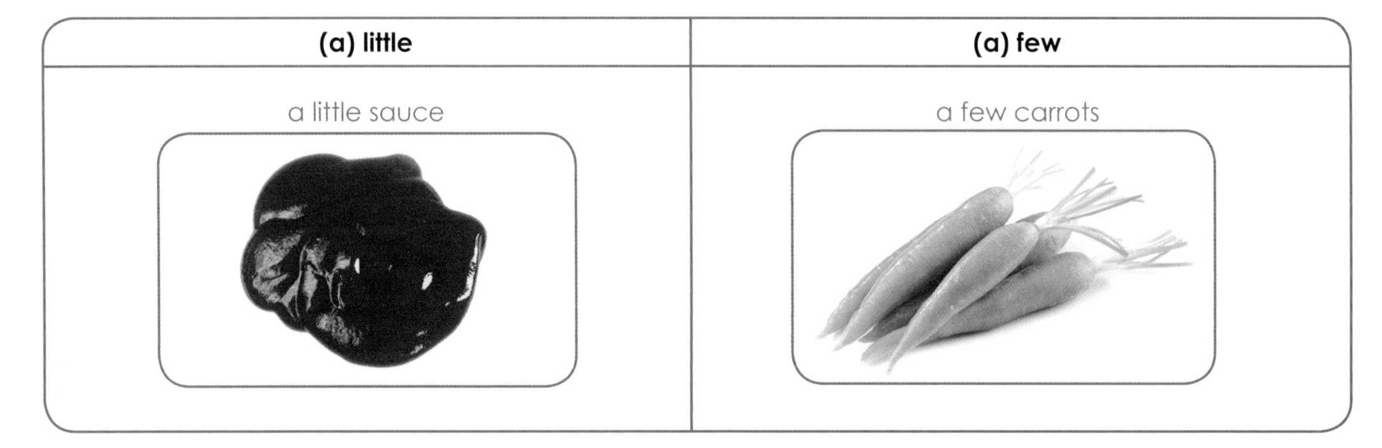

(a) little	(a) few
a little sauce	a few carrots

Using *little* and *few*

1. Use **a little** and **a few** as a factual description (meaning 'not much or many'):
   We've got a few carrots. (enough for dinner)

2. Use **little** and **few** without **a** to mean 'not enough':
   We have little sauce left. (we need to buy some more)

3. **Little** and **few** on their own are quite formal. In normal spoken English, you can express 'not enough' with **only a little** and **only a few**:
   We've only got a few carrots. (we need to buy some more)

4. In informal spoken English it's common to say **a bit of** or **a little bit of** instead of **a little**:
   We've only got a bit of oil in the cupboard.

Teaching ideas

Don't teach all at once, or it's too confusing; consider **much/many/a lot** and then **little/few** in a later lesson.

From a dialogue

Teach students a dialogue where amounts naturally occur (e.g. shopping or making a recipe):

Flatmate 1: We've got a lot of rice.
Flatmate 2: Do we have much oil?

no/some/most/all + of the?

Level:
Pre-Intermediate
Intermediate

What are they?

You can use **no, some, most** and **all** with or without **of the**:

most...	most of the...
Most people like music.	Most of the people in my block like music.

Don't use **of the** if you mean 'in the world'.	Use **of the** to specify a group of people or things.

Using *no/some/most/all*

1. These can only come before plural or uncountable nouns:
   I like most music; I like some of the songs on the CD.

2. Use **no** directly before a noun, but **none** in the phrase **none of the**… (not ~~no of the~~…):
   No instrument is easy to play; None of the music in the concert was very good.

3. You can say **all of the**… or just **all the**…:
   All of the students in my class play music or all the students in my class play music.

4. You can replace **the** with a different determiner, e.g. **my** or **those**:
   Most of my friends listen to music online.

5. You can replace **the** + noun with a pronoun:
   Some of us can sing.

Careful:

1. Don't use **the most** with a noun (e.g. ~~The most people can sing~~): **the most** is used to form a superlative adjective (see page 185).
2. Its normal to bring negatives to the start of the sentence: ~~I know none of the songs in the Top 10~~ →
   I don't know any of the songs in the Top 10.

Teaching ideas

From a discussion

Have students discuss people's habits and beliefs in their country. You can write prompts on the white-board to guide them (e.g. home/transport/hobbies). Have students feed back to the class and elicit the target language: Most people live in apartments. Most of the people in the countryside have a car.

From statistics, a graph or pie chart

Find some interesting results of a survey to elicit the language:

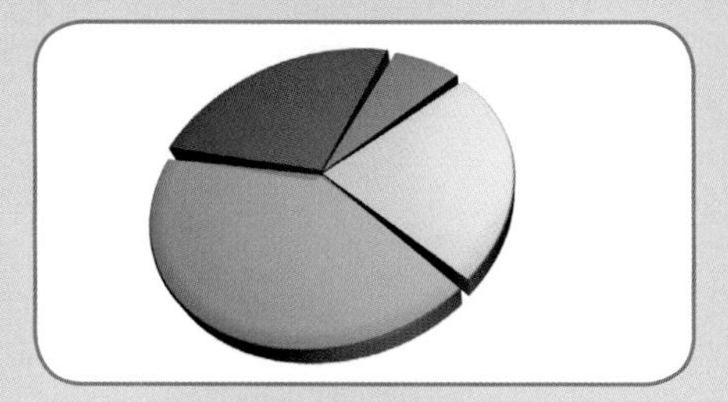

Almost no Japanese speak Swahili.

Pronunciation hint

- **none** and **some** have the vowel sound /ʌ/.
- **no** and **most** have the vowel sound /əʊ/.

2. Pronouns

These are words that replace nouns, like she (= my sister) or it (= the computer). This section looks at, for example, subject and object pronouns (the difference between I and me) and indirect objects (I gave her a present).

A Subjects and objects: *I* vs *me*?

Level:
Elementary
Pre-Intermediate

What are they?

- Pronouns replace a noun: My brother's a student → He's a student.

Subject	Object
I	me
you	you
he	him
she	her
it	it
we	us
they	them

Using subject and object pronouns

- In most cases, subjects come before a verb, and objects after: He (subject) likes me (object).

- You have to use an object pronoun after a preposition: I live with them (prepositional object).

(!) Sometimes both are possible:
After the verb **be**, use the object form. Some books recommend using the subject form, but this is very old-fashioned.
It's me.
It is I. (very old-fashioned)

- To answer the question **Who ...?** you can use a subject + auxiliary, or the object form (informal):
Who did their homework? - I did.
Who did their homework? - Me. (informal)

- After a comparative adjective and **than** (see page 183), you can use a subject + auxiliary, or the object form (informal):
She's smarter than I am.
She's smarter than me. (informal)

(!) Special cases:

1. The question word who can be a subject or object:
   Who (subject) went to class?
   Who (object) did you (subject) see?
   In very formal language there is an object form **whom**:
   Whom did you see?
2. You can use **you** or **one** (same meaning and formal) to mean 'people in general':
   You should practise speaking as much as possible. (= it's a good idea for everybody)
   One should practise speaking as much as possible. (formal)

3. You can use **they** and **them** to mean one person, if you don't know or don't want to say whether the person is male or female:
   Someone texted me and they want me to meet them after class.
4. You can use **they** to mean 'the Government' or 'the authorities':
   They're going to build a new school near here.
5. You can use **one** to replace the article **a + a singular countable noun**:
   - Can I borrow a dictionary? – Sorry, I haven't got one. (= a dictionary)
   (**It** replaces a noun with **the**: – Can I borrow the key?- Sorry, I haven't got it. (= the key)
   If there's an adjective before **one,** you also need **a**:
   - Can I borrow a dictionary? – Here you go. – Sorry, do you have a better one?

Teaching ideas

From a discussion

Have students discuss what they do together after class. Elicit examples of the target language:
I visit her. I go shopping with them.

From a text

Give students a text with a simple story about people. Have students underline all the pronouns and elicit their function:

Text: He phoned her and asked her for help.
Teacher: Which person did this? (mime using phone) Who did this? (mime answering phone)

Pronunciation hint

- Indefinite you is always unstressed and pronounced /jə/.

B Indirect objects

Level:
Pre-Intermediate
Intermediate

What are they?

Some verbs, like **give**, have two objects – usually a person who receives it (indirect object) and the thing you give (the direct object):

We gave her (indirect object) a present (direct object).

Using indirect objects

1. An indirect object pronoun can be a noun or a pronoun. Indirect object pronouns have the same form as direct object pronouns (e.g. her).

2. Usually the indirect object comes first:

subject	**verb**	**indirect object**	**direct object**
I	bought	my sister	a vase.
I	sent	her	a card.

3. You can put the indirect object second, but you need to use **to** or **for** before it:

Subject	**Verb**	**Direct Object**	**Indirect Object**
I	bought	a vase	for my sister.
I	sent	a card	to her.

You have to learn which verbs need **to** and which need **for**:

verbs with **to**	verbs with **for**
bring	book
email	buy
give	choose
hand	cook
lend	find
offer	get
owe	leave
pass	make
pay	order
send	pick
show	save
take	
teach	
tell	
write	

4. If there are two pronouns, put the indirect object second: ~~I gave her it~~ → I gave it to her.

Teaching ideas

Using visuals

Use visuals of two people and a thing to elicit the parts of the sentence. You can move them around to elicit variations on the target language:

She

gave

him

a present.

She

gave

a present

to him.

He

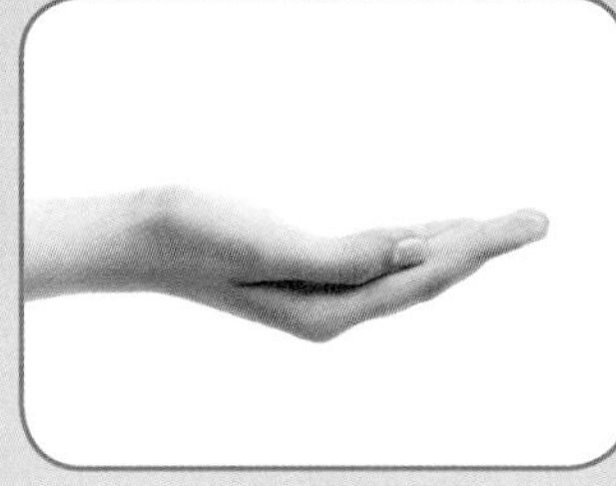
gave

her

a present.

Pronunciation hint

- **to** and **for** are usually unstressed and pronounced /tə/ and /fə/.

C *Somebody vs anybody?*

Level:
Pre-Intermediate
Intermediate

What are they?

You can use these to refer to people and things without saying exactly who or what they are.

	some-	**any-**
people	someone, somebody (inf)	anyone, anybody (inf)
things	something	anything
place	somewhere	anywhere
time	sometime	anytime

Using somebody and anybody

The difference is the same as between **some** and **any** (see page 169).

some-	**any-**
• In positive statements: Somebody rang before.	• In negative statements: I don't know anybody here. • In questions: Did anyone ring when I was out?
Also • In offers: Would you like something to eat? • In requests: Could you explain something to me? • In questions when we think the answer is yes: Has something happened? (you can see people look worried)	**Also** • After if: If anyone rings, tell them I'm out. • In statements to mean 'it doesn't matter what': Ask me anything you like.

Special cases:

1. It's normal to bring a negative to the start of a sentence: ~~I know nothing~~ → I don't know anything.
2. **Sometime** is one point in time (You should visit sometime next week); **sometimes** is more than one time (Sometimes I visit her on Saturdays).

3. **Somehow** and **anyhow** have different meanings. **Somehow** means 'I don't know how', especially when you think something's impossible: I didn't have a job but somehow I saved money. **Anyhow** (or **anyway**) is used in conversation to say 'the last point wasn't important': I didn't save much money. Anyhow, I survived.

Teaching ideas

From a dialogue

Teach students a dialogue in a real-life context where questions about availability, offers and requests naturally occur (such as a restaurant):

Would you like something to drink?
Do you have anything spicy?

D *Myself vs each other*

Level:
Pre-Intermediate
Intermediate

What are they?

Both of these show the subject and object are the same people.

myself, yourself...	each other
He cut himself.	They aren't talking to each other.
• This can refer to one person or more than one • Person A does something to Person A.	• This always refers to two or more people. • Person A does something to Person B, and Person B does something to Person A.

Reflexive pronouns change to match the subject:

subject	object
I	**myself**
you	**yourself**
he	**himself**
she	**herself**
it	**itself**
we	**ourselves**
they	**themselves**
one	**oneself**

Note: **Each other** never changes.

Using reflexive pronouns and *each other*

Common examples:

Verbs with reflexive pronouns	*Verbs with each other*
be hard on yourself	bump into each other
behave yourself	email each other
blame yourself	get each other something
cut yourself	give each other something
dry yourself	help each other
enjoy yourself	know each other
get yourself something	like each other
give yourself something	live with/near (etc) each other
help yourself	look at each other
hurt yourself	love each other
introduce yourself	send each other something
kill yourself	talk to each other
look at yourself in the mirror	text each other
make yourself something	work with each other
talk to yourself	write to each other

Special cases:

1. You can use reflexive pronouns after prepositions – I bought it for myself – except when it describes *where* – Are you carrying your passport on you?
2. Be Careful – English, unlike other languages, does not use **a reflexive** after dress, feel, relax, shave and wash, and does not generally use **each other** after marry and meet.
3. You can use reflexives for emphasis. Straight after a noun they mean 'in person': The principal herself spoke to me. At the end of the sentence they mean 'without any help': I made this myself.

Teaching ideas

From visuals

Use visuals showing someone's daily routine to elicit examples of verbs with and without reflexive pronouns:

He has a shave.

He makes himself breakfast.

Use visuals to tell a story about two people's relationship:

They used to work with each other.

Now they don't talk to each other.

Pronunciation hint

- Reflexive pronouns and **each other** are not normally stressed.

E There's ... vs It's ...

Level:
Pre-Intermediate
Intermediate

What are they?

There's...	It's...
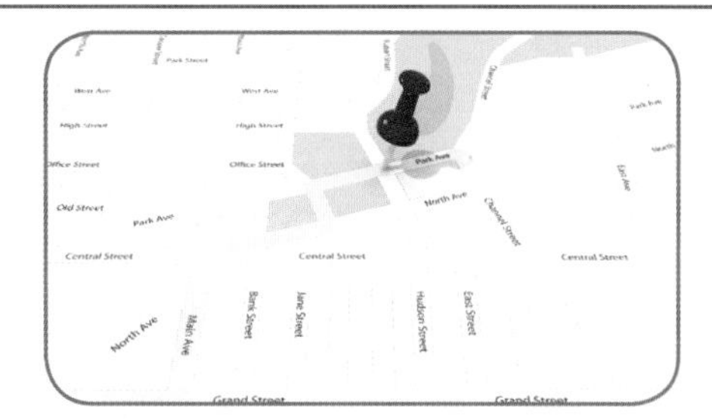	
There's a new restaurant in our street.	- What's that? – It's a restaurant.
• Use **There's**... with a noun to say something exists somewhere:	• Use **It's**... with a noun to answer **What's that?** • Otherwise use **It's**... with an adjective to describe something: It's expensive.

Using *There's*... and *It's*...

There's...

1. When you introduce something for the first time, don't start a sentence with **A**... – use **There's**...:
   ~~A good café is near here~~ → There's a good café near here.

2. After **There's**... you should specify where you mean, often with a preposition such as **in** or **on**:
   ~~There's a fast food place~~ → There's a fast food place on the corner.

3. **There's**... does not have the same meaning as the single word **there** (i.e. 'in the distance').
   There's... can introduce something very close: There's a menu here on the table.

4. You can use other forms of **be**: **There are** lots of places to eat around here / When I was growing up **there were** no restaurants here.

It's ...

1. Use **It's**... to talk about time, distance and weather. This it doesn't refer to anything concrete:
   It's seven o'clock.
   It's two kilometres to the nearest restaurant.
   It's windy. (with an adjective; compare There's a strong wind with a noun)
   It's twenty degrees.

2. Use **It's** + adjective + infinitive with to say how difficult, expensive etc doing something is.
   ~~I am easy to cook at home~~ → It's easy to cook at home.

Other adjectives and phrases you can use like this:

It's ...	**difficult** **possible** **impossible** **cheap** **expensive** **a good idea** **a bad idea**	to ...

3. Use **It's** ... to announce someone's arrival somewhere: It's me. / It's Gary!

Teaching ideas

From visuals

Show visuals of a country or city (perhaps the students' own town or city). Elicit the target language with questions such as:

Teacher: Tell us about Shanghai.
(Elicit:) There are lots of modern buildings here. There's a big river in the centre.

Teacher: What's that?
(Elicit:) It's a bridge.

3. Adjectives

Adjectives are words that describe nouns and pronouns (small, intelligent). This section looks at how adjectives are used, including adjective order (is it an old beautiful town or a beautiful old town?) and comparative adjectives (London's bigger than Leeds).

A Adjectives: order
B Adjectives: position
C Comparatives: *-er & more*
D Superlatives: *the -est & the most*
E *as... as*
F Ungradable adjectives/using *very/too/quite/enough*
G *-ing* or *-ed?*

A Adjectives: order

Level:
Pre-Intermediate
Intermediate

Using more than one adjective

1. It's rare to use more than two adjectives in a row.

2. Opinion goes before a fact: ~~That's a new nice jacket~~ → That's a nice new jacket.

3. Otherwise adjectives tend to go in this order:

size	**age**	**shape**	**colour**	**origin**	**material**	**purpose**	**NOUN**
big	old	round	black	German	plastic	reading	glasses

For example:

- ~~reading plastic glasses~~ → plastic reading glasses
- ~~plastic black glasses~~ → black plastic glasses

Using commas & *and*:

Before a noun	**After the verb 'be' etc**
• if there are two adjectives, don't use **and**: I bought an expensive new suit (unless the two adjectives are colours: I bought a green and orange shirt.) • if there are three adjectives, use **a comma** then **and**: They had Italian, French and Spanish shoes.	• if there are two adjectives, use **and**: The shoes are old and worn out.

Teaching ideas

It's a lot for students – and teachers – to remember these rules all at once! Consider teaching 'opinion before fact' first (see example below). Then add the other categories across a number of lessons – and, if possible, as they come up naturally in context.

From a situation or dialogue build

Set up a context where characters describe (and compare) a number of objects or people. A furniture or clothes shop, a lost property office, or a dating agency work well. Use pictures to elicit the language. For example:

Customer: I'm after a new leather armchair.
Shop assistant: Well, we have a beautiful German armchair right here.

Stick to short two - or three-adjective strings – anything longer can be quite contrived, not to mention the fact it's too hard to remember!

Once you've elicited the sentences and students are confident with them, write the categories (size, age etc) on the whiteboard in random order, and challenge students to work out the sequence.

From a text
Find a text with short descriptions (e.g. a catalogue or a travel brochure). First have students read for understanding, then ask them to underline any examples of two or more adjectives in a row. Get students to work out the rules.

B Adjectives: position

Level:
Intermediate
Upper Intermediate

Using adjectives with nouns

Most adjectives can go either before a noun, or after a linking verb like **be**:
That's a nice tie. (before a noun)
Your tie's nice. (after be)

A small number of adjectives can only go before a noun or only after a noun:

Before a noun	After be etc
• elder (e.g. brother or sister) • old (e.g. friend or school; meaning you've known them for a long time, or ex-) ~~My friend is old~~ → She's an old friend	• afraid • alive • alone • asleep • awake • ill • well ~~I woke up the asleep guy~~ → The guy was asleep, so I woke him up.

Teaching ideas

It's difficult to present this inductively. It may be more efficient simply to present the rules on the whiteboard, then give students a controlled practice activity (e.g. un-jumbling sentences or correcting errors):

Ill is sister feeling your ? → Is your sister feeling ill?
My brother's elder. → My brother's older/I have an elder brother.

Pronunciation hint

- Adjectives before a noun are not usually stressed: ~~He's my EL•der brother~~ → He's my elder BRO•ther (the noun is stressed).
- Adjectives after be are usually stressed: My bro•ther's ILL.

C Comparatives: -er and more

Level:
Elementary
Pre-Intermediate

What are they?

We use comparatives when comparing two people or things.

Meaning	Checking meaning
• We're talking about two things. • They're different.	• Are we talking about one thing? (No) Three things? (No) Two? (Yes) • Are they the same or different? (Different)

From

To form a comparative adjective, you need to add **-er** or use **more**.

adjectives	comparative forms
one syllable: small	**add -er:** smaller
two-syllable adjectives ending in **-y:** noisy	change **-y** to **-ier**: noisier
two or more syllables: modern	add **more**: more modern
Note: Some adjectives e.g. clever, narrow, quiet & simple	can be either: simpler or more simple
Irregular: good, bad, far, old	better, worse, further, elder (for family members)

The rules are very similar to superlative adjectives (see page 185).

Spelling Rules

• Adjectives that end in **-e:** large	just add **–r:** larger
• Adjectives ending in a **consonant + vowel + consonant**: big	double the consonant: bigger

Using comparative adjectives

1. The most common structure is:

subject	be*	comparative	than	object
Warsaw	's	smaller	than	Moscow.

*or other linking verb

2. After **than** you can use an object pronoun or a subject pronoun + **be** (more formal): He's older than me/I am.

3. You can replace the object with a different clause: Warsaw's smaller than I thought.

4. You can use comparatives as you would normal adjectives, before a single noun: I want to live in a more modern city.

5. You can use words to say how much bigger/ better etc something is: Tokyo's much busier than Kyoto.

informal/neutral	formal
a (little) bit quite a lot a lot, much	slightly somewhat considerably

(!) Special uses:

1. '**X-er and X-er**' (or '**more and more X**') means something's changing: Sydney's getting more and more crowded.
2. '**The X-er the better**' means it's what you want: The cheaper the better.

Teaching ideas

From visuals

Show the class visuals of two contrasting cities or countries and elicit the differences. Use a question that focuses the students on the topic (e.g. weather) rather than feeding them the adjectives (hot etc):

Teacher: Is the weather the same?
Students: No ... Singapore is hot ... Tokyo is cold ...
Teacher: How do you say that in one sentence?
(Elicit:) Singapore is hotter than Tokyo.

From contrasting information

Find two short texts that describe different places. The bullet-pointed facts in Wikipedia or *New Internationalist* work well. Elicit the differences from the information.

From a ranking activity

Have students in groups rank items from good to bad (e.g. brands of car, food, types of holiday) according to their opinions. Then elicit the language from their rankings:

Teacher: Why did you say Jaguar is number one, and Lada number five?
Student: Jaguars are fast ... Ladas are slow ...
Teacher: How do you say that in one sentence?
(Elicit:) Jaguars are faster than Ladas.
Teacher: A bit faster?
(Elicit:) No. Much faster. Jaguars are much faster than Ladas.

Pronunciation hint

- The ending **–ger** in younger, longer & stronger is pronounced /ŋgə/

D Superlatives: the -est and the most

Level:
Elementary
Pre-Intermediate

What are they?

We use superlative when comparing three or more people or things, and singling one out as special.

Meaning	Checking meaning
• We're talking about three or more things. • They're different. • One is special.	• Are we talking about one thing? (No) Two things? (No) Three or more? (Yes) • Are they the same or different? (Different) • Is one special? (Yes)

From

To form a superlative adjective, you need to use **the -est** or **the most**.

adjectives	superlative form
one syllable: old	add **the –est**: the oldest
two-syllable adjectives that end in -y: busy	add **the** and change **-y** to **-iest**: the busiest
two or more syllables: crowded	add **the most**: the most crowded
Nb Some adjectives e.g. clever, narrow, quiet & simple	can be either: the quietest or the most quiet
Irregular: good, bad, far, old	the best, the worst, the furthest, the eldest (for family members)

Spelling rules

• Adjectives that end in **-e**: nice	just add **–st:** the nicest
• Adjectives ending in a **consonant + vowel + consonant**: hot	double the consonant: the hottest

Using superlative adjectives

1. The most common structure is:

subject	**be***	**superlative**	**noun**	**in ...**
Jakarta	's	the largest	city	in Indonesia.

*or other linking verb

2. You don't need anything after the superlative if the context is clear: China's the biggest (e.g. we were comparing countries on a map).

3. You can use **of all** to mean the group of things you've been talking about: Hanoi was the most charming place of all (e.g. we were talking about cities on our trip).

4. You can use **easily** or **by far** before a superlative to say the difference is very large: Mexico City's by far the largest city in Mexico.

Special use:

To say something stands out in your life experience, you can use a superlative with **I've ever...** (present perfect simple, see page 225): It's the most beautiful place I've ever seen.

Teaching ideas

From visuals

Show the class visuals of three contrasting places or things (e.g. mobile phones, holiday destinations, places to eat) and ask them to discuss which one they like best. Use a topic you know your class likes arguing about. Write up some prompts to help the discussion along (e.g. price, size, style). Then ask students to give their opinions and to say why, in order to elicit superlatives:

Students: I like the black phone.
Teacher: Why?
Student: It's small.
Teacher: How do you say this one is small, but these two aren't so small?
(Elicit:) It's the smallest.

From a situation

Have students read resumes of three applicants for a job, or listen to three job interviews. Make sure they understand the reading or listening texts. Then ask students to decide who should get the job and why in order to elicit superlatives: She gave the best answers etc.

From a video

Show students a competition (e.g. a game show or a number of short Olympic events). First have students listen for understanding, then elicit who was the winner and why: She ran the fastest etc.

Pronunciation hint

• The ending **–gest** in youngest, longest & strongest is pronounced /ŋgəst/.

as...as

Level:
Pre-Intermediate

Meaning
Saying two things are the same in some way.

Meaning	Checking meaning
• We're talking about two things. • They're the same.	• Are we talking about one thing? (No) Three things? (No) Two? (Yes) • Are they the same or different? (The same)

Using as...as
1. The most common structure is:

subject	be*	as	adjective	as	object
Jacqueline	's	as	tall	as	her brother.

*or other linking verb

2. You can replace the object with a clause: Jacqueline's as tall as I remember.

3. In a negative sentence, you can use **not as...as** or **not so...as**: She's not so tall as her father.

Special uses:
1. You can quantify how similar or different two things are:
   half as tall as
   twice as tall as
   three times as tall as/taller than

2. To say something is a good idea, use **as** (adjective) **as possible**: I want to get as fit as possible.

3. To say 'very', use **as** (adjective) **as anything** (informal): She's as nice as anything.

Teaching ideas
Use the ideas for teaching comparatives (see page 183).

Pronunciation hint
- Stress the adjective and the personal pronoun/noun, but not **as** (pronounced /əz/): She's as WEAL•thy as HIM.

Ungradable adjectives; *using very/too/quite/enough*

Level:
Intermediate
Pre-Intermediate

What are they?
Gradable adjectives
With most adjectives (e.g. good) you can 'grade' their intensity by adding a grading adverb such as **fairly**, **rather** or **very**: The restaurant was very good.

Ungradable adjectives
However, you can't use a grading adverb with some adjectives, because they're already an extreme: The restaurant was ~~very~~ fantastic or they describe an absolute (e.g. something is either dead, or it's not): Your answer is ~~very~~ perfect.

Here are some common examples of gradable and ungradable adjectives:

gradable adjectives	ungradable objectives
good bad small hot cold	**Extreme adjectives:** fantastic, excellent, wonderful, terrific awful, terrible, ridiculous, enormous, gigantic, huge, tiny, minute, boiling, freezing **Absolute adjectives:** alive, dead, correct, incorrect, full, perfect, impossible

Using gradable adjectives with very/too/quite/enough

adverb		Nb:
more ↑	**very/really/too**	**too** means 'it's bad': I couldn't eat it, it was too salty.
	quite/pretty(informal)**/rather**(UK)	**rather** suggests the speaker didn't expect it: I'd heard bad reports but the meal was rather good.
	slightly/a little/a bit (informal)	**slightly/a little/a bit** often mean the speaker isn't happy: It was a bit expensive, so I didn't get it. **a bit** (or **not very**) can be used as a polite way of saying you really don't like something: It was a bit crowded, and the waiters weren't very polite.
less	**(not) … enough**	**(not)…enough** means you are (un)happy: The place was comfortable enough

Using ungradable adjectives with absolutely/really/quite

absolutely, **really** and **quite** are exceptions to the rule and can be used with ungradable adjectives:

adverb		Nb:
more ↑	**absolutely**, **really** (informal),	You can use **absolutely** and **really** to stress it was 100%: The meal was absolutely fantastic.
less	**quite (formal)**	**quite** has a different meaning with gradable and ungradable adjectives: The food is quite good (= not very) He's quite dead (= 100%)

* There are other adverbs you can use, but they only go with certain adjectives (e.g. The place was ~~totally~~ completely full) – you need to check a good dictionary. However, **absolutely** and **really** are possible in most cases.

Word order rules:

1. **very**, **pretty**, **rather** and **slightly** come after **a**: We had a very nice lunch.
2. **too**, **a little** and **a bit** with an adjective can't go before a noun: ~~It was a too hot curry~~ → The curry was too hot.
3. **enough** comes after an adjective: ~~It was enough sweet~~ → It was sweet enough.
4. quite goes before **a**: We had quite a good time.

Teaching ideas

From visuals or word prompts

Divide students into groups. Give each group a set of visuals related by topic (e.g. curry, pizza, ice cream) and ask them to brainstorm what is good and bad about each. Elicit opinions to the whole class:

Teacher: What about pizza?
Student: It's too nice.
Teacher: Do you mean you don't like it?
Student: No, I like it.
Teacher: So don't say too, say...
(Elicit:) Very.
Teacher: The whole sentence?
(Elicit:) Pizza's very nice.

From a situation or dialogue build

Set up a context where characters have to compare different things and make a choice, such as a shop. Use visuals to elicit the characters and what they'll say about the items, e.g. a large man and a small jumper: It's too small.

Pronunciation hint

- **really**, **very**, **too** and **quite** are usually stressed as well as the adjective: VE•ry GOOD
- **enough** is not usually stressed: not GOOD enough
- Use a wide voice range to show your strong feelings about something.

-ing or -ed?

Level:
Intermediate
Pre-Intermediate

What are they?

There are a lot of adjectives in English ending in **-ing** and **-ed** (all formed from verbs). They all express feelings and opinions.

-ing	-ed
This causes someone to feel something: The movie was boring.  **-ing** often describes a thing (you can use the mnemonic '**the thing is -ing**'). However, a person can cause a situation: He's boring because he never stops talking.  	This is how the person feels: I was bored. 

Using -ing and -ed

Some common examples:

It's.../He's.../I find it, him...	I'm.../I feel...
boring	bored
interesting	interested
tiring	tired
annoying	annoyed
depressing	depressed
disappointing	disappointed
exciting	excited

Teaching ideas

Teach example's that include **it's...** with **-ing** and **I feel...** with **-ed** to help students remember the difference in meaning.

From a situation

It's possible to elicit adjectives like 'bored' from visuals, but it's difficult to check students have understood the picture correctly (bored can look like tired etc!). Consider instead telling a story:

Anna went to a new French movie. It was three hours long. Nothing happened. The film was very... (elicit:) boring.

From the news

Have students watch or listen to short news items. Set them a comprehension task to ensure they understand. Then elicit their reactions using '**What do you think?**' and '**How do you feel?**':

Teacher: What do you think about the first story?
Student: It's very... shock...
Teacher: How do you say that?
(Elicit:) It's very shocking.
Teacher: How does it make you feel?
(Elicit:) I feel depressed.

Make sure the items are not too shocking or depressing for your students: finish with something light and interesting.

From discussion about feelings

Have students discuss important experiences in their lives (e.g. coming to a new country; going to university). Elicit the language using 'How did you feel?' and 'Why?':

Teacher: How did you feel during lectures?
Student: I was bored.
Teacher: Why?
Student: The subject was boring, not interesting.

Pronunciation hint

- To show emotion, use wide voice range.

. Adverbs

This section looks at adverbs in -ly which describe verbs (He plays piano beautifully) as well as adverbs that describe frequency (sometimes) and certainty (definitely).

A Describing actions: adverbs in *-ly*
B Comparative adverbs
C Superlative adverbs
D Frequency: *never, sometimes, always*
E Certainty: *maybe, probably, definitely*
F *still, already, yet, any more*

Describing actions: adverbs in *-ly*

Level: Pre-Intermediate

Form

Normally you just add **-ly** to an adjective.

adjective	adverb
beautiful bad	beautifully badly

Irregular adverbs:

	adjective	adverb
adjectives in **-ic** add **-ally**	fantastic	fantastically
adjectives in **-y** change **-y** to **-ily**	easy	easily
good becomes **well**	good	well
fast, hard & **late** don't add **-ly**	hard	hard

In informal US English, adverbs may drop the **-ly**: He sings real bad.

Using adverbs in *-ly*

1. Adverbs in **-ly** come at the end of the sentence.

subject	verb		adverb in **-ly**
Luis	sings	Karaoke	beautifully.

Other types of adverbs have different word order (see page 195 and page 196).

Meaning

Describing how someone does something.

Meaning	Checking meaning
• We're not describing a person. • We're describing how someone does something. • (e.g.) We're not describing Luis, we're describing his singing.	• Are we describing a person? (No) • Are we describing how they do something? (Yes) • (e.g.) Are we describing Luis, or his singing? (Singing)

Teaching ideas

From visuals

Use instructions or signs to elicit:

You have to drive slowly.

You need to drive carefully.

You can drive fast.

Just make sure as much of the vocabulary as possible is familiar.

Using mime

Mime actions (e.g. running on the spot, writing on the Whiteboard) to elicit the language:

Teacher: What am I doing?
(Elicit:) Running.
Teacher: How fast?
(Elicit:) Quickly.
Teacher: Can you say that in a sentence?
(Elicit:) You're running quickly.

You can check the meaning here: is it describing me, or the running?

From a text

Use a text containing instructions e.g. a recipe (Pour slowly...).

Pronunciation hint

- Adverbs are usually stressed: You have to drive SLOW•ly.

B Comparative adverbs

Level: Pre-Intermediate

Form

To form a comparative adverb, use **more** with the **-ly** form (see page 191).
She speaks more fluently than me.
Could you please speak more clearly?

1. Unlike adjectives, the number of syllables is irrelevant (see page 184).

2. The rules are very similar to superlative adverbs (see page 194).

(!) Some comparative adverbs use the same form as the comparative adjective:

adverb	comparative adverb
bad	worse (not more badly)
early	earlier
fast	faster
good	better
hard	harder
late	later
long	longer
soon	sooner

In informal English, three comparative adverbs have the same form as the comparative adjective:

	informal	**formal**
loudly quickly slowly	louder quicker slower	more loudly more quickly more slowly

Using comparative adverbs

1. The most common structure is:

subject	verb		comparative adverb	**than**	object
She	speaks	English	better	than	me.

2. After than you can use an object pronoun or a subject pronoun + auxiliary (more formal):
   She speaks better than me/I do.

3. You can use words to say how much more fluently etc someone does something: He writes much more accurately than me.

informal/neutral	**formal**
↑ a lot, much quite a lot a (little) bit	considerably somewhat slightly

Meaning

Comparing how two people do something.

Meaning	**Checking meaning**
• We're thinking about two people. • We're describing how they do something. • They do it differently	• Are we talking about one person? (No) Two people? (Yes) • Are we describing the people, or how they do something? (Do something) • Do they do it differently? (Yes)

They can also compare the way something is done at two different times: Can you speak more slowly?

Teaching ideas

Teach this after students are confident with adverbs in ***-ly***.

From a video

You need to see actions in progress to be able to compare adverbs.

Show the class a video of two contrasting performances (e.g. sport or music). Elicit the language. Focus on describing the activity, not the people:

Teacher: What are they doing?
(Elicit:): Running.
Teacher: Are they running in the same way?
(Elicit:) She's running faster.

Pronunciation hint

• Comparative adverbs are usually stressed: She's running FAST•er.

C Superlative adverbs

Level: Intermediate

Form

To form a comparative adverb, use **the most** with the **-ly** form (see page 191):
She sings the most beautifully.

1. Unlike adjectives, the number of syllables is irrelevant (see page 184).

2. The rules are very similar to comparative adverbs (see page 192).

(!) Some superlative adverbs use the same form as the superlative adjective:

adverb	superlative adverb
bad	the worst (not the most badly)
early	the earliest
fast	the fastest
good	the best
hard	the hardest
late	the latest
long	the longest
soon	the soonest

In informal English, three have the same form as the superlative adjective:

	informal	formal
loudly	**the loudest**	the most loudly
quickly	**the quickest**	the most quickly
slowly	**the slowest**	the most slowly

Using superlative adverbs

1. The most common structure is:

subject	verb		superlative adverb	of ...
She	plays	guitar	the best	of all the students.

2. You don't need anything after the superlative adverb if the context is clear: He plays the fastest (e.g. we were comparing players in a competition).

3. You can use of all to mean the people you've been talking about is clear: She plays violin the most sensitively of all (e.g. we were talking about a group of violin players).

4. People often avoid superlative adverbs, and rephrase the sentence, often with a superlative adjective (see page 185):
   He plays the most confidently → He's the most confident player.
   He's the most confident at piano.

5 You can use easily or by far before a superlative adverb to say the difference is very large: He can read music by far the best.

Meaning

Comparing how three or more people do something, and singling one out as special.

Meaning	Checking meaning
• We're thinking about three or more people. • We're comparing how they do something. • One is special.	• Are we talking about two people? (No) Three people? (Yes) • Are we describing the people, or how they do something? (Do something) • Is one special? (Yes)

Teaching ideas

Teach superlatives after students are confident with comparative adverbs.

From a video

Like with comparative adverbs, you need to see actions in progress to be able to single one out as special.

Show the class a video of three contrasting performances (e.g. sport or music). Elicit the best, worst, most interesting etc.

Teacher: Who's best?
(Elicit:): Performer X.
Teacher: Why?
(Elicit:): He sings the best.

Pronunciation hint

- Superlative adverbs are usually stressed: He sings the BEST.

D Frequency: never, sometimes, always

Level: Pre-Intermediate

What are they?

often	**always**
	normally, usually
	frequently (formal), **often**
↑	**sometimes**
	occasionally
	seldom, rarely (formal), **hardly ever** (inf)
not often	**never**

Using frequency adverbs

1. Frequency adverbs usually go:

- after an auxiliary:

subject	aux **have**	frequency adverb	past participle	
I	've	never	been	to Spain.

- after **be**:

subject	**be**	frequency adverb	
She	's	often	away.

- before other verbs:

subject	frequency adverb	verb	
He	never	takes	his family.

2. You can use normally, usually, sometimes and occasionally at the start of sentence to stress the time, especially to contrast with the current situation: Normally I travel for work, but this time I'm on holiday.

3. You can only use **always** and **never** at the start of a sentence in a command: Always remember to check your bag!

4. To specify exactly how often, you can use **every**... at the end of the sentence: I go abroad every June.

5. As these are habits, use present simple (see page 216), not continuous: ~~I'm often staying in Prague~~ → I often stay in Prague.

(!) Two informal expressions:
all the time = very often (at the end of the sentence): I go to Europe for work all the time.
now and then = sometimes (at the start of the sentence): Now and then I go by car, but I usually fly.

(!) You can use present continuous with always to mean a bad habit: She's always showing off about how much she travels.

Meaning

These answer the question *How often*?

Teaching ideas

From a dialogue build
Teach a dialogue where people naturally talk about frequency:

Doctor: How often do you drink?
Patient: Once a year.
Doctor: So... rarely.

Housemate 1: You never wash up!
Housemate 2: You always say that!

You can check the meaning of the adverbs as you go by writing them against a cline (as at the start of this section).

From a class survey
Have students conduct a class survey: How many times a week do you (do homework/play sport/eat out etc.)? Summarise the results on the board. Elicit the language from the results: People often eat out.

Pronunciation hint

- Adverbs of frequency are usually stressed: I've NE•ver been to SPAIN.

E Certainty: *maybe, probably, definitely*

Level:
Intermediate
Pre-Intermediate

What are they?

sure	certainly, definitely
↑	probably
not sure	maybe, perhaps

Using adverbs of certainty

1. Certainly, definitely and probably usually go:

- after an auxiliary (except **do**):

subject	aux **will**	adverb of certainty	bare infinitive	
The shops	will	probably	close	soon.

- after **be**:

subject	**be**	frequency adverb	
It	's	definitely	too expensive.

- before auxiliary **do**:

subject	adverb of certainty	aux **do**	**not**	bare infinitive	
I	certainly	do	n't	like	shopping.

- before other verbs:

subject	adverb of certainty	verb	
I	definitely	want	something cheaper.

2. **Maybe** and **perhaps** come at the start of the sentence: Maybe I'll use my credit card.

Meaning

These express how sure you are about something. There are other ways of expressing certainty (see modals page 164).

Teaching ideas

From visuals about the future

Use visuals to make predictions about the economy, technology, the environment etc (students will need to know will for predictions.) Make some predictions clearly more likely than others:

Maybe we'll live on the moon.

We definitely won't live on the sun.

From a dialogue: investigating a crime scene

Investigators discuss what possibly/probably/certainly happened. Make some deductions clearly more likely than others:

Investigator 1: They definitely didn't come through the door. It's locked.
Investigator 2: Maybe they climbed in through the window.

Pronunciation hint

- Adverbs of certainty are usually stressed: MAY•be we'll live on the MOON.

still, already, yet, any more

Level: Intermediate

What are they?

These four adverbs all refer to time, and whether something is finished or not.

Using *still, already, yet, any more*

1. You usually use:

- still with a state verb like be or present continuous (see page 231)
- already and yet with present perfect (see page 225)
- any more with present simple (see page 216)

2. still and already usually go:

- after an auxiliary:

subject	aux **have**	adverb	past participle	
I	've	already	changed	jobs.

- after be:

subject	**be**	adverb	
He	's	still	a hairdresser.

- before other verbs:

subject	adverb	verb	
I	still	like	my work.

3. yet and any more usually goes at the end of the sentence: Have you applied for the job yet?

Meaning

1. **still** means a situation hasn't changed (and possibly you want it to change):
I'm still working for the bank.
She's still looking for a new job.
I can't believe it – he's still in bed!
2. **already** means something happened sooner than you expected:
He's already got a promotion (e.g. only after a month)
3. **yet** (in questions and negative statements) means you think it should happen:
Has she found any work yet?
He hasn't finished university yet (e.g. after ten years)
4. **any more** (in negative statements) means a situation was true in the past, but not now:
I don't work here any more.

Teaching ideas

Don't teach still, already, yet, and any more all at once – they're too similar and confusing. Teach them as they come up naturally in context.

Pronunciation hint

- Still often has sentence stress: She's STILL looking for a JOB. The others are usually unstressed.
- any more is stressed on MORE, not any.

. Questions

The two essential question types this section covers are yes/no questions (Are you Japanese?) and wh- questions, which ask for information (Where's the café?).

A *Yes/no* questions
B *Wh-* questions
C Subject & object questions
D Embedded questions
E Tag questions

Yes/no questions

Level:
Elementary
Pre-Intermediate

What are they?

These are questions that ask for a yes or no answer.

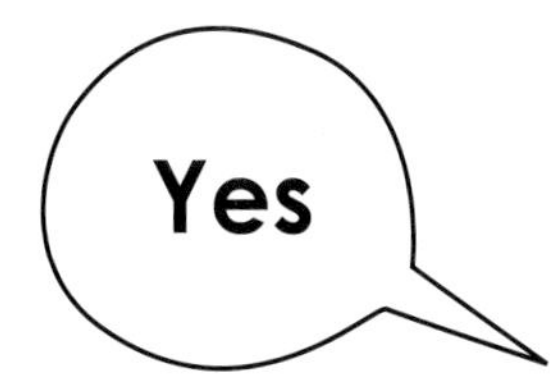

Form

- Questions with be invert the subject and verb:

Statement

subject	**be**	
He	's	a teacher.

Question

be	subject	
Is	**he**	**a teacher?**

- Questions with an auxiliary (be/have or a modal verb) invert the subject and auxiliary:

Statement

subject	aux **have**	past participle	
He	's	taught	overseas.

Question

aux **have**	subject	past participle	
Has	**he**	**taught**	**overseas?**

- Other verbs need to add the auxiliary do and the main verb becomes a bare infinitive:

Statement

subject	verb (present simple)	
He	knows	a lot about grammar.

Question

aux **do**	subject	bare infinitive	
Does	**he**	**know**	**a lot about grammar?**

Using yes/no questions

1. It's polite to use a 'short answer' (just repeat the subject + be or auxiliary; negatives add -n't):
   - Is he a teacher? – ~~Yes.~~ → Yes, he is.
   - Has he taught overseas? – ~~No.~~ → No, he hasn't.

2. It's very unusual to repeat the whole question when you answer:
   - Is he a teacher? – ~~Yes, he is a teacher.~~ → Yes, he is.

(!) Negative questions:

1. You can use a negative question by inserting -n't after be or the auxiliary:
   Is she our teacher? → Isn't she our teacher?
   Have you done your homework? → Haven't you done your homework?
2. Negative questions are not very common. This can either show you're annoyed, or you expect the listener to agree with you:
   Didn't you set your alarm? (= I'm waiting and I'm annoyed)
   Didn't you go to conversation class yesterday? (= I remember you said you planned to)
3. Answers to negative questions say what happened; they're not agreeing or disagreeing with the question:
   - Haven't you finished yet? - **Yes, I have.** (= I have finished) - No, I haven't. (= I haven't finished)

(!) Special cases:

1. When you offer something, you can just say the thing, with rising intonation: Coffee? Something to eat?
2. In informal language, if a question contains a verb in the present continuous or present perfect, you can just say the **V+ing** or past participle with rising intonation: Coming? (= Are you coming?) Finished? (= Have you finished?).

Teaching ideas

You'll be teaching questions from the very start of an elementary course, first as expressions that students need to memorise. Later, introduce each type of question one at a time and analyse the structure on the whiteboard.

From answer prompts

Put a picture of a person on the whiteboard along with prompts for the questions and answers. For example:

Japan / ✓ Is she from Japan? Yes, she is.
a student / ✗ Is she a student? No, she isn't.
a car / ✗ Does she have a car? No, she doesn't.

From a dialogue build
Similarly, you can teach a social dialogue where two people ask and answer yes/no questions:

you / French? - Are you French? – Yes, I am.
here / before? / first time – Have you been here before? – No, I haven't, it's my first time.

Pronunciation hint

- Yes/no questions usually have rising intonation on the last stress: Are you a ↗TEA•cher?
- the ending **-n't** is pronounced /ənt/: isn't = /ˈɪ•zənt/

B *Wh-* questions

Level:
Elementary
Pre-Intermediate

What are they?

These are questions that ask for information.

7.30 at the station.

They need a question word (usually beginning with wh-):

what which who	where when why how	how much how many how long how often

Form

Wh- questions usually start with the question word. Otherwise they follow the same rules as yes/no questions.

- Questions with be invert the subject and verb:

Statement

subject	**be**	
That	's	a ticket.

Question

question word	**be**	
What	**'s**	**that?**

- Questions with an auxiliary (be/have or a modal verb) invert the subject and auxiliary:

Statement

subject	aux **have**	past participle	
She	's	travelled	in Asia.

Question

question word	aux **have**	subject	past participle
Where	**'s**	**she**	**travelled?**

- Other verbs need to add the auxiliary do (but see page 203 subject/object questions)

Statement

subject	verb (past simple)	
He	worked	in South America.

Question

question word	aux **be**	subject	bare infinitive
Where	**did**	**she**	**work?**

Using wh- questions

1. In the answer, it's unusual to repeat the whole question. You only need to give the relevant information:
   - How long do you want to stay? – ~~I want to stay about two days.~~ → **About two days.**

2. If there's a preposition, put it at the end of the question:
   I carried my passport in my bag. → **What did you carry your passport in?**
   I travelled with Peter. → **Who did you travel with?**
   Some books recommend putting the preposition first, and using whom instead of who: With whom did you travel? However, this is very formal and rare in modern spoken English.

3. What, which, how much and how many can stand alone, or go before a noun:
   How much was the ticket?
   How much time do we have before the flight?

4. Which asks for a choice from a small number; what is from all possible choices:
   Which city do you want to see most, Tokyo or Beijing?
   What cities in Asia do you want to see?

Special cases:

1. How come…? (informal) with statement, not question, word order, means 'why?': How come the train's late?
2. What… for? (informal) asks why someone did something, especially if you think it wasn't a good idea: What did you spend all your money for?

Teaching ideas

Students need to learn questions from the very start of their English study. First they'll simply need to memorise them but later you can analyse the structure on the board.

From answer prompts
Put a picture of a person on the whiteboard along with some answers. Elicit the questions. For example:

22. How old is he?
France. Where's he from?
He's a student. What does he do?

From a visual
Show the class an intriguing visual. Elicit what questions they'd like to ask:

What is it?
Where is it?
How old is it?

Pronunciation hint
- **wh-** questions usuallly have falling intonation on the last stress: Where's the STA•tion?
- You can stress the question word, and use rising intonation to show surprise: WHERE are you living?

C Subject & object questions

Level:
Pre-Intermediate

What are they?
The way you make a who and what question depends on whether you're asking about the subject or the object.

subject	verb	object
John	cooked	dinner.

subject question	object question
- Who cooked dinner? - John.	- What did John cook? - Dinner.

Form

You don't use auxiliary do in subject questions, unlike object questions (see Wh- questions page 201).

Statement

subject	verb (past simple)	
Aneta	made	breakfast.

Question

question word	verb (past simple)	
Who	made	breakfast?

Not: ~~Who did make breakfast?~~

Statement

subject	verb (past simple)
The toast	burned.

Question

question word	verb (past simple)
What	burned?

Teaching ideas

From a story

Tell students a simple story, but miss out key information (by coughing or making a noise) to elicit subject and object questions as you go:

Teacher: I was walking down the street and saw (cough).
(Elicit:) Who did you see?
Teacher: (cough) came up to me …
(Elicit:) Who came up to you?

From visuals

Use pictures of historical events to elicit questions:

Teacher: They invaded England.
(Elicit:) Who invaded England?

Pronunciation hint

- Subject and object questions usually have falling intonation on the last stress: Who bought the ↘GRO•ce•ries?

Embedded questions: *Do you know where the bus stop is?*

Level:
Pre-Intermediate
Intermediate

What are they?

This is a question inside another question or statement:

Where's the bus stop? → Do you know where the bus stop is?
Is the bus on time? → I don't know if the bus is on time.

Form

The way you form an embedded question depends on whether it's a *yes/no* or *wh-* question (see page 199 and page 201).

Yes/no question	Wh- question
• start with if (or whether – formal) • use normal statement word order • don't use auxiliary do; use normal endings on the main verb	• start with the question word • use normal statement word order • don't use auxiliary do; use normal endings on the main verb
Is the station near here? → Do you know if the station's near here? (not ~~Do you know is the station near here?~~)	Where's the taxi rank? → Could you show me where the taxi rank is? (not ~~Could you show me where is the taxi rank?~~)
Do the trains run after twelve? → Can you tell me if the trains run after twelve? (not: ~~Can you tell me do the trains run after twelve?~~)	When does this depart? → Could you tell me when this departs? (not ~~Could you tell me when does this depart?~~)

Don't use a question mark if the question is embedded inside a statement: I'm not sure where the taxi rank is.

Using embedded questions

1. Embedded questions are more polite than direct questions. To ask a stranger for information, it's normal to say Excuse me and then ask an embedded question.

2. Here are common ways to start embedded questions:

Questions	Statements
Do you know...? Can you tell me...? Could you tell me...? Could you show me...? Could you explain...?	I don't know... I don't understand... I wonder... I'm not sure... It doesn't say... I can't remember...

3. To ask for help doing something, use how to + infinitive: Do you know how to buy a ticket?

4. You use the same rules for reporting a question: She asked me where the station was. You need to follow the same rules for changing the subject, tense and time expression as for reported speech (see page 210).

Teaching ideas

From a dialogue

Teach a dialogue where a visitor is trying to find their way around a strange city. Teach at least one question and one statement.:

Excuse / can / tell / where /station?
Don't / understand / how / buy / tickets

Sure / that / direction
I / show

Elicit: Excuse me, can you tell me where the station is?
Sure, it's that direction.
I don't understand how to buy tickets.
I'll show you.

Pronunciation hint

- Intonation depends on the purpose of the entire sentence, not the embedded question.
- Yes/no questions such as Could you tell me… usually have rising intonation.
- Statements such as I don't know… usually have falling intonation.

E Tag questions

Level:
Pre-Intermediate
Intermediate
Upper Intermediate

What are they?

These are short yes/no questions at the end of a sentence: Our class starts at ten, doesn't it?

Form

1. If the statement is positive, the tag question is usually negative. If the statement is negative, the tag question is usually positive:
   We have a test tomorrow, don't we? We don't have any homework, do we?
   The most common tag questions are positive + negative.

2. The verb in the tag question is the same one you use in a yes/no question (see page 199): followed by -n't if negative.

- Statements with be use be in the tag question:

subject	**be**		**be**	**not**	subject
This	is	our classroom,	is	n't	it?

- Statements with an auxiliary (be/have or a modal verb) use the auxiliary in the tag question:

subject	aux **can**		aux **can**	**not**	subject
We	can	go in,	can	't	we?

- Statements with other verbs use the auxiliary do in the tag question:

subject	verb (past simple)		aux **do** (past)	**not**	subject
I	passed	the test,	did	n't	I?

3. The tag question am + not + I becomes aren't I: I'm on time, aren't I?

Using tag questions

1. Use a tag question with falling intonation when you're sure of the answer, and you want someone to agree with you. In reality it's not a question at all (and some people write it without a question mark). It's useful for making small talk.:
   Room 36 is down here, isn't it? (= I seem to remember)

2. Use a tag question with falling intonation to make small talk. This is not really a question at all (and some people write it without a question mark); you know the answer, and want the other person to agree with you.
   It's noisy in here, isn't it.
   That test was hard, wasn't it.

3. Use a negative statement with a positive question tag when you expect it not to be true:
   She's not the teacher, is she? (e.g. she looks too young)

4. It's polite to answer a tag question with a short answer (see page 199):
   - Our class is here, isn't it? – Actually no, it isn't, it's in the next room.

5. You should generally agree with the 'small talk' tag question, or you'll seem antisocial:
   - (It's) hot, isn't it. – Yes, it is, absolutely.

6. Use You couldn't + infinitive + could you? to make a friendly, informal request:
   You couldn't turn down the music could you? – Sure.

Teaching ideas

Tag questions are very confusing so teach one tense or use at a time.

CONFIRMING INFORMATION - From a situation

Set up a context where people require genuine information (for example, a student visiting the school office). Elicit the questions from prompts:

(you / English student)
You're an English student, aren't you?

SMALL TALK - From a situation

Use a context (such as a party) where it is very clear people are making small talk and not asking for information. Elicit the questions from prompts:

(Good music)
Good music, isn't it.

Pronunciation hint

- Use rising intonation to ask a genuine question.
- Use falling intonation to make small talk, when you expect the listener to agree.

6. Building Sentences

This section looks a range of ways of building sentences, including the use of conjunctions (but/although), and how we report what someone says (She said, "I'm ready" → She said she was ready).

A Conjunctions and prepositions: (*although/because* vs *in spite of/because of*)?

Level:
Intermediate
Upper Intermediate

What are they?

These words have similar meanings, but you need to use them differently in a sentence.

Conjunctions	Prepositions
because	because of, due to
although, even though	in spite of, despite

Using conjunctions and prepositions

1. Conjunctions need a clause after them (ie a subject and verb):
   I stayed home because I had a test.
   Even though the teacher's good I don't understand the subject.

2. Prepositions just need a noun (or V+ing), or a pronoun:
   I stayed home because of the test.
   Despite the teacher I don't understand the subject.

3. All of these prepositions are quite formal. In spoken English, it's more common to use because and even though.

4. You can use a preposition plus the fact that to turn it into a conjunction: Few students passed in spite of the fact that they had attended revision classes. (This is very formal).

Meaning

1. Because and because of/due to tell us why: He always achieves good marks due to his excellent study habits.
2. Although/even though and in spite of/despite contrast two situations in a surprising way: Although she's normally a good student, she went badly this year. (we expect her to go well)

(!) Although has a similar meaning to but, but they start different parts of the sentence:

	situation 1		**situation 2**
Although	I studied		I didn't pass.
	I studied	but	I didn't pass.

- Don't use although and but in the same sentence: Although I studied ~~but~~ I didn't pass.
- In informal English you can say though (instead of although and but) at the end of the sentence: I studied. I didn't pass though.

Teaching ideas

Don't teach all at once; it's better to teach two contrasting items at one time (e.g. because vs although, or because vs because of).

From visuals

Show pictures of causes and effects (e.g. the environment, lifestyles, the economy) to elicit the language:

It flooded because of the rain.

From a text

Give students a text describing causes and effects that contains the target language. First have them read for understanding, then ask them to underline any examples of language that tells us why something happened. Check the meaning:

In text: There was no food because of the bad weather.

Teacher: What happened first, no food or bad weather? (Bad weather) Are they connected? (Yes)

Pronunciation hint

- Because is unstressed, and reduced to /kəz/ in informal speech.

B Purpose: *to/for/so that*

Level:
Pre-Intermediate
Intermediate

What are they?

These all describe why people do things.

Using to/for/so that

1. To (or in order to, formal) with infinitive describes why someone does something.
   I went to the supermarket to buy some milk.

2. If your purpose is to help someone else do something, you need to use so (that) + different subject + can/could:
   I gave my husband some money so he could buy the groceries.

3. You can use so (that) + subject + can/could with the same subject, when there's a serious purpose.
   She rang the supermarket so she could arrange a time for a job interview.

4. You need to use for with a noun.
   I got some tablets for my cold.

ⓘ Special cases

1. You can use for + V+ing to describe the purpose of a thing: – What's this for? – It's for weighing things.
2. You can use to + infinitive after enough time/enough money/a chance to say whether you could do something: I didn't have enough time to buy everything.

ⓘ Don't use for + V+ing to show why a person does something; use to instead: ~~I rode to the shops for buying some things~~ → I rode to the shops to buy some things.

Teaching ideas

From visuals

Show people leaving different destinations. Ask: Why did she go (to the shop)?

Elicit: To buy bread.

From a class survey

Have students conduct a class survey: 'Why did you (come to the UK/learn English etc)?' Don't expect students to know the target language (they can use because). Elicit the language from the results: She came to the UK to do an English course.

Pronunciation hint

- to and for are usually unstressed and pronounced /tə/ and /fə/.

C Reported speech

Level:
Pre-Intermediate
Intermediate
Upper Intermediate

What is it?

When you report what someone says, the pronoun and the verb often changes:

→

Using reported speech

Direct speech		Reported Speech
• present tense "I play football." • past tense "I was in the team." • **will / can / must** "I'll join the gym."	→	• past tense She said she <u>played</u> football. • past perfect She said she<u>'d been</u> in the team. • **would / could / had to** She said she<u>'d join</u> the gym.

1. The tense change also applies to auxiliary verbs (so, for example, present perfect becomes past perfect):
   "We've won the competition!" → She said they'd won the competition.

2. The pronoun changes (usually to **he, she** or **they**)

3. Any reference to time and place may change, depending on context: "I'm playing today" → She said she was playing that day. Common changes are:

Direct speech		Reported Speech
now last week this week next week yesterday today tomorrow here	→	**then** **the week before** **that week** **the week after** **the day before** **that day** **the day after** **there**

This won't change if the speech is being reported, for example, on the same day: (on 1 December) "I have a match today" → (also on 1 December) **She said she had a match today.**

Special cases:
1. You can keep the original tense if the statement is still or always true: "I play a lot" → She said she played a lot OR She said she plays a lot.
2. You must change the tense if the situation is finished: (on 1 December) "I have practice tonight" → (on 10 December) She said she had practice that night.
3. Change the tense to show you don't believe someone. For example, if a player says, "I'm too ill to play", then later you see them looking fine, walking down the street: He said he was too ill to play!
4. When you are reporting a request or commend, you can use tell/ask someone to do something: The coach said, "Run faster!" → The coach told us to run faster.
5. To report a question, see page 199.

Don't use a comma (,) or quotation marks (" ") in reported speech. (~~He said, "He was good at soccer".~~)

Teaching ideas

From a situation

Set up a story where people are likely to report what someone said (e.g. a famous person and a reporter, or friends gossiping).

Like the pictures above, use a visual to show someone speaking, and write exactly what they say on the whiteboard. Then show a visual of someone reporting what they said to another person. Elicit the reported speech forms.

From an email

Have students read an email for understanding, then have them report back to the class what it said:

In email: I'm coming back to China soon.
Elicit: He said he was coming back to China soon.

D Relative clauses 1: overview

Level:
Pre-Intermediate
Intermediate

What are they?

A relative clause tells us more about a noun. It comes straight after the noun it's describing:
The man is from Russia. He lives above me. → The man who lives above me is from Russia.

Form

1. Use who or that for people, and that or which (formal) for things:
   The woman is nice. She works next door. → The woman who/that works next door is nice.
   The house is really beautiful. It is for sale. → The house that/which is for sale is really beautiful.

2. You can leave out who, that or which, but only if it's the object of the relative clause:
   The man is from Russia. He (subject) lives above me. → The man who lives above me is from Russia.
   The man is from Russia. I met him (object) last week. → The man who I met last week is from Russia.
   OR The man I met last week is from Russia.

3. If there's a preposition, it comes at the end of the relative clause. You can leave out who, that or which (because it's an object):
   The house is really beautiful. I live near it. → The house that/which I live near is really beautiful.
   OR The house I live near is really beautiful.

Using relative clauses

1. In very formal language, if it's an object, use whom instead of who:
   The man is from Russia. I met him (object) last week. → The man whom I met last week is from Russia.

2. In formal language, put a preposition first, followed by whom or which:
   The house is really beautiful. I live near it. → The house near which I live is really beautiful.

3. Use whose to replace a possessive (his, her etc)
   The man is from Russia. His wife is Italian. → The man whose wife is Italian is from Russia.

4. Use where to mean that... in or at:
   That's my apartment. I live in it. → That's the apartment that I live in.
   OR That's the apartment where I live.

(!)
1. Don't add an extra object:
   The man is from Russia. I met him (object) last week. → The man who I met ~~him~~ last week is from Russia.
   OR The man I met ~~him~~ last week is from Russia.
2. Don't forget the preposition:
   The man is from Russia. I talked **to** him. → The man that I talked to is from Russia.
   OR The man I talked to is from Russia.

E Relative clauses 2: defining & non-defining clauses

Level:
Intermediate

What are they?

Some relative clauses give us essential information about a noun (called 'defining'); others simply give us interesting extra information (called 'non-defining').

defining	non-defining
A geologist is a person who studies rocks. If you leave out who studies rocks the sentence becomes meaningless.	**My father, who was a geologist, worked in Australia.** You could omit who was a geologist and still have an informative sentence.

Form

1. Defining relative clauses never have commas. Non-defining relative clauses always have commas at both ends.

2. Non-defining relative clauses always use who (never ~~that~~) for people and which (never ~~that~~) for things.

3. In a non-defining relative clause, you can't omit who or which, even if it's an object: ~~My girlfriend, I met in Thailand, is also a student~~ → My girlfriend, who I met in Thailand, is also a student.

Using defining and non-defining relative clauses

1. Defining relative clauses are often used for dictionary-style definitions: A computer is a device that processes information.

2. Non-defining relative clauses are often used for describing historical events or painting a detailed picture of a person or place: Lenin, who was born in 1870, was leader of the Russian Revolution.

3. Defining relative clauses are common in speaking and writing. Non-defining relative clauses are only common in writing.

You can use which in a non-defining relative clause to refer to all of the previous information, not just the noun: Her mother used to work on a farm, which surprised me.

F Relative clauses 3: *-ing* & *-ed* clauses

Level: Upper Intermediate

What are they?

If the relative clause contains:

- aux be + V+ing (present continuous, see page 231) or
- aux be + past participle (passive, see page 244)

you can omit the relative pronoun (who, that, which) and be:

be + V + ing	**be** + past participle
I know the student who is sitting over there → I know the student sitting over there.	That's the book that is recommended by the school. → That's the book recommended by the school.

Using *-ing* & *-ed* clauses

1. It's very common to omit subject and auxiliary be before V+ing in speaking and writing: Do you know the guy drinking coffee over there? (= who is drinking); Who's the woman talking to our teacher? (= who is talking)

2. It's not common to use passive in speaking, so the second sort of sentence is more common in writing. In speaking, you could say That's the book the school recommends.

The past participle without the relative pronoun and be can be confusing, as it look like a past tense (e.g. All subjects taught here are free). If you see two verbs (i.e. taught and are), it is likely the first is a relative clause with the pronoun and auxiliary omitted:
All subjects that are taught here are free.

Teaching ideas

DEFINING RELATIVE CLAUSES

From visuals

Show visuals of people and things. Elicit what they are, and what they do. Make notes on the whiteboard. Then put the information together in one sentence:

A mechanic is a person who fixes cars.

A car is a small vehicle that has four wheels.

From a text

Find examples in dictionary definitions. Check the meaning: If you take it out, is the sentence still OK? (No).

NON-DEFINING RELATIVE CLAUSES

From a text

Find an example of a non-defining clause (often texts about history, people and places). Check the meaning: If you take it out, is the sentence still OK? (Yes).

Using the whiteboard

Start with skeleton sentences, and add more information: Tokyo is a big city → Tokyo, which has a population of 12 million, is a big city.

Pronunciation hints

- There is no pause before a defining relative clause.
- There must be a pause before and after a non-defining relative clause.

. Verbs

erbs can be actions (take, buy, eat) or states (be, have, like). Verbs are the most complex area English for learners, and for this reason this section systematically treats each use of each verb ucture, as described earlier.

A Simple
Present simple 1: a habit
Present simple 2: a current situation
Present simple 3: a timetabled future event
Past simple 1: a past action
Past simple 2: a past situation

B Perfect
Present perfect simple 1: a past event that's still important
Present perfect simple 2: a situation that started in the past and is still true
Present perfect simple 3: a life experience
Past perfect simple: a past event that happened before another event

C Continuous
Present continuous 1: a short activity happening now
Present continuous 2: a temporary situation
Present continuous 3: a change we're in the middle of
Present continuous 4: a future arrangement
Past continuous 1: the background to a story
Past continuous 2: a past activity happening around a point in time

D Perfect + continuous
Present perfect continuous 1: a past activity we now see a result of
Present perfect continuous 2: a situation that started in the past and is still true
Past perfect continuous: a situation that continued up to a point in time in the past

E Passives
Present simple passive: a process
Past simple passive: an historical event

F Future forms
will 1: a decision made now
will 2: a prediction
going to 1: a plan
going to 2: a prediction from evidence
Future continuous: a future activity happening around a point in time
Future perfect: a future action finished before a point in time

G Conditionals
Zero conditional: a general truth
1st conditional: thinking about future possibilities
2nd conditional: imagining a different now
3rd conditional: imagining a different past

H Wishes
wish + past: wanting now to be different
wish + past perfect: wanting the past to be different

I Modals
Obligation (present & past)
Deduction (present & past)

J Special verbs
Phrasal verbs
used to
get used to or be used to?
have or have got?
Verb patterns: gerund or infinitive?
have/get something done; make/let someone do something

A Present simple 1

Level: Elementary

a habit

Form

positive

subject		present simple	
I		wear	a tie to work.
He	always	comes	to work in a t-shirt.

negative

subject	aux **do** (present)	**not**	bare infinitive	
I	do	n't	smoke.	
He	does	n't	wash	his hair.

question

aux **do** (present)	subject	bare infinitive	
Do	you	shave	every day?

short answer

yes	subject	aux **do** (pres)
Yes,	I	do.

no	subject	aux **do** (pres)	**not**
No,	I	do	n't.

1. Present simple normally looks the same as the base form.
2. Third person singular adds **-s** (e.g. I work, she works).
3. Sometimes the spelling changes (e.g. I go, she goes).
4. Some verbs are irregular (e.g. I have, she has). – see page 317.
5. You can use a time adverb to show how often (e.g. every day/always). – see page 195.

For more on questions in the present simple, see page 195: **present simple 2.**

Meaning

A habit. (Compare page 231 **present continuous 1.**)

Meaning	Checking meaning
• We're talking about the past, now & the future. • It happens more than once.	• Are we talking about the past, now and the future? (Yes.) • Does it happen one time, or many times? (many times)

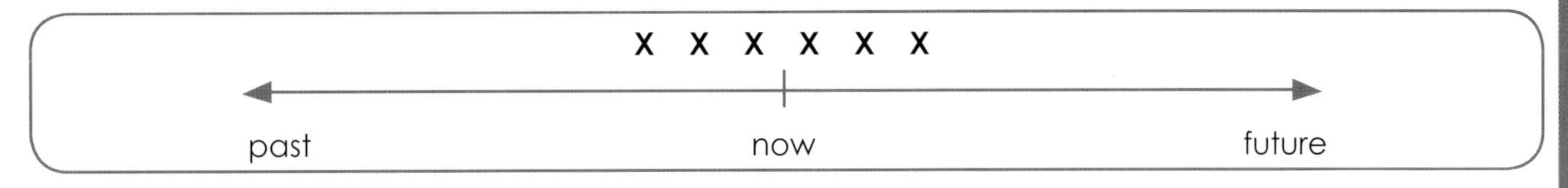

Anticipated problems and solutions

Problems	Solutions
Students omit third person **-s**.	Draw **-s** in the corner of the whiteboard and point to it to remind them.
Students omit auxiliary do in questions and negatives.	Write do in the corner of the whiteboard and point to it to remind them.

Teaching ideas

From a situation or dialogue build

The situation needs to show that it's a habitual action.

- Expressions like every morning, every Tuesday make this clear.
- Daily routines (work, housework, free time) work well:

What does he do every day?

He gets up at 7.

He has breakfast at 8.

- Bad habits are particularly memorable! (She smokes two packets a day etc.)
- You can also compare cultural habits in different countries.

From a text

Use a text that describes:

- a person's day-to-day life (e.g. a celebrity or politician)
- the responsibilities of a particular job.

First have students read for understanding, then underline all the verbs. Check meaning and analyse the form on the whiteboard.

A Present simple 2

Level: Elementary

a current situation

Form

positive

subject	present simple	
I	live	in the country.
My wife	works	on a farm.

negative

subject	aux **do** (present)	**not**	bare infinitive	
I	do	n't	like	the city.
She	does	n't	want	a different job.

question

aux **do** (present)	subject	bare infinitive	
Do	you	have	a tractor?

short answer

yes	subject	aux **do** (pres)
Yes,	I	do.

no	subject	aux **do** (pres)	**not**
No,	I	do	n't.

For basic rules on forming present simple, see page 216 **present simple 1.**

1. The verb be and modals (can, must etc – see page 264) do not use do in negatives. They add not after the verb: I'm not a farmer. (~~I don't be a farmer~~.)

2. The verb be and modals do not use do in questions. They invert the subject and verb: Can they sell their rice? (~~Do they can sell their rice?~~)

Meaning

A current situation. (Compare page 231 **present continuous.**)

Meaning	Checking meaning
• We're talking about the past, now & the future. • It's true for a long time.	• Are we talking about the past, now and the future? (Yes). • Is it a short time, or a long time? (A long time).

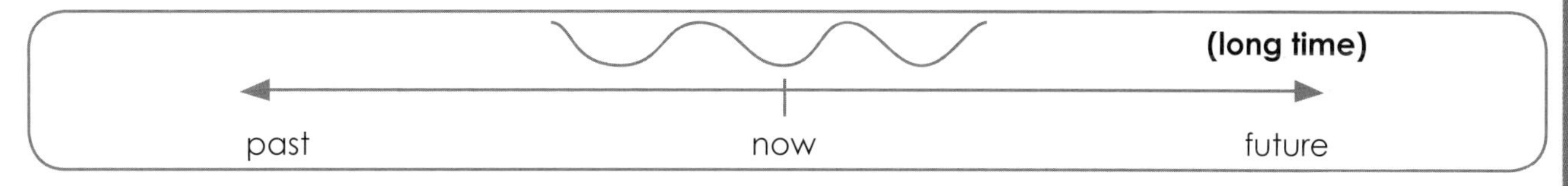

Anticipated problems and solutions

Problems	Solutions
Students omit third person **-s**.	Draw **-s** in the corner of the whiteboard and point to it to remind them.
Students omit auxiliary do in questions and negatives.	Write do in the corner of the whiteboard and point to it to remind them.

Teaching ideas

From a situation

Use visuals to elicit people's situations:

He lives in Russia.

She lives in Japan.

Clarify this is true for a long time.

From a text or visuals

Use pictures or information from the Web to contrast different things that are always true (e.g. countries and cities). Elicit facts: Monaco is small, China has a big population.

You can also describe animals or people, but be aware you will probably elicit habits as well as states

e.g. Elephants live in Asia and Africa (state); They eat grass (habit). Both are present simple however (see page 216 present simple 1).

A Present simple 3

Level:
Pre-Intermediate
Intermediate

a timetabled future event

Form

positive

subject	present simple	future time expression
My flight	leaves	in thirty minutes.

question

when/what time	aux **do**	subject	bare infinitive
What time	does	the next train	arrive?

1. A very common verb used with this meaning is be: My flight's at six.

2. Remember a question with be does not need auxiliary do: When's your train? (see page 218 **present simple 2**).

Meaning

A timetabled future event. (Compare other ways to talk about the future: page 247 **will**, page 250 **going to**, page 235 **present continuous 4**.)

Meaning	Checking meaning
• We're talking about the future. • You can see it on a timetable.	• Are we talking about now, or the future? (The future). • Is it on a timetable? (Yes).

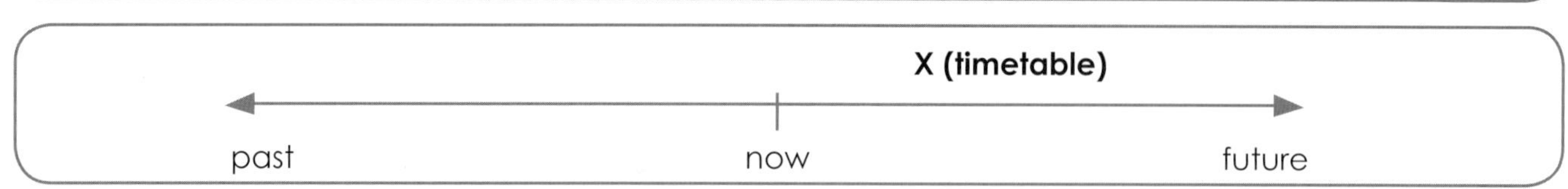

Anticipated problems and solutions

Problems	Solutions
It's called present tense, so students think we're talking about now.	Contrast form and meaning explicitly: 'What's the tense called?' (Present simple) 'Is he talking about now?' (No). 'Is he talking about the future?' (Yes).
Students think they should always use *will* to talk about the future	Remind students there are different ways to talk about the future. Ask: for timetables do we use *will*? (No)

Teaching ideas

From a situation or dialogue build

Use a situation that makes it clear someone will catch transport in the near future (a PA talking to a manager, someone running). Elicit the language from prompts:

why / hurry?

train / leave / 10 minutes

- Why are you hurrying?
- My train leaves in ten minutes!

From a text

Find an official schedule (e.g. in the newspaper) and ask students to describe the program:

> Exhibition
> doors open 9am
> official welcome 10am

The exhibition opens at 9am. The official welcome is at 10am.

A Past simple 1

Level:
Elementary

a past action

Form

positive

subject	past simple		(past time expression)
I	looked	round the mall	yesterday.
I	bought	a few things	

negative

subject	aux **do** (past)	**not**	bare infinitive	
I	did	n't	spend	much.

question

aux **do** (past)	subject	bare infinitive	
Did	you	go	with anyone?

short answer

yes	subject	aux **do** (past)
Yes,	I	did.

no	subject	aux **do** (past)	not
No,	I	did	n't.

1. Past simple normally adds **-ed** to the infinitive.

2. Sometimes the spelling changes (e.g. study → I studied).

3. Many verbs are irregular (e.g. buy → I bought) – see page 317.

4. Like present simple (see page 218 **present simple 2**), be and modals do not use do in negatives and questions: I couldn't find anything I wanted (~~I did not can find anything I wanted~~).

5. You can use a past time expression to say exactly when something happened (e.g. yesterday, a week ago).

Meaning

- A past action. (Compare other ways to talk about the past: page 225 **present perfect**, page 229 **past perfect**, page 237 **past continuous.**)
- A rule of thumb: if you're talking about the past, use past simple, unless there's a good reason to use another verb structure.
- Past simple can also describe something that happened more than once, or a habit: Last week I went to the mall twice (see page 272 **used to**).

Meaning	Checking meaning
• It happened in the past. • We know when it happened.	• Are we talking about now, or the past? (The past). • Do we know when it happened? (Yes).

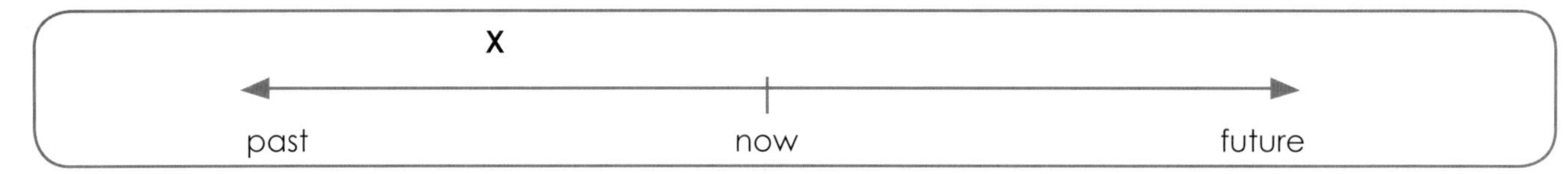

Anticipated problems and solutions

Problems	Solutions
Students forget to use past tense.	Write 'past' in the corner of the whiteboard and point to it to remind them. Ask: are we talking about now, or the past? (The past) So what tense do we use? (Past tense).
Students overuse other past structures (especially past perfect).	Reassure students that past simple is usually correct to talk about the past. The main problem is forgetting to use past tense at all.

Teaching ideas

From a situation or dialogue build

When you set the context, first clarify this is set in the past (e.g. use a date or a time, and ask, 'Are we talking about now or the past?'). Tell a story using pictures – the funnier or more memorable the better!

My terrible holiday

From a text
Use a comic strip or a story on video. Again, establish first that we are talking about the past, and then elicit what happened.

A Past simple 2

Level: Pre-Intermediate

a past situation

Form

positive

(past time expression)	subject	past simple	
In the 1990's	I	lived	in Japan.
	I	worked	there for five years

negative

subject	aux **do** (past)	**not**	bare infinitive	
I	did	n't	speak	any Japanese.

question

aux **do** (past)	subject	bare infinitive	
Did	you	like	it?

short answer

Yes	subject	aux **do** (past)
Yes	I	did.

No	subject	aux **do** (past)	not
No,	I	did	n't.

For basic rules on forming past simple, see page 221 **past simple 1**.

1. You can use for (not since) to show a period of time e.g. I worked there for five years. (Compare present perfect simple 2 where for and since are both possible, with different meanings.)

Meaning

- A past situation. (Compare page 237 **past continuous** and page 272 **used to**)

Meaning	Checking meaning
• It happened in the past. • We know when it happened. • It was true for a period of time.	• Are we talking about now, or the past? (The past). • Do we know when it happened? (Yes). • Was it a short time, or a long time? (A long time).

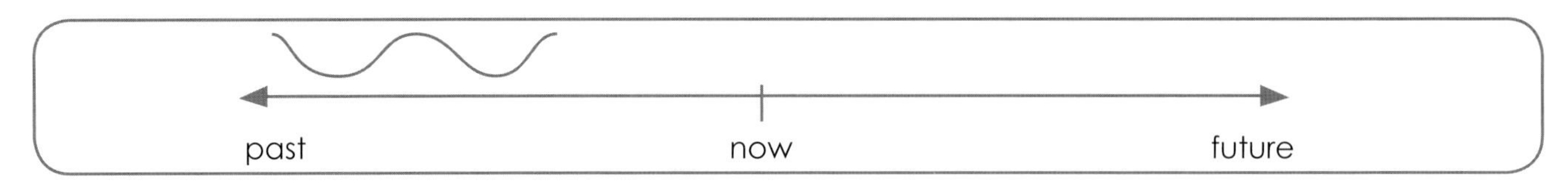

Anticipated problems and solutions

Problems	Solutions
Students forget to use past tense.	Write 'past' in the corner of the whiteboard and point to it to remind them. Ask: are we talking about now, or the past? (The past) So what tense do we use? (Past tense).

Teaching ideas

From a situation or dialogue build

Set a context where someone talking now talking about how their life was different in the past. You can use contrasting pictures of the person now and then:

Five years ago I...

2006

lived at home

From a text

Find an interesting text about the world in the past. This may include a mixture of states and habits (e.g. People lived in the wild and ate nuts and berries).

Have students read for understanding, and then focus on the past simple verbs: check the meaning, and analyse the form.

B Present perfect simple 1

Level:
Pre-Intermediate
Intermediate

a past event that's still important

Form

positive

subject	aux **have** (present)	past participle	
I	've	lost	my passport.

negative

subject	aux **have** (present)	**not**	past participle	
I	have	n't	told	the embassy.

question

aux **have** (present)	subject	past participle	
Have	you	checked	your room?

short answer

Yes	subject	aux **have** (present)
Yes	I	have.

No	subject	aux **have** (present)	not
No,	I	have	n't.

1. You can't specify the time: I've lost my passport ~~an hour ago.~~
2. Use **just** to mean 'a very short time ago': Sorry, I've just found it.
3. Use **yet** at the end of questions and negatives to mean 'you should': Have you reported it yet?
4. Many US speakers use past simple for this meaning: I lost my passport!

Meaning

A past event that's still important. (Compare page 221 **past simple.**)

Meaning	**Checking meaning**
• It happened in the past. • It's still important.	• Did it happen in the past? (Yes). • Is it still important? (Yes).

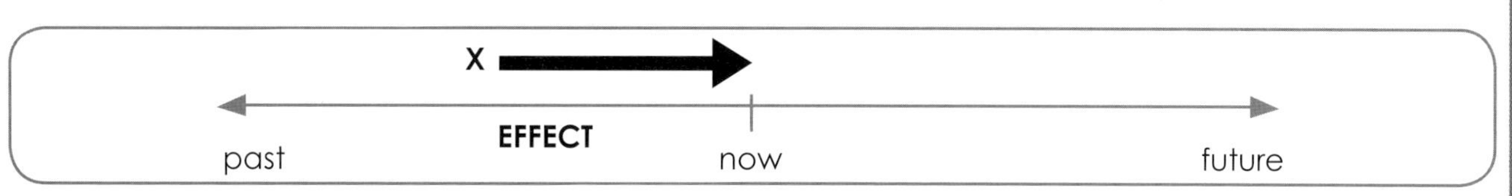

Anticipated problems and solutions

Problems	Solutions
Students specify the time e.g. I've lost my money yesterday.	Check: can you say the time? (No)
Students believe the key meaning is locating in time (it's 'recent') rather than the effect on now.	Use an example where the event is possibly not recent, but the effect is still important e.g. 'I can't travel because my passport's expired'.

Teaching ideas (see Lesson Plans page 115)

From a situation or dialogue build

The situation should clearly show that someone is talking about a past even that's still important. Conveying news (e.g. an accident, a crime, a prize) works well:

From a text

Use a newspaper or news broadcast – stories often begin with present perfect with this meaning (The Prime Minister has announced...). Make sure students understand the text, and then focus on the example(s) of present perfect. Check the meaning and analyse the form.

B Present perfect simple 2

Level:
Pre-Intermediate
Intermediate

a situation that started in the past and is still true

Form

positive

subject	aux **have** (present)	past participle		time expression with **for/since**
I	've	lived	here	for a year.
She	's	worked	with me	since April 1st.

negative

subject	aux **have** (present)	**not**	past participle		time expression with **for/since**
I	have	n't	been	here	for long.

question

how long	aux **have** (present)	subject	past participle	
How long	have	you	been	in China?

1. With this meaning you need use for (to describe a period, e.g. for a year) or since (to describe a point in time when the situation started, e.g. since April 1).

Meaning

A situation that started in the past and is still true. (page 241 **present perfect continuous 2** can be used interchangeably with this meaning.)

Meaning	**Checking meaning**
• The situation started in the past. • It's still true.	• When did it start? (In the past). • Is it still true? (Yes).

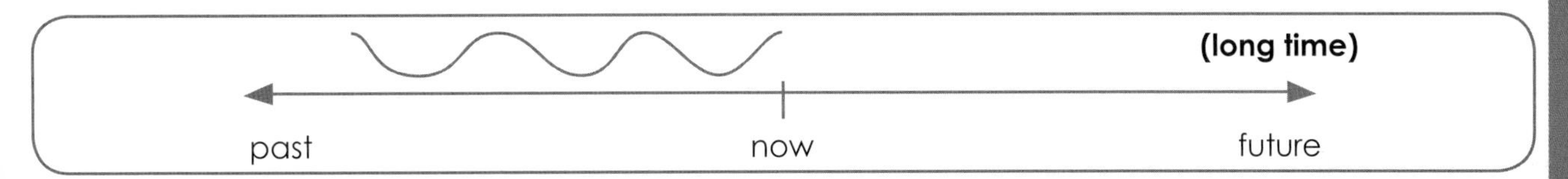

Anticipated problems and solutions

Problems	**Solutions**
As the situation is still true, students use present tense.	Use a timeline to clarify that if a situation started in the past and is still true, and you give the length of time, you must use present perfect.
Students confuse for and since.	Use a timeline to show for tells us a period of time, and since tell us a point in time.
Students have been taught they should use present perfect continuous with this meaning.	Reassure students present perfect simple and continuous are interchangeable, but only with this meaning.

Teaching ideas

From a situation

Use visuals of someone doing something your students can relate to (e.g. studying; learning English; living abroad) now and in the past to elicit the target structure:

2001 started English

now still learning

Elicit: He's studied English since 2001.

From a text
Information about countries, companies and organisations often proudly states how long they've done something (e.g. We have been the market leader for over twenty years). Check meaning with students: When did they start? Are they still in operation? Then analyse the form including the preposition for or since.

B Present perfect simple 3

Level:
Pre-Intermediate
Intermediate

a life experience

Have you ever used a database?

Form

positive

subject	aux **have** (present)	past participle	
I	've	worked	all over the world.

negative

subject	aux **have** (present)	**not**	**ever**	past participle	
I	have	n't	ever	studied	IT.

question

aux **have** (present)	subject	**ever**	past participle	
Have	you	ever	used	a database?

short answer

yes	subject	aux **have** (present)
Yes,	I	have.

no	subject	aux **have** (present)	**not**
No,	I	have	n't.

1. This use of present perfect means 'in your life', and you cannot specify when the experience happened (e.g. - Have you ever used a database ~~last year~~?).
2. This use of present perfect often starts a topic. Then, to discuss it in more detail, use past simple: Have you ever used a database? (present perfect) - What was it? Did you find it useful? (past simple).
3. I've been to... means I went and came back (e.g. I've been to Korea).

Meaning
A life experience. (Compare page 221 **past simple**.)

Meaning	Checking meaning
• We're talking about the past. • It's an important life experience. • We don't need to know exactly when it happened.	• Are we talking about now, or the past? (The past). • Is it a life experience? Is it important in their life? (Yes). • Do we need to know exactly when it happened? (No).

Anticipated problems and solutions

Problems	Solutions
As this happened in the past, students want to use past tense.	Tell them that is normally correct. This use of present perfect is only to describe important experiences in someone's life.
Students specify the time e.g. I've been to China in 2008.	Check: can you say the time? (No)

Teaching ideas
From a situation or dialogue build
Use a situation where people don't know each other, and are likely to talk about their life experiences (e.g. a job interview, the first day in class, or a first date). If there are follow up questions, they'll normally be in past simple:

travel? Have you ever travelled?
where / go? Where did you go?
who / go / with? Who did you go with?

B Past perfect simple

a past event that happened before another past event

Level:
Intermediate
Pre-Intermediate

Form
positive

subject	past simple	subject	aux **have** (past)	past participle	
I	saw	someone	had	robbed	my house.

Meaning

Meaning	**Checking meaning**
• We're talking about the past. • Two things happened. • The event in the past perfect happened first.	• Are we talking about now, or the past? (The past). • How many things happened? (Two). • Which happened first? (e.g. 'Me seeing it, or the robbery?') ('The robbery').

A past event that happened before another past event. (Compare page 221 **past simple**.)

1. Past perfect is normally used to tell a story in the wrong order. It clarifies what happened first.
2. You can use past perfect for dramatic effect – start with an interesting situation, and work backwards. (e.g. - My car was a wreck! - Why? - Someone had...).
3. If you use a word like before to show the order of events, you do not need to use past perfect (e.g. Before I got home someone robbed my house).

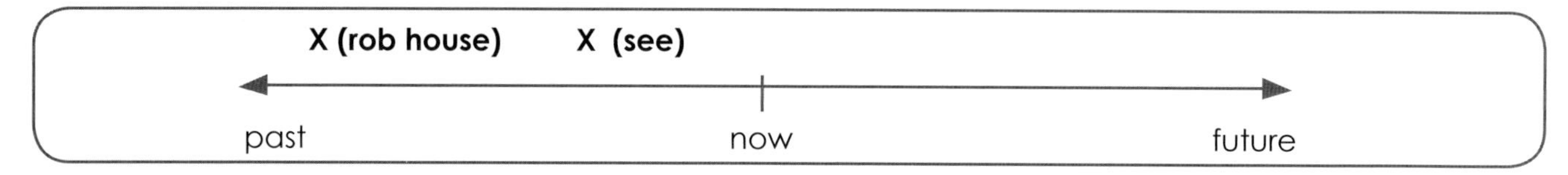

Anticipated problems and solutions

Problems	**Solutions**
Students overuse past perfect instead of past simple.	Explain to students that you only use past perfect if the events are in the wrong order, and even then, it is unnecessary if there is a word like before that makes the sequence clear.

Teaching ideas
From a situation or dialogue build

Use pictures to elicit a story, first in the correct order (all in past simple), then reverse the order of two steps (to elicit past perfect):

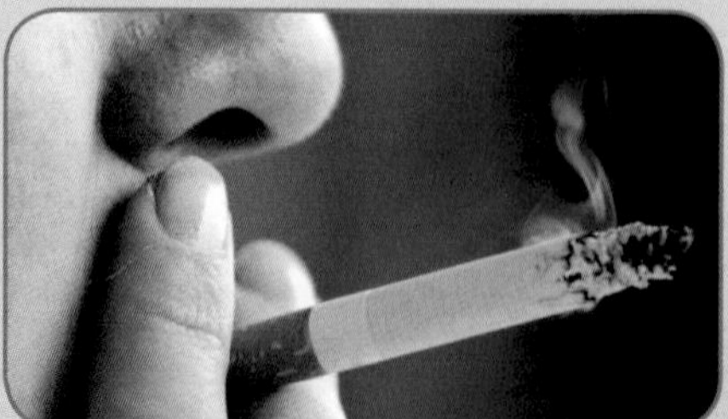

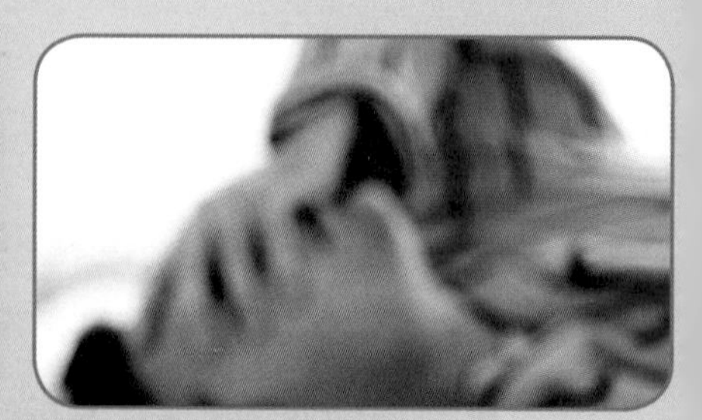

Teacher: On Sunday he felt bad because...
(Elicit:)... he'd drunk too much and smoked too much.

From a text

Use a story or narrative that includes past perfect, and have students work out the order of events.

Present continuous 1

Level: Elementary

a short activity happening right now

Form

positive

subject	aux **be** (present)		Verb + ing
I	'm	just	looking.

negative

subject	aux **be** (present)	**not**	Verb + ing	
I	'm	not	looking	for anything.

question

aux **be** (present)	subject	verb + ing	
Are	you	waiting	for someone?

short answer

Yes	subject	aux **be** (present)
Yes	I	am.

No	subject	aux **be** (present)	not
No,	I	'm	not.

1. Some verbs are not usually used in the continuous e.g. I like cars (not ~~I'm liking cars~~).
   The main ones are:
   feelings (hate, like, love, prefer, want, wish);
   mental states (believe, feel, know, mean, remember, understand);
   senses (hear, see, sound, smell, taste);
   and several others (belong, contain, include).

2. The verbs have and think have a different meaning in the simple and continuous:
   I think that's good. (opinion)
   Wait a minute, I'm thinking. (activity happening now)
   I have a car. (possession)
   I'm having lunch. (activity happening now)

3. The verb be is only used in the continuous to describe temporary bad behaviour
   e.g. He's being an idiot.

Meaning

A short activity happening right now. (Compare page 216 **present simple**.)

Meaning	Checking meaning
• We're talking about now. • It's a short activity.	• Are we talking about now, or the past? (Now). • Is it a short time or a long time? (A short time).

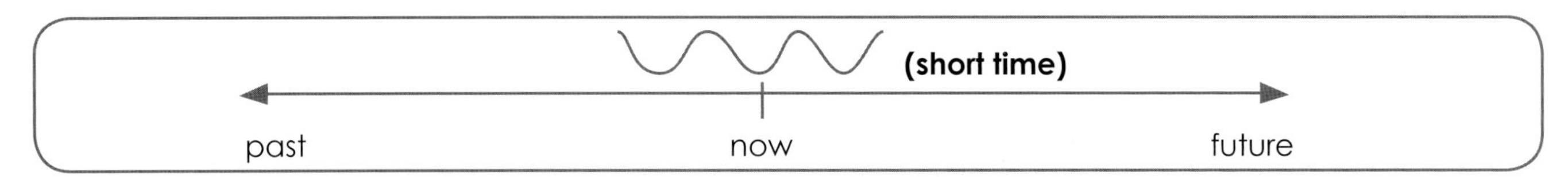

Anticipated problems and solutions

Problems	Solutions
Students use continuous to describe a habit.	Ask: are we talking about one time, or many times? (One time) So for many times, what tense do you use? (Present simple)
Students use a verb that cannot be used in the continuous e.g. I am being very happy.	Ask: which verbs don't we use in continuous?
Students omit the auxiliary e.g. I going home.	Highlight the fact there must be a word between the subject and **V+ing**. Elicit it is the verb be.

Teaching ideas

From visuals or video

Use visuals of people clearly in the middle of temporary actions – or freeze-frame a video. Elicit what is happening. Unusual and mysterious actions work well, so you can clarify this is a short one-off action and not a habit:

Teacher: What is she doing?

(Elicit:) I think she's reading.

C Present continuous 2

a temporary situation

Level:
Elementary
Pre-Intermediate

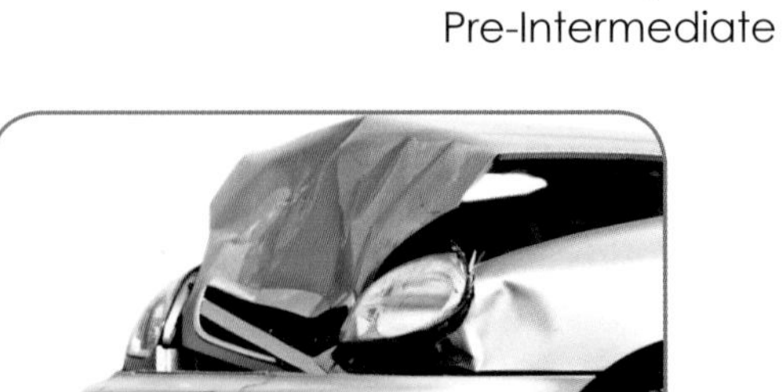

Form

positive

	subject	aux **be** (present)	Verb + ing	
This week	I	am	walking	to work.

negative

subject	aux **be** (present)	**not**	Verb + ing	
I	'm	not	driving	because of the accident.

question

aux **be** (present)	subject	verb + ing	
Are	you	staying	close by?

short answer

Yes	subject	aux **be** (present)
Yes	I	am.

No	subject	aux **be** (present)	not
No,	I	'm	not.

1. This use often has a time expression (e.g. this week or today) to show it's temporary.
2. It's often contrasted with the usual situation (in present simple): Normally I drive, but this week...
3. Even though present continuous is not normally used for habits (see page 216 present simple), with the meaning of 'temporary situation' it can be a repeated action This week I'm walking to work as well as a state This week I'm staying with my parents.

Meaning

A temporary (and perhaps unusual) situation. (Compare page 216 **present simple**)

Meaning	**Checking meaning**
• We're talking about now. • It's temporary.	• Are we talking about now, or the past? (Now). • Is it the normal situation? (No.) / Will it last a long time or a short time? (A short time).

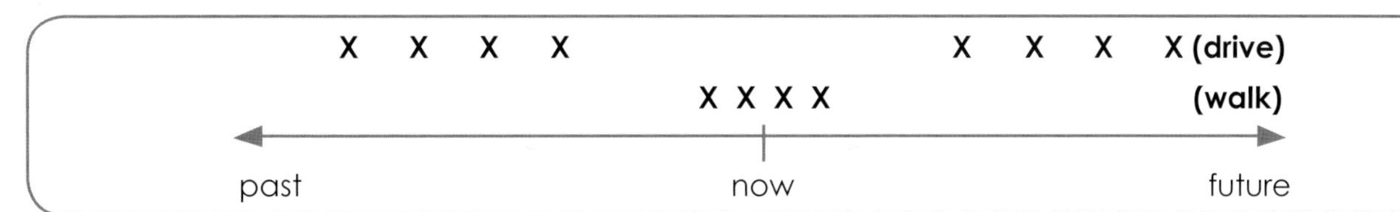

Anticipated problems and solutions

Problems	**Solutions**
Students use continuous to describe an ongoing state.	Contrast use of present simple (for normal situation) vs continuous (for a short, temporary action) e.g. I usually live with my family, but this week I'm staying with a friend.
Students omit the auxiliary e.g. I going home.	Highlight the fact there must be a word between the subject and V+ing. Elicit it is the verb be.

Teaching ideas

From visuals or video

Use visuals of people clearly in the middle of temporary actions – or freeze-frame a video. Elicit what is happening. Unusual and mysterious actions work well, so you can clarify this is a short one-off action and not a habit:

C Present continuous 3

Level:
Pre-Intermediate
Intermediate

a change we're in the middle of

Form

positive

subject	aux **be** (present)	verb + ing	
I	'm	getting	cold.
The temperature	's	dropping.	

negative

subject	aux **be** (present)	**not**	verb + ing	
That wind	is	n't	dying	down.

question

aux **be** (present)	subject	verb +ing	
Is	it	becoming	chilly?

short answer

yes	subject	aux **be** (present)
Yes,	it	is.

no	subject	aux **be** (present)	**not**
No,	it	is	n't.

Meaning

A change we're in the middle of. (Compare page 216 **present simple**.)

Meaning	**Checking meaning**
• We're talking about now. • We're in the middle of a change.	• Are we talking about now, or the past? (Now). • Are we in the middle of a change? (Yes).

Anticipated problems and solutions

Problems	Solutions
Students use present simple e.g. Prices now go up.	Ask: is there a change? (Yes) Are we in the middle of it? (Yes) So do you use simple or continuous? (Continuous).

Teaching ideas

From a text

You can use graphs (that show the past, now and projections) to elicit the structure.

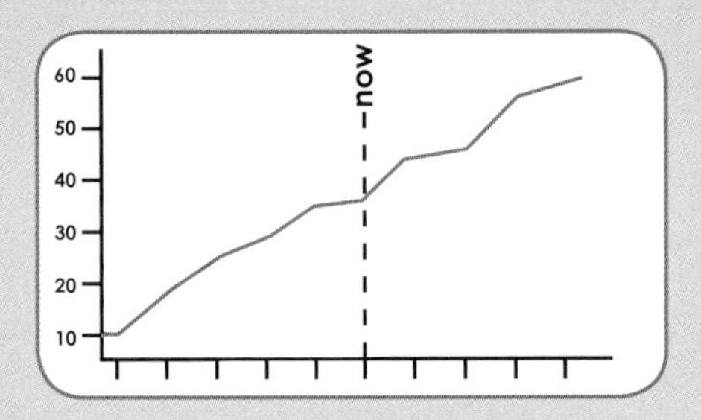

Teacher: Tell me about the price now.
(Elicit:) It's increasing.

C Present continuous 4

a future arrangement

Level:
Pre-Intermediate
Intermediate

Form

positive

subject	aux **be** (present)	verb + ing		future time expression
I	'm	meeting	someone	in half an hour
We	're	having	lunch	at one.

negative

subject	aux **be** (present)	**not**	verb + ing	
He	is	n't	bringing	anyone.

question

aux **be** (present)	subject	verb +ing	
Are	you	doing	anything afterwards?

short answer

yes	subject	aux **be** (present)
Yes,	we	are.

no	subject	aux **be** (present)	**not**
No,	we	're	not.

1. This future meaning of present continuous must have a future time expression (e.g. in half an hour), or the future time must be very clear from context; otherwise the listener will think you're talking about now. (See page 231 **present continuous 1**.)

Meaning

A future arrangement. (Compare other ways to talk about the future: page 247 **will**, page 250 **going to**, page 216 **present simple**.)

Meaning	**Checking meaning**
• We're talking about the future. • Two people have already agreed to do it.	• Are we talking about now, or the future? (The future). • How many people know about it – one or two? (Two). • Have they already agreed to do it? (Yes).

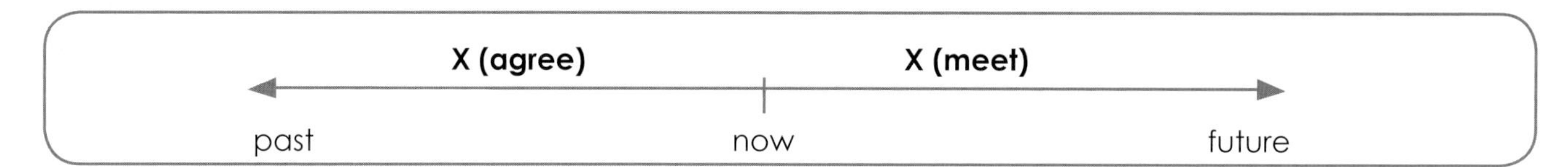

Anticipated problems and solutions

Problems	**Solutions**
It's called present tense, so students think we're talking about now.	Contrast form and meaning explicitly: 'What's the tense called?' (Present continuous). 'Is he talking about now?' (No). 'Is he talking about the future?' (Yes).
Students think they should always use will to talk about the future.	Remind students there are different ways to talk about the future. Ask: if two people have an arrangement, do we use will? (No).
Students confuse going to (for one person's plans) with present continuous (for arrangements between people).	Contrast two examples, e.g. in a restaurant context: 1. I'm angry. I'm going to talk to the manager! How many people know? (One). Does the manager know? (No). Is it just my plan? (Yes) / In 2, how many people know? (Two). 2. I'm seeing the manager at three. Does the manager know? (Yes) Did we arrange it? (Yes).

Teaching ideas

From a situation or dialogue build

The situation shows someone making arrangements for the future, e.g. for work or a social event:

free / Saturday?	sorry / work	sorry / see family
- Are you free Saturday?	- Sorry, I'm working.	- Sorry, I'm seeing my family.

C Past continuous 1

the background to a story

Level:
Pre-Intermediate
Intermediate

Form

positive

subject	aux **be** (past)	verb + ing		subject	past simple	
I	was	walking	down the street and	I	saw	a guy break into a car.

negative

subject	aux **be** (past)	**not**	verb + ing		subject	past simple	
I	was	n't	looking	carefully so	I	didn't see	his face.

question

question word	aux **be** (past)	subject	verb + ing
Where	were	you	going?

- This sets the scene for a story. Use past simple for all the subsequent events.

Meaning

The background to a story. (Compare page 221 **past simple**.)

Meaning	Checking meaning
• We're talking about the past. • It describes a situation. • It begins a story.	• Are we talking about now, or the past? (The past). • Was it a short event, or a situation? (A situation). • Is it the beginning of a story? (Yes).

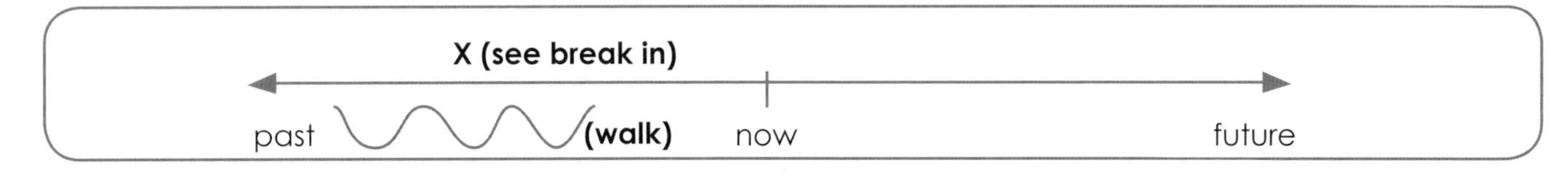

Anticipated problems and solutions

Problems	Solutions
Students are confused by what is a 'longer' and 'shorter' action.	Explain past continuous starts a story. It paints a picture of the scene. Then everything that happens is normally in past simple.

Teaching ideas

From a situation or dialogue build

Use visuals to elicit a story, the first setting a scene:

She was going on holiday.

When she lost her handbag.

She found it in lost property.

From a text

Most jokes and anecdotes start with a past continuous. Have students read for understanding (and hopefully get the joke!). Then contrast the meaning and form of the past continuous and past simple.

C Past continuous 2

a past activity happening around a point in time

Level:
Pre-Intermediate
Intermediate

Form

positive

subject	aux **be** (past)	verb + ing		point in time
I	was	driving	home	at six.

negative

subject	aux **be** (past)	**not**	verb + ing	point in time
I	was	n't	working	at six.

1. The point in time needs to be stated or clear from context.

Meaning

A past activity happening around a short event or point in time. (Compare page 221 **past simple**.)

Meaning	Checking meaning
• We're talking about the past. • We first think about a point in time. • Something starts before that point in time, and finishes after it.	• Are we talking about now, or the past? (The past). • Are we thinking about a point in time? (Yes). • Did something else happen? (Yes) When did it start? (Before 6) When did it finish? (After 6).

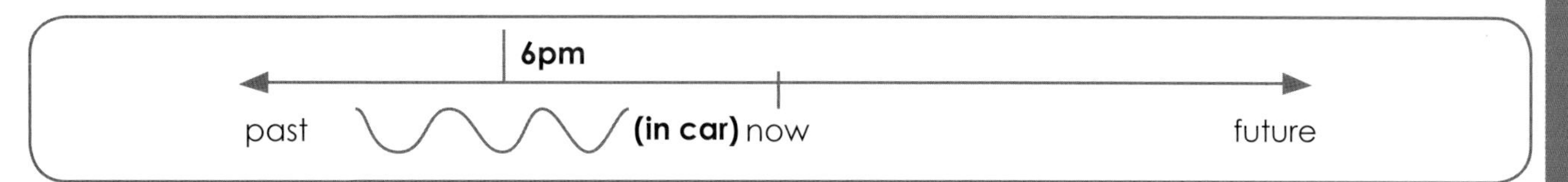

Anticipated problems and solutions

Problems	Solutions
Students ask for the difference between past simple and past continuous.	Clarify past continuous starts before the point in time. Past simple starts at the point in time: I drove home at six.

Teaching ideas

From a situation or dialogue build

Use a context where someone is checking up on everything someone did all day (suspicious parent and child?):

12.00? (have / lunch) 1.00? (study / library)

- What were you doing at twelve?
- I was having lunch.

- What were you doing at one?
- I was studying in the library.

D Present perfect continuous 1

Level: Intermediate

a past activity we now see a result of

Form

positive

subject	aux **have** (present)	**been**	verb + ing	
I	've	been	running	round all day.

negative

subject	aux **have** (present)	**not**	**been**	verb + ing	
I	have	n't	been	sitting.	around the house.

question

aux **have** (present)	subject	**been**	verb + ing	
have	you	been	looking	for a present for me?

short answer

yes	subject	aux **have** (present)
Yes,	I	have.

no	subject	aux **have** (present)	**not**
No,	I	have	n't.

- You can't specify the time (e.g. Have you been looking for a present ~~at three o'clock?~~).

Meaning

A past activity we now see a result of. (Compare page 225 **present perfect simple**.)

Meaning	Checking meaning
• We're talking about the past. • We can now see the result of the activity. • We don't need to know if the activity is finished.	• Did this happen in the past? (Yes). • Can we see a result now? (Yes). • Do we need to know if they finished what they were doing? (No).

Anticipated problems and solutions

Problems	Solutions
Students confuse this with present perfect simple, as both describe something in the past that has an effect on now	Clarify we are not interested in whether the action is finished, just the effect. (For example, you ask someone why they are covered in oil, and they say 'I've been working on the car'. We don't want to know if the repair work is finished – just why there's so much oil.)

Teaching ideas

From a situation or dialogue build

Use visuals to show people in a certain state (e.g. covered in paint, tired etc) and elicit why:

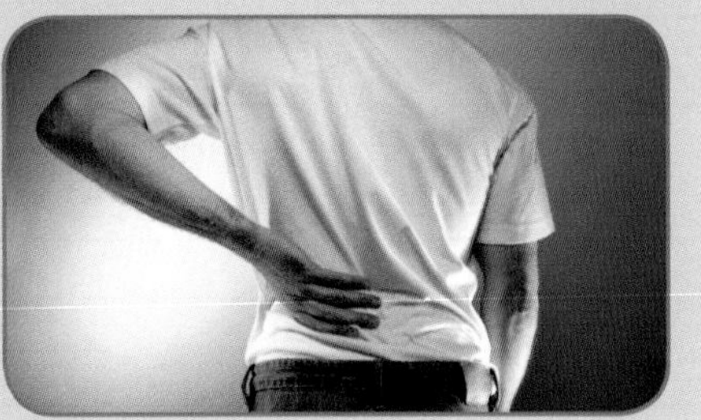

Teacher: What's he been doing?
(Elicit:) He's been working too hard. / He's been lifting things. / She's been walking in the rain.

D Present perfect continuous 2

Level: Intermediate

a situation that started in the past and is still true

Form

positive

subject	aux **have** (present)	**been**	verb + ing		time expression with **for/since**
I	've	been	studying	English	for sixty years.

negative

subject	aux **have** (present)	**not**	**been**	verb + ing		time expression with **for/since**
I	have	n't	been	going	to class	for all that time.

question

aux **have** (present)	subject	**been**	verb + ing		time expression with **for/since**
Have	you	been	studying	English	for sixty years?

short answer

Yes	subject	aux **have** (present)
Yes	I	have.

No	subject	aux **have** (present)	not
No,	I	have	n't.

- With this meaning you **must** use for (to describe a period, e.g. for a year) or since (to describe a point in time when the situation started, e.g. since April 1).

Meaning

A situation that started in the past and is still true. (Page 226 **present perfect simple 2** can be used interchangeably with this meaning. Continuous is more common in spoken language.)

Meaning	Checking meaning
• The situation started in the past. • It's still true.	• When did it start? (In the past) • Is it still true? (Yes)

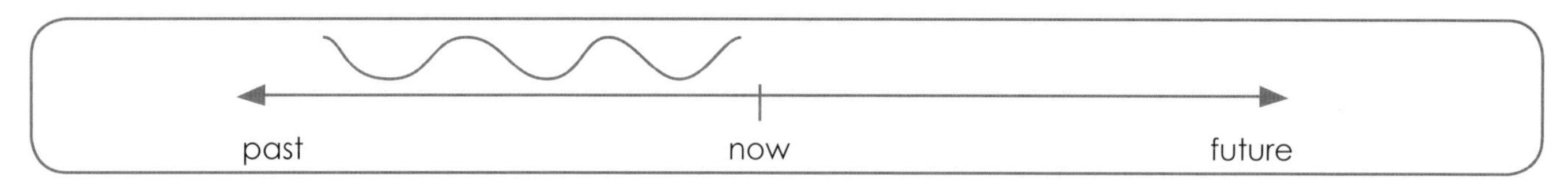

Anticipated problems and solutions

Problems	Solutions
As the situation is still true, students use present tense.	Use a timeline to clarify that if a situation started in the past and is still true, and you give the length of time, you must use present perfect (simple or continuous).
Students confuse for and since.	Use a timeline to show for tells us a period of time, and since tell us a point in time.
Students have been taught they should use present perfect simple with this meaning.	Reassure students present perfect simple and continuous are interchangeable, but only with this meaning.

Teaching ideas

See page 226 present perfect simple 2.

D Past perfect continuous

Level:
Intermediate
Upper Intermediate

a situation that continued up to a point in time in the past

When I got there she'd been waiting for an hour!

Form

positive

subject	aux **have** (past)	**been**	verb + ing	time expression with **for/since**
She	'd	been	waiting	for an hour.

negative

subject	aux **have** (past)	**been**	verb + ing	time expression with **for/since**
She	'd	been	waiting	for an hour.

question

aux **have** (past)	subject	**been**	verb + ing	
Had	you	been	trying	to call her?

short answer

Yes	subject	aux **have** (past)
Yes	I	had.

No	subject	aux **have** (past)	not
No,	I	had	n't.

- With this meaning you usually use for (to describe a period, e.g. for an hour) or since (to describe a point in time when the situation started, e.g. since 9am).

Meaning

A situation that continued up to a point in the past. We first think about the point in time (When I got there...), and then look back on the situation before it. (Otherwise just use past simple for a situation in the past: She waited for an hour but no-one came.)

Meaning	Checking meaning
• We're talking about the past. • We first think about a point in time. • We look back from that point in time to the situation before it.	• Are we talking about now or the past? (The past). • Are we thinking about a point in time? (Yes). • Are we looking forward or back from that point in time? (Back).

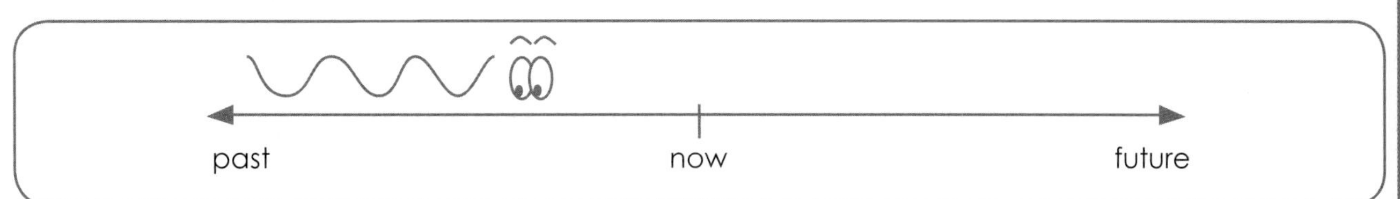

Anticipated problems and solutions

Problems	Solutions
Students are confused why past simple and past perfect continuous both talk about a situation in the past.	Use a timeline to clarify that we start with a point in time in the past, and then look back.
Students confuse for and since.	Use a timeline to show for tells us a period of time, and since tell us a point in time.

Teaching ideas

Using the whiteboard

Reveal a story, starting with a dramatic past event, and then look back from that event to see why it was a problem. For example:

Yesterday 5pm
(Elicit:) They had a fight.
Teacher: Why?

Yesterday 9am – 5pm
(Elicit:) He'd been drinking for eight hours.

E Present simple passive

Level:
Pre-Intermediate
Intermediate

a process

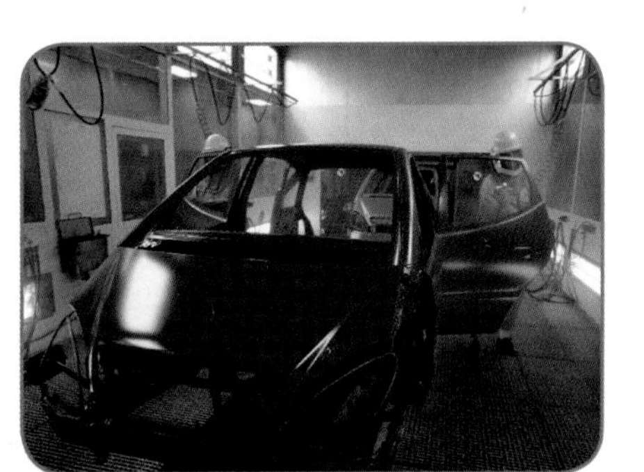

1. First the body's painted.

2. The seats are fitted later.

Form

positive

	subject	aux **be** (present)	past participle	
First	the body	's	painted.	
	The seats	are	fitted	later.

1. You can use sequence words first/next/later/finally etc to make the sequence clear.

2. Only transitive verbs can be used in the passive.

3. A passive sentence cannot have an object.

4. Passive is usually formal. In informal, spoken language it's more common to say First they paint the body etc.

Meaning

Describing a process.

Meaning	**Checking meaning**
• It happens every time. • We don't need to know who does it.	• Does it happen once or every time? (Every time). • Do we need to know who does it? (No).

Anticipated problems and solutions

Problems	Solutions
Students have learnt they should use by with a passive, and it is important who does it.	Clarify that this is a different use of passive – it's not a process. See page 245 **past simple passive.**
Students use an intransitive verb e.g. The cars are go into the factory.	Check: is go transitive or intransitive? (Intransitive.) So can we use *go*? (No.)
Students use an object e.g. The cars are fitted the seats.	Reanalyse the structure on the whiteboard: is there an object? (No.) Elicit the correct sentence.

Teaching ideas

From a situation

Use visuals of a process (industry, agriculture, services etc) to elicit the stages, eg:

1

2

3...

To make paper, first the trees are cut down...

Just make sure as much of the vocabulary as possible is familiar.

From a text

Use a brochure, textbook or website that describes a process. Make sure students understand the text (possibly by drawing the process or sequencing pictures), then analyse the meaning and form of the passive structures.

E Past simple passive

Level:
Pre-Intermediate
Intermediate

An historical event

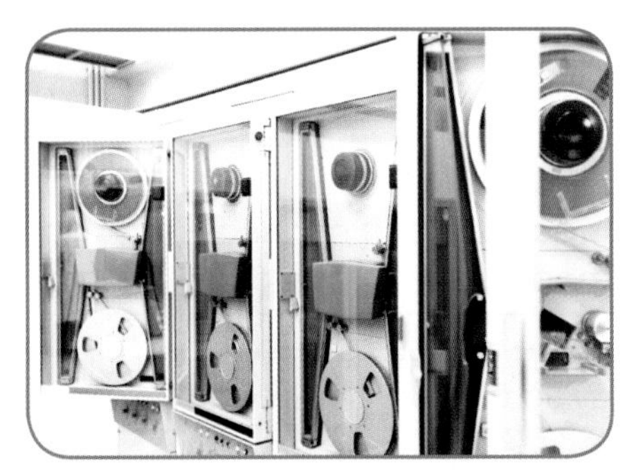

Computers were invented in the 1930's.

Form

positive

subject	aux **be** (past)	past participle	past time expression
Computers	were	invented	in the 1930's.

negative

subject	aux **be** (past)	**not**	past participle	
The Internet	was	n't	designed	for the public.

question

aux **be** (past)	subject	past participle	past time expression
Were	the first PCs	intorduced	in the 1970's?

short answer

yes	subject	aux **be** (past)
Yes,	they	were.

no	subject	aux **be** (past)	**not**
No,	they	were	n't.

1. You can use time expressions (e.g. in 1970) to specify the time.

2. You can, but it is not essential, to use by to describe who did it. By puts particular emphasis on the person.

3. Only transitive verbs can be used in the passive.

4. A passive sentence cannot have an object.

5. Passive is usually formal. In informal, spoken language it's more common to say They invented computers in the 1930's etc.

Meaning

Often used to describe an historical event. (Compare page 221 **past simple.**) It is also used in many written genres (such as newspaper articles and reports) to describe other past events.

Meaning	**Checking meaning**
• We're talking about the past. • It's an important event in history. • We don't need to know who did it (if by is not used).	• Are we talking about now, or the past? (The past). • Are we talking about something important in history? (Yes). • Do we need to know who did it? (No).

Anticipated problems and solutions

Problems	**Solutions**
Students have learnt they should use by with a passive.	Clarify *by* is possible, but not necessary, and only if it is really important to say who did it.
Students use an intransitive verb e.g. PCs were come on the market in the 1970's.	Check: is *come* transitive or intransitive? (Intransitive) So can we use *come*? (No)
Students are not sure whether to use a passive if a verb can be transitive or intransitive (e.g. increase).	Passive suggests the event was under someone's control, e.g. Prices were increased (the Government did it) vs Prices increased (we don't know why).

Teaching ideas

From a situation or dialogue build

Elicit examples of famous historical events and inventions using visuals:

1860s?, 1903?, 1908?

I think cars were invented in 1903.

From a text

Most texts about history (of countries, products, organisations etc) have many examples of past perfect. After students read for understanding, elicit examples of the passive. Clarify the meaning and form on the whiteboard.

F *will* future 1

Level: Pre-Intermediate

a decision made now

Form

positive

subject	aux **will**	bare infinitive	
I	'll	have	a hamburger.

negative

subject	aux **will**	**not**	bare infinitive	
I	wo	n't	drink	anything.

question

aux **will**	subject	bare infinitive	
Will	you	order	any dessert?

short answer

Yes	subject	aux **will**
Yes	I	will.

No	subject	aux **will**	not
No,	I	wo	n't.

1. Will is always contracted with this meaning: ~~I will have a hamburger~~ → I'll have a hamburger.

2. Since the speaker is deciding now, I'll is often combined with I think and maybe.

3. Some books call any will future 'future simple'.

Meaning

Describing a decision made now. It's often used to offer help, as soon as you see a problem: I'll carry those for you. (Compare page 250 going to **future 1: a plan**)

Meaning	**Checking meaning**
• We're talking about the future. • The speaker's deciding now.	• Are we talking about now, or the future? (Future). • Is she deciding now, or before now? (Now).

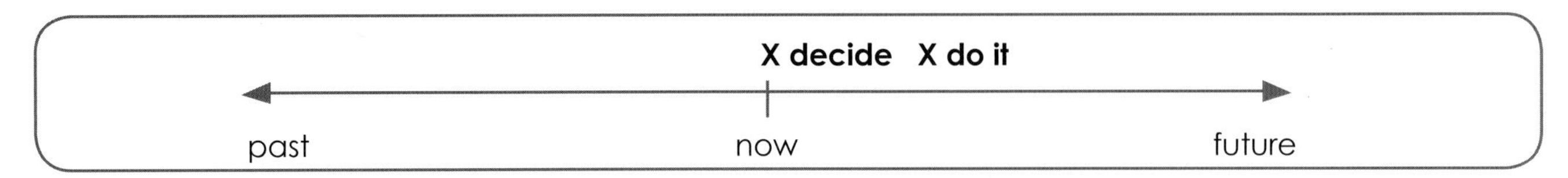

Anticipated problems and solutions

Problems	**Solutions**
Students are confused by the range of structures they can use to talk about the future.	Avoid telling students they're interchangeable. Teach each use in a clear context, and check meaning thoroughly.
Students don't contract will (e.g. they say I will instead of I'll).	Model and drill the contractions in a meaningful context.
Students do not use bare infinitive after will (e.g. ~~I will having a drink~~).	Analyse the structure on the whiteboard, and elicit the correct form.

Teaching ideas

From a situation or dialogue build

Use a context that shows a decision is being made now, and is not a long-standing plan: e.g. choosing what to eat in a restaurant, or deciding what to do when something unexpected happens:

bus / cancelled

don't / worry / give / lift

Student: The bus is cancelled
Driver: Don't worry, I'll give you a lift!

F will future 2

Level:
Pre-Intermediate

a prediction

Form

positive

subject	aux **will**		bare infinitive
Prices	will	probably	rise.

negative

subject	aux **will**	**not**	bare infinitive	
Costs	wo	n't	increase	in the near future.

question

aux **will**	subject	bare infinitive	
Will	the government	raise	taxes?

short answer

Yes	subject	aux **will**
Yes	it	will.

No	subject	aux **will**	not
No,	it	wo	n't.

1. Will is often used with adverbs of certainty (probably etc): see page 196.

2. Will does not need to be contracted with this meaning.

3. Some books call will future 'future simple'.

Meaning

Describing a prediction. It is not a plan; it is outside the speaker's control. It may only be the speaker's opinion with no strong evidence. (Compare going to **future 2: a prediction from evidence**)

Meaning	Checking meaning
• We're talking about the future. • It's not a plan. • It's the speaker's opinion about the future.	• Are we talking about now, or the future? (Future). • Is it a plan? (No). • Is it her opinion about the future? (Yes).

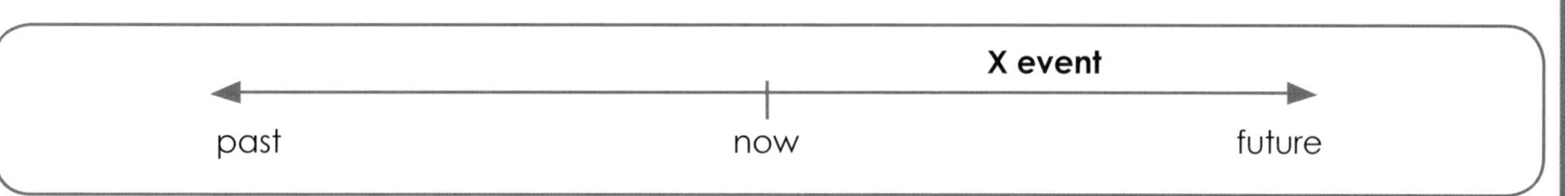

Anticipated problems and solutions

Problems	Solutions
Students do not use bare infinitive after will (e.g. ~~They will rising~~).	Analyse the structure on the whiteboard, and elicit the correct form.

Teaching ideas

From visuals about the future

Use visuals to make predictions about the economy, technology, the environment etc, and elicit the target language.

Cities will be more crowded.

Maybe we won't drive cars.

From a text

Many newspaper and current affairs magazines make predictions using will. Have students read for understanding, then ask them to underline any examples of will. Check the meaning and analyse the structure.

F going to future 1

Level:
Elementary
Pre-Intermediate

a plan

I'm going to work for a bank.

Form

positive

subject	aux **be** (present)	**going to**	infinitve	
I	'm	going to	work	for a bank.

negative

subject	aux **be** (present)	**not**	**going to**	infinitive	
I	'm	not	going to	do	a Master's degree.

question

aux **be** (present)	subject	**going to**	infinitive	
Are	you	going to	live	with your parents?

short answer

Yes	subject	aux **be** (present)
Yes	I	am.

No	subject	aux **be** (present)	not
No,	I	'm	not.

1. Although be am going (etc.) is the continuous form of go, it's better to teach going to as a special phrase as it has a different meaning (a plan vs an arrangement, see present continuous 4 page 235).

2. I'm going to go (to work tomorrow) is often reduced to I'm going (to work tomorrow).

Meaning

Describing a plan. The speaker decided to do this before now. (Compare page 247 **will future 1: a decision made now**)

Meaning	**Checking meaning**
• We're talking about the future. • It's a plan. • The speaker decided to do this before now.	• Are we talking about now, or the future? (Future) • Is it a plan? (Yes). • Did she decide now, or before now? (Before now).

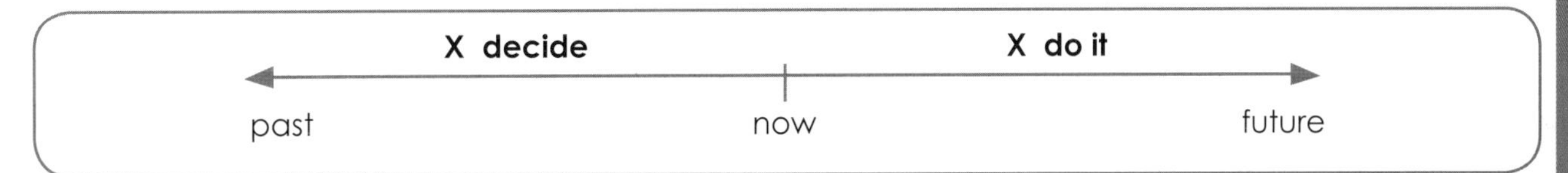

Anticipated problems and solutions

Problems	**Solutions**
Students think they should always use will to talk about the future.	Remind students there are different ways to talk about the future. Ask: 'to talk about a plan, do we use will?' (No).
going to looks like present continuous, so can be confused with an arrangement to go somewhere.	Avoid teaching going to with travel plans, so the meaning isn't confused.

Teaching ideas

From a situation

Use a context that makes it clear the person has already decided, possibly by showing two stages:

31 Dec
Teacher: She writes 'get fit, eat healthy' in her diary. Why?
(Elicit:) Plans for next year.

1 Jan
Teacher: What does she tell her friend?
(Elicit:) I'm going to get fit.

F going to future 2

Level: Pre-Intermediate

a prediction from evidence

It's going to pour!

Form

positive

subject	aux **be** (present)	**going to**	infinitive
It	's	going to	pour!

negative

subject	aux **be** (present)	**not**	**going to**	infinitive	
We	're	not	going to	get	home on time.

question

aux **be** (present)	subject	**going to**	infinitive	
Are	the roads	going to	be	okay?

short answer

yes	subject	aux **be** (present)
Yes,	they	are.

no	subject	aux **be** (present)	**not**
No,	they	're	not.

- Words that make a prediction less certain (like probably) are more common with will (see page 249).

Meaning

A prediction from evidence. The future event may have already started (e.g. there are dark clouds and it's starting to rain) or you can see something is very likely to happen: That glass is going to fall off! (it's right on the edge of the table) or We're going to be late! (because the traffic isn't moving). (Compare page 249 ***will* future 2: a prediction**)

Meaning	Checking meaning
• We're talking about the future. • We can see something that makes us certain.	• Are we talking about now, or the future? (Future) • Can we see something that makes us certain? (Yes)

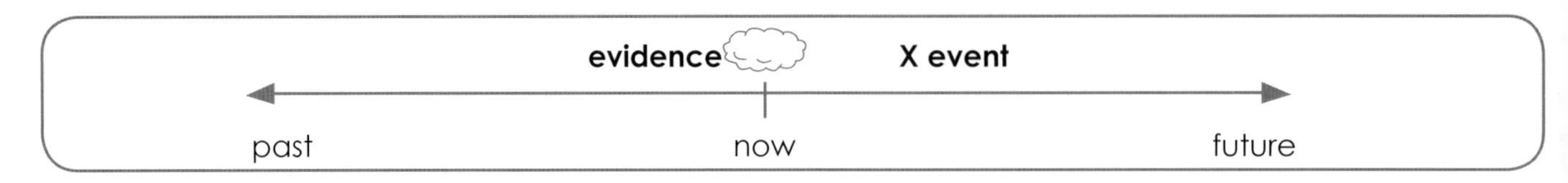

Anticipated problems and solutions

Problems	Solutions
Students think they should always use will to talk about the future.	Remind students there are different ways to talk about the future. Ask: 'To talk about a plan, do we use will?' (No).

Teaching ideas

From visuals

Use visuals to show a very likely future events to elicit the target language:

Brazil: 3
Colombia: 0
Brazil's going to win!

They're going to fall over!

F future continuous

Level: Intermediate

a future activity happening around a point in time

Form

positive

subject	aux **will**	**be**	verb + ing		point in time
We	'll	be	having	dinner	at eight.

negative

subject	aux **will**	**not**	**be**	verb + ing		point in time
We	wo	n't	be	listening	for the phone	then.

The point in time needs to be stated or clear from context.

Meaning

A future activity happening around a point in time. (Compare page 235 **present continuous 4** which is an arrangement to do something exactly at a point in time.)

Meaning	Checking meaning
• We're talking about future. • We first think about a point in time. • Something starts before that point in time, and finishes after it.	• Are we talking about now, or the future? (The future). • Are we thinking about a point in time? (Yes). • Did something else happen? (Yes) When does it start? (Before 8) When does it finish? (After 8).

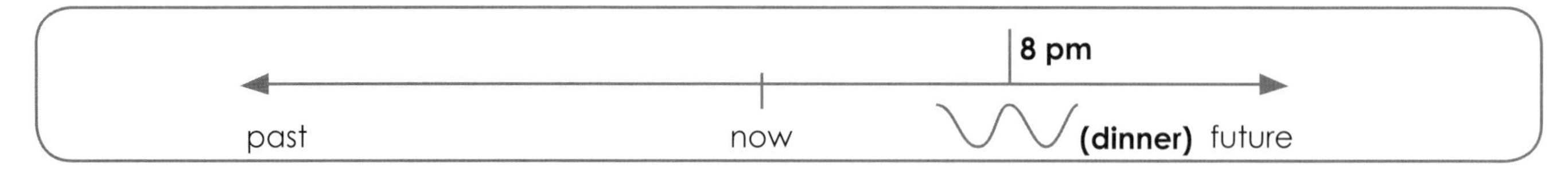

Anticipated problems and solutions

Problems	Solutions
Students ask for the difference between will and future continuous.	Clarify future continuous starts before the point in time. Will starts at the point in time: We'll have dinner at eight means dinner starts at 8.00.

Teaching ideas

Using a dialogue

Teach a dialogue where someone is trying to arrange to telephone or visit (perhaps a salesperson), and the other is trying to put them off:

3.00?

- How about three?
- Sorry, I'll be washing my hair.

4.00?

- How about four?
- Sorry, I'll be walking the dog.

Using the whiteboard

Write up random future days and times on the whiteboard. Ask students to imagine their situation at this time. Elicit the target language:

Teacher: What will you be doing at 4pm tomorrow, Anita?
(Elicit:) I'll be sitting here in this class!

F future perfect

Level: Intermediate

a future action finished before a point in time

Form

positive

subject	aux **will**	**have**	past participle		time expression with **by**
I	'll	have	finished	the report	by six.

negative

subject	aux **will**	**not**	**have**	past participle		time expression with **by**
I	wo	n't	have	copied	it	by then.

question

aux **will**	subject	**have**	past participle	
Will	you	have	checked	it for errors?

short answer

Yes	subject	aux **will**	**have**
Yes	I	will	have.

No	subject	aux **will**	**not**	**have**
No,	I	wo	n't	have.

1. With this meaning you usually use by to specify the deadline.

2. You can use *will* future (instead of future perfect) with by and still be understood. However future perfect emphasises that something should be finished before this deadline.

Meaning

A future action finished before a point in time. We first think about the point in time – a deadline (by six) – and then look back to see if something's finished.

Meaning	**Checking meaning**
• We're talking about the future. • We first think about a point in time. • We want to know if something is finished before the point in time.	• Are we talking about now or the future? (The future). • Are we thinking about a point in time? (Yes). • Do we want to know if something is finished? (Yes) When? (Before the point in time).

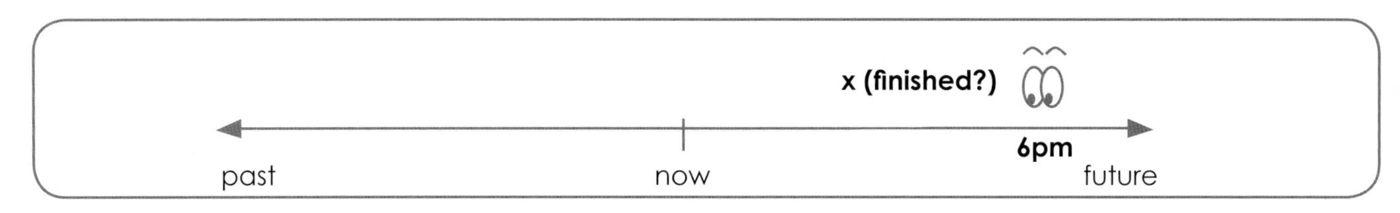

Anticipated problems and solutions

Problems	Solutions
Students are unclear when to use will and future perfect for a future event.	Ask: which one means it's important that we finish something? (Future perfect).
Students ask if there is a difference between by and before.	Explain by is for deadlines. By six means we finish something before or exactly at six, but not after.

Teaching ideas

Using a dialogue build

Consider teaching a dialogue between a boss and employee, or teacher and student, asking when things will be finished:

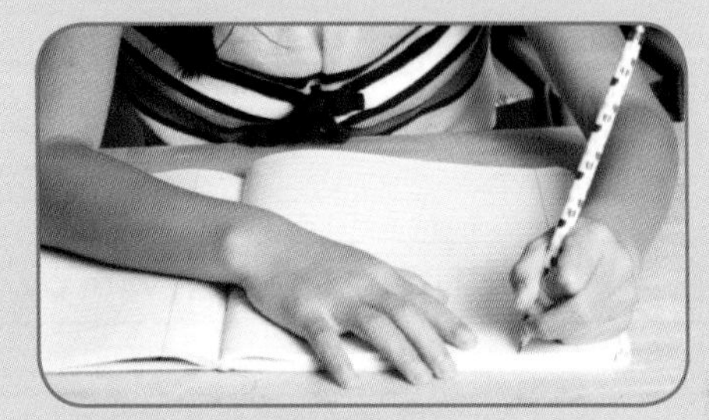

write / tomorrow

I will have written my homework by tomorrow.

clean / 2.00

We will have cleaned the classroom by two o'clock.

G zero conditional

Level:
Pre-Intermediate
Intermediate

a general truth

If you heat water to 100 degrees, it boils.

Form

Positive

If	subject	verb (present simple)		subject	verb (present simple)
If	you	heat	water to 100 degrees	it	boils.

Negative

If	subject	aux **do** (present)	**not**	bare infinitive		subject	verb (present simple)
If	trees	do	n't	get	enough water	they	die.

- While there's no strict rule, most people use a comma after a long if clause.

Meaning
Describing

Meaning	Checking meaning
• The two events are connected. • It's always true. • (It's a scientific fact.)	• Are the two events connected? (Yes). • Is it always true? (Yes). • Is it an opinion or a scientific fact? (A fact).

Anticipated problems and solutions

Problems	Solutions
Students have learned they must use will or would in a conditional.	Tell students that is often true. Clarify the meaning of the zero conditional: Is it true every time? (Yes.)

Teaching ideas
From visuals
Show students groups of pictures and elicit the relationship between them:

If you heat ice,

it melts.

From a text
Use a popular science article that has at least one example of zero conditional to describe scientific truths/causes and effects. First have students read for understanding. Then elicit the example(s) to the whiteboard. Check the meaning (using the questions above) and analyse the structure.

1st conditional

Level:
Pre-Intermediate

thinking about future possibilities

Form

Positive

If	subject	verb (present simple)		infinitive	will	bare infinitive	
If	it	's	sunny	we	'll	have	a picnic.

Negative

If	subject	verb (present simple)	subject	will	**not**	bare infinitive
If	it	rains	we	wo	n't	go.

1. While there's no strict rule, most people use a comma after a long if clause.

Meaning

Thinking about future possibilities; weighing up plans for the future

Meaning	Checking meaning
• The two events are connected. • We're talking about the future. • It is possible.	• Are the two events connected? (Yes). • Are we talking about the future? (Yes). • Is it possible or impossible? (Possible).

Anticipated problems and solutions

Problems	Solutions
Students use will in the if clause (If it will be sunny...).	Analyse the structure. Highlight there is no will after if.
Students use this to talk about an imaginary now.	Use an example that is clearly in the future (e.g. On Saturday...). Ask: Are we talking about the **future**? (Yes).

Teaching ideas

From a situation

Set a context where people are clearly weighing up options for the future (e.g. planning a weekend trip, or discussing potential changes to a city). Use pairs of pictures to elicit the target language:

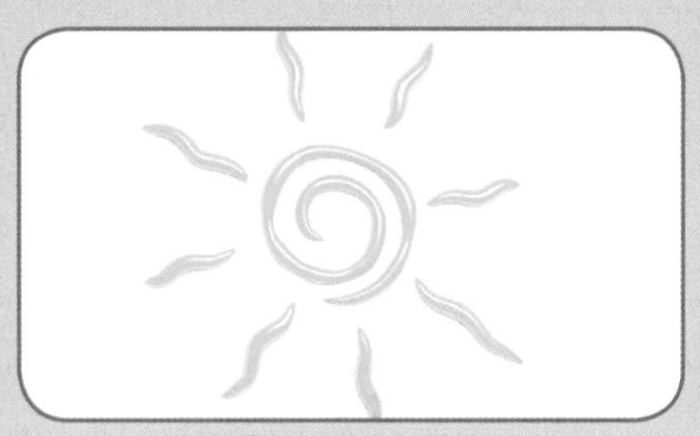

If it's sunny

we'll have a picnic.

If it rains

we'll see a movie.

2nd conditional

imagining a different now

Level:
Pre-Intermediate
Intermediate

If I had lots of money, I'd buy a mansion.

Form

Positive

If	subject	verb (past simple)		subject	**would**	bare infinitive	
If	I	had	lots of money	I	'd	buy	a mansion.

Negative

If	subject	aux **do** (past)	**not**	bare infinitive		subject	**would**	
If	I	did	n't	need	to work	I	'd	travel.

1. While there's no strict rule, most people use a comma after a long if clause.

Meaning

Imagining a different now. It's often used for dreams and wishes for something not likely to happen.

Meaning	**Checking meaning**
• The two events are connected. • He's thinking about now. • He's imagining life is different. • The situation is not real.	• Are the two events connected? (Yes). • Is he thinking about now or the future? (Now). • Is he imagining life is different? (Yes). • Is the situation real? (No).

Anticipated problems and solutions

Problems	**Solutions**
Students use would in the if clause (If I would have lots of money…).	Analyse the structure. Highlight there is no would after if.
Students use this to talk about possible future.	Use an example that is clearly impossible (e.g. If I was a child again…). Ask: Is it real? (No.)
Some students may have learnt they must use were not was.	Clarify that were is correct but formal. Check the level of formality: 'Is this a formal or informal situation?'

Teaching ideas

From a situation

Use a context that makes it clear the statement is imaginary (e.g. If I was Prime Minister). Use prompts to elicit the target language:

(lower) taxes If I was Prime Minister I'd lower taxes.
(sell) the banks If I was Prime Minister I'd sell the banks.

From a text

Some songs use second conditional with this meaning: If I were a rich man, If I was a carpenter.

Have students listen for understanding. Then elicit the lines starting with **If**… Check the meaning and analyse the structure on the whiteboard.

3rd conditional

Level:
Intermediate
Upper Intermediate

imagining a different past

Form

Positive

If	subject	aux **have** (past)	past participle		subject	**would**	**have**	past participle	
If	I	'd	got up	earlier	I	would	've	made	the train!

Negative

If	subject	aux **have** (past)	**not**	past participle		subject	**would**	**have**	past participle
If	I	had	n't	stayed up	late,	I	would	've	woken up!

- While there's no strict rule, most people use a comma after a long if clause.

Meaning

Imagining a different past. It's often used to express regrets. (Compare page 263 *wish* + past perfect.)

Meaning	Checking meaning
• The two events are connected. • He's thinking about the past. • He's imagining something different happened.	• Are the two events connected? (Yes.) • Is he thinking about now or the past? (The past.) • Is he thinking about what really happened? (No.) Something different? (Yes.)

Anticipated problems and solutions

Problems	Solutions
Students struggle to produce this structure because it's complicated.	Focus on written controlled practice first. Give students alternatives to convey this meaning (e.g. *wish* + past perfect - see page 263).
Students use this to talk about an imaginary now.	Use an example that is clearly in the past (e.g. Yesterday …'). Ask: Are we talking about the now? (No). Are we talking about the past? (Yes).

Teaching ideas

From a situation

Use visuals to show a bad situation now, because of some mistake they made.

Teacher: What's her problem? (Elicit:) She's late.
Teacher: Why? (Elicit:) She didn't set her alarm.
Teacher: So if she'd set her alarm … (Elicit:) If she'd set her alarm she would've come on time.

From a text

Use an article which features people expressing regrets for a bad situation (e.g. politicians after losing an election, someone after an accident). The target language does not need to be in the text, but you can elicit: What is the person thinking?

Try to end a lesson talking about regrets on a high note – for example, students brainstorm how we could help this person in a bad situation, or prevent this sort of accident in the future

H wish + past simple

Level: Intermediate

Wanting now to be different.

I wish I had a better job!

Form
positive

subject	**wish** (present)	subject	past simple	
I	wish	I	had	a better job.
My husband	wishes	I	was	happier.

negative

			past simple (negative)			
subject	**wish** (present)	subject	aux **do** (past)	**not**	bare infinitive	
I	wish	I	did	n't	work	here.

1. The two subjects can be different (e.g. My husband… I) – the rule doesn't change.

2. After I/he/she/it use was in informal language and were in formal language.

3. For past simple forms see page 221 past simple.

Meaning
Wanting now to be different.

Meaning	Checking meaning
• She wants something different. • She's thinking about now.	• Does she want something different? (Yes) • Is she thinking about now or the past? (Now)

Anticipated problems and solutions

Problems	Solutions
The past tense verb makes students think we're talking about the past.	Contrast form and meaning explicitly: 'What's the tense called?' (Past) 'Is she thinking about the past?' (No) 'Is she thinking about now?' (Yes)
Some students may have learnt they **must** use *were* not *was*. (See also page 259 **2nd conditional**.).	Clarify that *were* is correct but formal. Check the level of formality: 'Is this a formal or informal situation?'
Students try to use this structure to wish other people luck for the future, e.g. 'I wish your exam went well tomorrow'.	Ensure the context is clear so students are confident with the meaning of *wish* + past simple. (To wish people success, use hope + present simple, e.g. 'I hope your exam goes well tomorrow' – consider teaching this in a different lesson.)

Teaching ideas
From a situation
The situation should clearly show that someone is in a bad situation, and they're dreaming of a better life.

For example:
Is he happy? Why/Why not?

He talks to a friend about life. What does he say?

I wish I had a sports car.

H wish + past perfect

Wanting the past to be different.

Level:
Intermediate
Upper Intermediate

I wish I'd caught the train!

Form

positive

subject	**wish** (present)	subject	aux **have** (past)	past participle	
I	wish	I	'd	caught	the train.
My wife	wishes	we	'd	bought	a newer car.

negative

			past perfect (negative)			
subject	**wish** (present)	subject	aux **have** (past)	**not**	bare infinitive	
I	wish	I	had	n't	driven	today.

1. The two subjects can be different (e.g. My wife ... we) – the rule doesn't change.

Meaning

Wanting the past to be different. It's often used to talk about regrets.

Meaning	**Checking meaning**
• He wants something different. • He's thinking about the past. • He made a bad decision.	• Does he want something different? (Yes) • Is he thinking about now or the past? (Past) • Did he make a good or a bad decision? (Bad)

Anticipated problems and solutions

Problems	**Solutions**
Because the speaker's in a bad situation now, students may think he's talking about a decision now.	Ask: 'Is he talking about his situation now, or his decision?' (His decision) 'When did he decide to do this?' (In the past) 'Is he thinking about now or the past?' (Past)

Teaching ideas

From a situation

Use visuals to show a bad situation now, and the person regretting something they did that contributed to it. Elicit the language:

I wish I'd brought an umbrella!

When you teach the language of regret, it's good to end on a positive note: e.g. what can the people do now to make their situation better?

Modals 1: obligation

Level:
Elementary
Pre-Intermediate
Intermediate

What are they?

Modals are often used to talk about rules and advice.

have to vs must

Form

Present

subject	aux **have** (present)	**to**	infinitive	
She	has	to	start	work at six

subject	**must**	bare infinitive	
You	must	leave	right now.

Past

subject	aux **have** (past)	**to**	infinitive	
She	had	to	go	to work early yesterday.

Have to has the past form **had to**. **Must** has no past form to talk about obligation.

Using have to and must

1. Both **have to** and **must** are used to talk about rules. They mean someone has no choice.

2. **Have to** is more common. It means obligation comes from outside, either from rules and regulations (You have to get a visa) or physical reality (We have to get off at this stop).

3. **Must** means obligation comes from a person's authority (You must stop smoking). This use is disappearing in modern English. To make someone do something, you can just use an imperative (~~You must sit down~~ → Sit down!). **Must** is most common in:

   - friendly invitations: You must come to dinner.
   - reminding yourself to do something: I must finish my homework.
   - written instructions: All cars must turn left.

4. There are other ways to talk about obligation without using modals (e.g. We're required to wear a uniform).

should

Form

Present

subject	**should**	bare infinitive	
You	should	look	for a better job.

Past

subject	**should**	**have**	past participle	
You	should	have	quit	yesterday.

Using should

1. **Should** is used to give advice. The speaker thinks doing something a good idea, but the listener has a choice whether to do it or not.
2. There are other ways to give advice without using modals (e.g. Why don't you look around for a different job?).

can

Form

Present

subject	**can**	bare infinitive	
You	can	wear	what you want.

Past

subject	**could**	bare infinitive	
We	could	take	coffee breaks.

Using can

1. **Can** can be used to talk about permission. The listener is free to do something they want to do.

2. Some books recommend **may** for this meaning. However may sounds quite formal and old fashioned.

3. There are other ways to talk about permission without using modals (e.g. We're allowed to wear anything).

don't have to/can't/mustn't/shouldn't

Form

Present

subject	aux **do** (present)	**not**	**have**	**to**	infinitive	
He	does	n't	have	to	work	so hard.

subject	**can**	**not**	bare infinitive
I	can	n't	quit.

subject	**must**	**not**	bare infinitive	
We	must	n't	get	to work late.

subject	**should**	**not**	bare infinitive	
You	should	n't	stay	in that job.

Past

subject	aux **do** (past)	**not**	**have**	**to**	infinitive	
He	did	n't	have	to	accept	that position.

subject	**could**	**not**	bare infinitive	
We	could	n't	take	any breaks.

subject	**should**	**not**	**have**	past participle	
You	should	n't	have	taken	that job.

- Mustn't has no past form to talk about obligation.

Using don't have to/can't/mustn't/shouldn't

1. **Don't have to** is not the opposite of have to. It means the listener has a choice to do, or not to do, something: You don't have to wear a tie (= if they want to, it's OK, and if they don't want to, it's OK).

2. **Can't** is the negative version of have to. It describes there is obligation not to do something: You can't come to work dressed like that (= it is wrong, and the listener has no choice).

3. **Mustn't**, like **must**, comes from speaker's authority. Again, it is rare in modern English. You can use a negative imperative to stop someone doing something: Don't leave yet.

4. **Shouldn't**, like **should**, is used to give advice. The speaker thinks it's a good idea not to do something, but the listener has a choice whether to do it or not.

Teaching ideas

Modals are confusing for students. The meanings are quite similar and may overlap with words in the students' first language.

It's recommended to teach two or three contrasting items at one time (e.g. have to/can/can't). Also, teach one tense at a time, present before past.

MUST/HAVE TO/CAN
From visuals
Use signs or written rules to elicit:

You can't smoke.

You have to drive on the left.

Use prompts to elicit what students know about different jobs:

Prompts: UNIFORM / SMOKE
They have to wear a uniform.
They can smoke in their free time

Using the whiteboard
Write 'school rules' on the whiteboard (and possible prompts to elicit a range of modals e.g. hours / sport etc). Elicit the target language:

We have to be on time.
We don't have to play sport

SHOULD
From a situation or dialogue build
Use visuals or drawings to set up a situation where someone will give advice (e.g. doctor and patient; friends talking about relationships). Use prompts to elicit sentences or a dialogue that contain the target language:

Doctor: You shouldn't smoke.
Patient: Yes I know but it's difficult.
Doctor: You should try nicotine patches.

(!) Pronunciation hint

- have to = /'hæftə/ , has to = /'hæstə/
- shouldn't = /'ʃʊdənt/, should've = /'ʃʊdəv/, shouldn't have = /'ʃʊdəntəv/
- mustn't = /'mʌsənt/, mustn't have = /'mʌsəntəv/

Modals 2: deduction

Level:
Intermediate
Upper Intermediate

What are they?
Modals can be used to say how sure we are about something.

must

Form
Present

subject	**must**	bare infinitive	
You	must	be	tired.

Past

subject	**must**	**have**	past participle	
They	must	have	forced	the lock.

Using must
This means you are sure something is true.

could/might

Form
Present

subject	**could**	bare infinitive	
That	could	be	rain.

subject	**might**	bare infinitive	
He	might	be	tired.

Could looks like a past tense, but here we're talking about a situation now. You can't use **can** with this meaning.

Past

subject	**could**	**have**	past participle	
They	could	have	had	an accident.

subject	**might**	**have**	past participle	
She	might	have	missed	the train.

Using could/might

This means you feel it is possible something is true, but do not feel strongly yes or no.

can't

Form

Present

subject	**can**	**not**	bare infinitive	
You	can	't	like	this music!

Past

subject	**can**	**not**	**have**	past participle	
You	can	't	have	bought	it yourself!

In informal English couldn't/couldn't have is also possible: He couldn't have got my message.

Using can't

This means you are very sure something is not true.

Teaching ideas

PRESENT MODALS

From visuals

Show students unclear or mysterious visuals that they need to interpret:

- It must be in Europe
- It could be Poland

Using real objects
Students work out what mystery objects are by feeling them in a bag.

PAST MODALS
Students discuss an unsolved mystery (e.g. crop circles, the Mary Celeste) or a crime scene. Try to make the likelihood obvious so the choice of modal is meaningful:

- They must have wanted to rob the place.
- They might have been drunk.

Pronunciation hint
Note the pronunciation:
- could've = /ˈkʊdəv/
- couldn't have = /ˈkʊdəntəv/

J Phrasal verbs

Level:
Pre-Intermediate
Intermediate

What are they?
1. Phrasal verbs are verbs with two (very occasionally three) parts: I **stood up**, I **ran into** a friend, we **came up with** a good idea.

2. Sometimes they have a literal meaning: I **picked up** the rubbish. Sometimes they don't: I **cleaned up** my room.

3. Phrasal verbs are very common in spoken English.

4. Phrasal verbs are informal. If there is a one-word synonym it will normally be more formal:

informal	formal
go in	enter
send out	distribute
pull down	demolish

Using phrasal verbs
There are three main types.
1. **no object**
   Yesterday I **went out**.
   You can't add an object: ~~I went out the street~~.

2. **with a preposition**
   I **ran into** a friend.
   You can't separate the verb and the preposition: ~~I ran a friend into~~. Some books call them 'inseparable phrasal verbs'.

3. **with a particle**
   I **asked** her **out**.

   - The object can come before or after a particle: I asked her friend out or I asked out her friend. Some books call them 'separable phrasal verbs'.
   - However, if it's a pronoun, it has to come before the particle: ~~I asked out her~~ → I asked her out.
   - If it's a long object, put it after the particle: not ~~I asked the woman I met yesterday out~~ → I asked out the woman I met yesterday.

keep something on must always have the object before on: ~~Keep on your shoes~~ → Keep your shoe on.

A good dictionary shows you what type of phrasal verb it is:
- **come in** means there is no object (it may also say vi 'intransitive verb')
- **run into SB** means the object must always come after **into**
- **bring STH up** means the object can go either before or after **up**

Meaning
1. You may be able to work out phrasal verbs yourself, but it's worth checking in a dictionary, as they are unpredictable.
2. Here are some common non-literal uses:

away – returning something to its normal place
clear STH away
put STH away
pack STH away

down – getting less
slow down
turn STH down
cut down on STH

in – entering
invite SB in
drop in on SB
join in

on and off – talking about clothes & transport
put STH on
take STH off
try STH on
have STH on
get on STH
get off STH

out – removing
leave STH out
pick STH out
cross STH out
or completing a process
sell out
work STH out
sort STH out

up – finishing
drink up
eat up
fill STH up
clean up
use STH up

Teaching ideas

Don't teach more than a handful of new phrasal verbs at once – the meanings are very difficult to remember and easy to confuse. The main problem for students is remembering – what a particular phrasal verb means, and which group it belongs to.

Consider dealing with phrasal verbs as they come up in context. You can get students to keep a vocab book, or keep a chart on the wall, divided into three (corresponding to the three types of phrasal verbs). Add new examples as you come across them in class.

From a reading or listening text

To analyse the three structures: find or script a short text that has examples of the different types. Have students listen for understanding. Then elicit the relevant sentences to the whiteboard, and get students to work out the rules for the three different rules.

Pronunciation hint

- particles are stressed: I got UP / I looked the word UP.
- prepositions are unstressed: I ran up the STREET.

J used to

Level:
Pre-Intermediate
Intermediate

Form

Positive

subject	**used**	**to**	infinitive	
I	used	to	play	sport.

Negative

subject	aux **do** (past)	**not**	**use**	**to**	infinitive
I	did	n't	use	to	swim.

Question

aux **do** (past)	**subject**	**use**	**to**	infinitive
Did	you	use	to	exercise?

Short answer

yes	subject	aux **do** (past)
Yes,	I	did.

no	subject	aux **do** (past)	**not**
No,	I	did	n't.

Using used to

1. **Used to** refers to a general period: We used to play sport at school
2. If you want to say how long it lasted, or how many times you did it, use past simple (see page 221): ~~I used to play in the basketball team for three years~~ → I played in the basketball team for three years.
3. There is no present form of **used to**; to talk about current habits and states, use present simple (see page 216): I play basketball.
4. **Would** can also be used to express past habits (but not states): Every morning we'd start class at 9.

(!) **Used to** is easy confused with **get used to/be used to** (see page 273).

Meaning

A past habit or state that no longer happens.

Meaning	Checking meaning
• We're talking about the past. • It happened more than once OR for a long time. • It's not happening now.	• Are we talking about now or the past? (The past) • Did it happen once or many times? (Many times) OR for a long time? (A long time) • Is it still happening? (No)

Teaching ideas

From a situation

Use a context where someone would naturally speak about past habits (e.g. someone who used to be unfit and is now showing off about being healthy; a grandparent looking at a photo album with their grandchild):

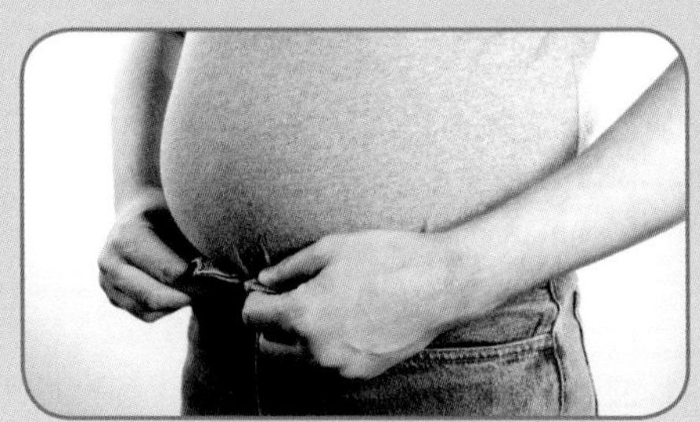
BEFORE

NOW

(Elicit:) I used to eat chips.

(Elicit:) I didn't use to exercise.

Pronunciation hint

- **used to** is pronounced /ˈjuːstə/.
- the s is pronounced /s/ not /z/, unlike in the verb use (e.g. a hammer).

J get used to or be used to?

Level: Intermediate

What are they?

These talk about a strange or difficult situation becoming easier.

get used to	be used to
I slowly got used to the warm weather.	Now I'm used to it.
I became more comfortable over time.	Now I'm completely comfortable.

(!) **Get used to/be used** to is easy confused with **used to** for past habits and states (see page 272).

Form

- **get used to**

	subject	**get** (any tense)	**used**	**to**	verb+ing/noun
When I was there	I	got	used	to	hot food.

- **be used to**

	subject	**be** (any tense)	**used**	**to**	verb+ing/noun	
Now	I	'm	used	to	eating	fish for breakfast.

Teaching ideas

From a situation

Use a context students will relate to where clearly someone might be uncomfortable at first (e.g. a new country). Use visuals as in the example above to elicit the change from feeling bad to feeling OK.

Pronunciation hint

- The pronunciation /ˈjuːstə/ is the same as used to (for past habits – see page 272) although the meaning is different.

J have or have got?

Level: Pre-Intermediate

What are they?

Have and **have got** mean exactly the same thing ('something is mine'). **Have got** is less formal, and common in spoken English.

have	**have got**
He has a new car.	He's got a new car.

Using have and have got

- **have**

Positive

subject	**have** (present)	object
I	have	a motorbike.

Negative

subject	aux **do** (present)	**not**	**have** (bare infinitive)	object
I	do	n't	have	a car.

Question

aux **do** (present)	subject	**have** (bare infinitive)	object
Do	you	have	a bike?

Short answer

Yes	subject	aux **do** (present)
Yes	I	do.

No	subject	aux **do** (present)	not
No,	I	do	n't.

1. To form the past tense, just turn **have** (pres) or auxiliary **do** to into past:
   I have a motorbike → **I had a motorbike.**
   I don't have a car → **I didn't have a car.**
   Do you have a bike → **Did you have a bike?**

2. With this meaning, don't use **have** in the continuous:
   ~~She's having a Mercedes~~ → She has a Mercedes.

3. **Have** is also used as an auxiliary in verb structures (e.g. present perfect **I've finished**) which has nothing to do with this meaning. The negative and question forms are quite different.

- **have got**

Positive

subject	aux **have** (present)	**got**	object
I	've	got	a new computer.

Negative

subject	aux **have** (present)	**not**	**got**	object
I	have	n't	got	a laptop.

Question

aux **have** (present)	subject	**got**	object
Have	you	got	a printer?

Short answer

Yes	subject	aux **have** (present)
Yes	I	have.

No	subject	aux **have** (present)	not
No,	I	have	n't.

There is no past form of **have got**; just use **had**: ~~Last year I had got a computer~~ → Last year I had a computer.

Meaning

1. Use **have** and **have got** to describe:
   - things that are yours
     We've got a house.
     My parents didn't have much money.

- friends and family
  She's got two brothers.
  Do you have any friends in the UK?
- features of a place or a thing (you can also use **There is** ... with a similar meaning; see page 180)
  My car's got central locking.
  London has lots of beautiful buildings.

2. You can also use **have**, but not **have got**, in set expressions that describe activities:
   have... breakfast/lunch/dinner
   a drink/a beer/a cup of tea/a cup of coffee
   a shower/a bath
   a rest/a holiday
   a look/a think/a listen
   fun/trouble/a good time/a bad time

   In these phrases, you can use **have** in the continuous. Use present simple for a habit, and present continuous for a short activity happening now:
   I have a cup of tea every morning. (habit)
   Give me five minutes, I'm having a cup of tea. (activity happening now)

(!) **Have** is also an auxiliary to form verb structures (e.g. Present Perfect see page 225). The meaning and the form are completely different.

Teaching ideas

Students will learn have from early elementary.

From visuals

Consider visuals of very different people to elicit questions and answers:

Teacher: Tell us about him.
(Elicit:) He has an expensive car.
Teacher: Good. What's another way to say 'he has'?
(Elicit:) He's got.

Pronunciation hint

- Have and has in questions are usually reduced to /həv/ and /həz/.

J Verb patterns: gerund or infinitive?

Level:
Pre-Intermediate
Intermediate
Upper Intermediate

Form

If you put two verbs together, the second verb will be one of three forms.

+ verb-ing	+ to + infinitive	+ bare infinitive
I enjoy speak**ing** English.	I want **to study** more.	I can **read** quite well.

Normally, the choice has nothing to do with meaning. Unfortunately you have to learn which verbs need what (or look them up in a good learner's dictionary).

1. Verbs + **V-ing**

admit	dislike	give up	practise
can't help	enjoy	imagine	recommend
consider	feel like	keep (= continue)	regret
discuss	finish	miss	suggest

2. Verbs + **to** + infinitive

agree
ask
attempt
can't afford
can't wait
decide
hope
intend
manage
mean
need
plan
pretend
promise
refuse
want
would like

3. Verbs + bare infinitive

can
should
must
will
would

Using verb patterns

1. Some verbs can be followed by either **V-ing** or **to** + infinitive: **I like cooking** or **I like to cook.**

begin
continue
like
love
prefer
start
try

- There's generally no difference in meaning.
- It's good style to avoid **V-ing** + **V-ing**:
  ~~It's beginning raining~~ → It's beginning to rain.
- **like to** and **love to** are more common in UK than US English. They have a slightly different meaning of 'a habit I do regularly':
  Every Sunday we like to meet in the library.

2. Stop and remember can be followed by either **V-ing** or **to** + infinitive, with different meanings.

Example	Meaning
She stopped talking to me.	She was talking, then was silent.
She stopped to talk to me.	She was walking, then stood still, and started talking.
I remember meeting him.	The picture of the first time I met him is in my head now.
I remembered to meet him yesterday.	It was in my diary, and I did it.

3. If a verb has a preposition after it, you must use V-ing:
~~I apologise for be late~~ → I apologise for being late.

advise against
apologise for
feel like
look forward to
plan on
succeed in
talk about
think about
wonder about
worry about

To after **look forward** is a preposition, not part of an infinitive:
~~I'm looking forward to meet you~~ → I'm looking forward to meeting you.

4. Some verbs with **to** + infinitive can, or must, have an object as well:
I want to practise English or I want him to practise English.
~~I encouraged to join a class~~ → I encouraged her to join a class.

- Don't use ~~I want that he practises English~~ etc.
- Verbs that can have object + **to**:

ask
expect
help
want
would like
would prefer

- Verbs that must have object + to:

advise
encourage
force
get
invite
order
remind
tell
warn

- **suggest** and **recommend** need subject + **should**, not **to**:
~~I suggested him to study more~~ → I suggested he should study more.

Teaching ideas

Students simply need to memorise what pattern any particular verb belongs to. Only teach a few examples at once to show the different rules.

Deal with more examples as they come up in context. You can keep a chart on the wall (or have students keep a vocabulary book) and add new examples as students come across them in class.

From a reading or listening text

Find or script a short text that has examples of the different types. Have students read or listen for understanding. Then elicit the sentences to the whiteboard, and get students to work out the rules.

Pronunciation hint

- to is usually unstressed and reduced to /tə/.

J have/get something done; make/let someone do something

Level:
Pre-Intermediate
Intermediate

What are they?

These all describe the speaker's influence on someone else's action: asking them to do something, or saying they must or don't need to do something.

have/get	make	let
The teacher had her car fixed.	She made us stay late.	She let us bring drinks into class.
She paid for a service.	We had no choice.	We had a choice.

1. **Get** is informal; **have** is more common in UK English.
2. There are other ways to talk about rules: **We're supposed to do homework; we're not allowed to eat in class.**

Form

have/get

subject	**have/got**	object	past participle
The teacher	had	her hair	cut.
They	got	the computers	fixed.

The form of **have/get** sentences is quite similar to **passives** (see page 244).

make/let (positive)

subject	**make/let**	object	bare infinitive	
The teacher	made	us	memorise	dialogues.
She	let	us	leave	early.

make/let (negative)

subject	aux **do**	**not**	**make/let**	object	bare infinitive	
The teacher	does	n't	make	us	do	homework.
She	did	n't	let	us	eat	in class.

Teaching ideas

HAVE SOMETHING DONE

From a situation

Use a context where someone needs to arrange lots of repairs and improvements (e.g. before a party/ before their parents visit etc). Elicit the story (to add some dramatic tension – a huge party and only a day to go!) and then use visuals to elicit the language:

He had the house painted.

He had the grass cut.

MAKE/LET

From a discussion

Have students brainstorm in groups what school life is like. Are they free? What are the rules? Then elicit the language using prompts such as The school .../The teachers ...

Students: We have to do homework.
Teacher: So the teachers ...
(Elicit:) ... make us do homework.

Appendix

Introduction
These appendices provide an ongoing useful source of instant information – come to this section to clarify anything you're unclear about!

1. TEFL A-Z
TEFL A-Z covers all the common terms associated with language teaching.

2. Grammar A-Z
Grammar A-Z is the place to go to clear up any grammar question.

3. Common Irregular Verbs
This is a list of common irregular verbs

4. Spelling Rules
This lists English spelling rules that students need to know (e.g. why you need a double t in getting). It also looks at key differences between UK and US spelling.

5. Useful Resources
Here you'll find recommendations for print and online resources to find information, teaching ideas and opportunities for work.

6. Phonemic Symbols and Abbreviations
These are the symbols you can use to represent sounds, so you can highlight particular sounds on the board, and students can record pronunciation of words. There's also a list of abbreviations we use as shorthand in this book.

5.1 TEFL A-Z

accuracy — Producing written or spoken language without **errors**. It's often contrasted with **fluency**; when a learner focuses on being accurate, their speech can slow down.

active vocabulary — Vocabulary learners can produce when they speak and write. It's generally much smaller than their **passive vocabulary**, which is words they can understand when they read or listen.

activity book — A book containing activities for the classroom. These often include handouts that you can legally photocopy.

advanced — See **proficiency level**.

affect — Emotions. Affect plays probably the most important part in a language learner's success. Trust in the teacher, and a supportive classroom environment, have huge affective benefits. Anxiety and boredom of course have the opposite result.

agent — Also called a recruiter, a person or company that arranges teaching work. While there are unquestionably effective agents in the TEFL world, do consider carefully what value an agent will add when it's generally easy to contact schools directly.

aim, lesson — What students will take from a lesson. Consider phrasing an aim from the point of view of the students, for example 'to learn and practise 'used to' for past habits'. Students tend to like a clear aim as it makes the class seem organised and purposeful. You can write the lesson aim in the corner of the whiteboard before you start a class.

anticipated problems and solutions — What you think students will have trouble with in a lesson, and, if so, how you're going to help them. This could be problems they'll have doing an activity (a good solution is often to **demonstrate** rather than explain), or difficulties they'll have with the **form**, **meaning** or **pronunciation** of the **target language**.

approach — A fairly general set of principles describing a way of teaching. For example, the **communicative approach** suggests students need to practise speaking in class. The **lexical approach** believes vocabulary should be the primary focus in the classroom. Approach contrasts with the more dogmatic **method**.

application letter — A letter sent to an employer, together with a resume, to apply for a job.

applied linguistics	The study of how language is used in the real world. Applied linguistics includes **SLA** (Second Language Acquisition), which investigates how languages are taught and learned.
aptitude	How 'good' someone is at languages. It's a controversial term, because there are so many factors (**affect**, **motivation**, input, opportunities for practice etc) that influence someone's success. Some people confuse aptitude with intelligence, but there's little to no correlation between intelligence and the ability to speak a language.
ARC	Authentic use, restricted use, clarification. These are proposed by the author Scrivener as an alternative to the **PPP** model. He suggests the times when explanation, accuracy practice and fluency practice happen in a lesson need to be fluid, and not occur in a set order.
assessment	Any measurement of a person's language ability. We tend to associate assessment with formal tests where students receive a mark (summative assessment). However, informal assessment - such as regular quizzes - can help students gauge their own progress. Also, it's very helpful if assessment can give students feedback to help them improve (formative assessment).
assimilation	When a sound changes because it's next to another sound. It makes sound combinations easier to say. English has backwards assimilation: an initial sound changes to 'get ready' for the second sound. For example, if you say 'ten girls' at natural speed, the /n/ sound becomes /ŋ/, which is more similar to the following sound /g/.
attitude	A learner's beliefs about English as well as their own learning. Believing English is the language of imperialism, for example, can create a negative attitude towards learning English.
audiolingualism	A teaching **method** popular for several decades after World War Two. It suggested that we learn language through performing habit-forming exercises. Drilling, a by-product of audiolingualism, is still frequently used today.
audio-visual aids	Teaching aids such as CD and DVD players, **OHPs**, **visualisers**, **flash cards** etc.
aural	Related to listening.
aural learners	See **learning style**.
authentic task	A task that replicates real use of language outside the classroom, for example making a phone call, writing an email or filling in a form.

authentic text
Material in English from the real world, not a textbook. Authentic texts can be written (e.g. newspapers, brochures, signs) or spoken (e.g. songs, TV programs, recordings of real conversations). Using authentic texts can motivate students as it bridges the gulf between the classroom and the real world.

backchaining
A modelling and drilling technique, starting from the end of the sentence, and building up the sentence backwards to the start (e.g. *you - are you - How are you?*). Its benefit is that, since the segments are meaningless, students focus solely on accurate pronunciation. In particular, it can help students notice and produce word-linking, for example in the sentence 'She's an artist': /ˈnaːtəst/ → /zəˈnaːtəst/ → /ʃiːzəˈnaːtəst/.

beating the stress
Using your hand to show stress when you **model** a word or sentence. The visual cue helps students produce more natural rhythm.

bilingual
A person who speaks two languages at a high level of **proficiency**.

blacklist
A website where teachers post negative comments about their employers

brainstorming
A group activity where students contribute ideas to a topic to generate ideas.

burden
How much students can cope with. One useful rule is never to teach 'new + new': if you're teaching new grammar, don't use examples that have new vocabulary in them. As a TEFL teacher, it's a useful exercise to attend a foreign language class yourself, to appreciate just how much is too much.

CALL
Computer Assisted Language Learning. From the days when computers were rare. Now generally called **TELL**.

CEF
The Common European Framework. Emerging as the standard measurement of language proficiency. Unlike **IELTS**, it is not a test, but a set of descriptions of six language levels, ranging from A1 (basic) to C2 (proficient).

CELTA
The Certificate in English Language Teaching to Adults, run by Cambridge University.

CELTYL
The Certificate in English Language Teaching to Young Learners. An extension to the CELTA.

certificate
The entry level of qualification for the TEFL industry. Some schools and accreditation bodies also require a degree (but generally in any field).

chain school
A school with a number of branches in different cities or countries. While some teachers dislike their uniformity (and may compare them to a well-known fast-food chain), chain schools tend to be professionally managed and resourced, and can provide a structured environment for a new teacher.

checking meaning
Also called concept checking. Making sure that students understand what some new vocabulary, functional language or grammar actually means. It's best to avoid asking 'do you understand?' as students will tend to say yes, even if they don't. Instead, ask a question that will prove to you that they understand. For example, when teaching 'sunglasses', you can ask 'Do you wear them inside or outside?'.

choral
Speaking together as a whole class (e.g. when **drilling**).

chunk
Several words that are commonly used together, for example 'Would you like to' or 'Excuse me'. It's very useful for students to learn chunks, and not always break sentences down to word level.

classroom management
What a teacher does in class to make the lesson work. This includes setting up activities, instructing, and using techniques to teach. We often associate classroom management with behaviour management, but this is only one small part. The biggest challenge for TEFL teachers is the fact the students may not understand the language they use when they instruct or teach.

cline
A scale used to clarify relative meanings of words. For example, a cline showing frequency would have 'never' at one end, 'sometimes' in the middle and 'always' at the other end.

cloze
A reading text with words removed that the student needs to fill in. These are often used to assess students; it has been argued clozes provide a good measurement of a learner's overall proficiency, although the jury is out on this.

collocation
Frequently used combinations of words, such as 'make the bed' or 'strong coffee'. Although we could possibly use alternatives and be understood ('fix the bed', 'heavy coffee'), they would sound strange to most speakers. Students need to learn collocations, and not just single words, in order to sound natural.

colloquialism
An informal word or phrase, often associated with a particular **variety**: I'm crook, mate, feeling rough as guts today!

communicative classroom
A classroom where students frequently interact and speak with other students.

communicative language teaching
An approach to language teaching which suggests the primary goal of learners is to communicate. It promotes language for social interaction, such as **functional language**, rather than a theoretical understanding of grammar. The typical model for a communicative lesson is **PPP**, where, after learning a new piece of language, students practise it in a range of interactive activities.

comprehensible input
Language that's understandable to students. It's been argued that, for students to progress, they need to be constantly exposed to input just slightly more challenging than their current level.

concept checking
Same as **checking meaning**.

consonant
A sound where the air from the lungs is stopped or restricted. Compare **vowel**.

consonant cluster
A group of two or more consonants together, such as /sp/ at the start of 'spin'. Some English clusters are not possible in other languages, and your students may simplify clusters as a result (e.g. /spɪn/ may become /pɪn/ or /sɪn/).

content words
Words that have meaning. This is generally used to refer to nouns, verbs, adjectives and adverbs. Compare **function words**.

context
A term used in **communicative methodology** to mean a life-like situation a teacher sets up in class in order to teach new language. For example, to teach restaurant language, the teacher might bring in a checked tablecloth and a menu and ask 'Where are we?'. She might then draw someone with a pad, and someone sitting down, and ask 'Who are these people?'. The aim is for the language to mean something in reality; when she teaches the sentence 'Are you ready to order?', students should immediately understand what this means, who would say it, and where.

contraction
The dropping of a sound, shown by an apostrophe in writing: do not → don't. Contractions are normal in spoken English but avoided in formal writing.

controlled practice
A practice activity where options are dictated by the teacher or the material, and students have no choice. Typical controlled practice activities are drilling and substitution exercises. It's generally used as the first stage of practice after learning new language. The aim is for students to become confident and correct with the **target language** before moving onto **free practice**.

coursebook
The main book used in class, often decided by the school. Coursebooks generally have a grammatical or topic-based syllabus. They usually teach all four **macroskills** along with grammar, vocabulary and functional language.

cue cards
Cards with pictures or word prompts on them to guide student speaking. Often used for **controlled practice**.

deductive teaching A teaching technique where the teacher tells students a rule, and the students practise it. This contrasts with inductive teaching, where students work out rules themselves. While the latter is arguably much more engaging, deductive teaching is very efficient for simple rules (e.g. 'Use 'an' before a vowel sound').

DELTA The Diploma in English Language Teaching to Adults, run by Cambridge University.

demonstrate To show students how to do an activity, rather than explain using words only. We all know when learning to do something complex (such as play a board game) we need to see it, not just hear about it.

detail (read/listen for) See **gist**.

dialect See **variety**.

dialogue build A technique to teach spoken language, especially **functional** language. The teacher sets a **context**, conveying the place and the people using pictures on the WB. She then uses visual **prompts** to **elicit** different lines of dialogue between the characters. The students repeat each line and then later practise the dialogue in pairs.

dictation A technique where one person reads out a short text and the other(s) write it down. While dictation has a bad name, it actually gives students integrated practice in listening, writing, grammar and vocabulary. Traditionally teachers have read the text, but student-student dictation is of course also possible.

diphthong A combination of two **vowel** sounds such as /aʊ/ in 'house'.

diploma A higher level of qualification than a **certificate**; generally only required for positions of responsibility in a school.

direct method A teaching method developed in the late 19th century as a reaction to **grammar translation**. The main principle was that a language should be learned 'directly', without another language in the classroom, just as a child learns a language. This greatly influenced modern-day assumptions about 'English-only classrooms' and has helped reduce time spent on grammatical explanations.

DOS Director of Studies. The term generally used for the top academic position in a language school. DOSs, in reality, spend a lot of time on operational tasks such as organising staffing and classrooms.

drill To have students repeat a word or sentence after the teacher **models** it. This is part of **controlled practice**.

EAP English for Academic Purposes. Teaching students who are at, or are planning to study at, college or university.

echo
Repeating what a student says. Teachers often do this to encourage students, but students are often confused by the purpose – is the teacher correcting me, telling me the language was correct, or telling me they agree with my ideas? Also, echoing tends to stop students talking. Instead, a teacher can use a question like 'Why?' to encourage the student to say more.

eclectic approach
The approach of many teachers and textbooks that does not rely on one 'right' method to teach a language.

EFL
English as a Foreign Language; generally used to mean English for work or study, for learners who do not intend to live permanently in an English-speaking country.

elementary
See **proficiency level**.

eliciting
Asking students to tell you, rather than you telling them. Teachers can elicit by asking questions ('What's this?') or by using prompts and realia. Eliciting helps students to engage with a lesson rather than simply being passive recipients. It also shows the teacher what students understand and can produce.

elision
Where sounds are dropped in natural spoken language. **Consonant clusters** are often simplified to make speech more efficient: the /d/ at the end of 'and' almost always disappears ('you and me' is pronounced /juːwən'miː/). However, while this is a feature of native speaker language, students will still sound natural if they don't use elision.

ELT
English Language Teaching, an umbrella term for TEFL, TESOL, ESL etc.

error
A mistake a learner makes in speaking or writing, caused by their language knowledge (not a slip caused by tiredness).

error correction
While many teachers worry that correcting errors will inhibit students, a number of studies show that students want much more correction than they receive. It is the teacher's decision, of course, when during a lesson this is appropriate. It could be suggested correction is most useful during controlled practice, when students need to be confident they've mastered any new language before moving onto free practice. So as not to inhibit fluency during free practice, teachers often wait till after the activity to discuss errors with the whole class.

ESA
Engage, Study, Activate. Like **ARC**, it is an alternative to **PPP**, proposed by Harmer. He suggested these components of a lesson could occur at any time and not in a strict sequence.

ESL
English as a Second Language; generally used to mean English taught to immigrants.

ESOL (especially Cambridge ESOL)	English as a Second or Other Language (used mainly in the UK); any English teaching to non-native speakers.
ESP	English for Specific Purposes; teaching students in a specialised field, such as engineering or health.
facilitator	A role a teacher can play to encourage student practice and interaction, rather than dominating from the front of the class.
false friend	An incorrect translation of a word or phrase from a student's own language. This is an example of **transfer**. For example, a German student may use the word 'box' to mean 'hi-fi system', as this is the word used in German.
feedback	Any sort of 'giving information back'. It's often used to mean a verbal or non-verbal response from a teacher to a student (e.g. to praise, or correct an error). Giving sincere positive feedback and constructive advice is very important for students' **motivation**. Also, feedback can be used to mean 'answers to an activity'. After students complete an exercise with right and wrong answers, they will need to know what the answers are. Many teachers recommend, in this situation, first having students compare answers, and then quickly eliciting the answers to the whiteboard so students can see them (you may want to avoid the traditional technique of asking one question at a time to each student in sequence; it's not very clear, it takes ages, and students tune out).
flashcards	Cards used by the teacher with word or picture prompts on them to elicit language from students. The main problem is when flashcards aren't large enough! You can either make bigger cards, or get students to stand close to you.
fluency	Speaking at natural speed, without unnatural pauses. This is contrasted with **accuracy**, as learners often produce more errors as their speed increases.
focus on form	Drawing students' attention to grammatical form. This may happen at any stage during a lesson. For example, in a lesson on 'invitations', the teacher notices that students are saying 'Would you like go to the cinema', so she asks, 'What do you need after Would you like?' (To.) There is some evidence helping students 'notice' grammar as it comes up in context is very helpful for their **accuracy**.
foreigner talk	The weird language native speakers use when they speak to learners: YOU – GO – NOW – YES? Teachers often resort to foreigner talk to help students understand the lesson. However, students quickly feel patronised, and it does little for their language development. However, it is still important to **grade** your language.

form, meaning and pronunciation	Three key elements of any new language students learn. You can suggest that all three need equal attention. Traditionally, grammar teaching focused on the structure, but students may not have known what the language meant, or how to say it.
formal	See **register**.
formative assessment	See **assessment**.
fossilised	An error that has become a habit in a learner's language. Someone who has been pronouncing /θ/ as /s/ for twenty years is unlikely ever to change, no matter how much **feedback** they get.
free practice	Practice where students participate in more lifelike, meaningful interaction. This conventionally occurs at the end of a PPP lesson. Unlike **controlled practice**, which aims to build **accuracy**, free practice aims to build **fluency**. Typical free practice activities include role plays, simulations, debates and games directed by the students.
function words	Words whose grammatical function is considered more important than their meaning. The term is often used to describe articles, prepositions and conjunctions. This contrasts with **content words**.
functional language	The language that people use to do things in the real world. Typical examples are making requests, introducing people, asking for permission or asking for advice. Functions are often taught as phrasebook-like formulaic **chunks**: for example, to make requests, students can learn ''Can you please…', 'Could you please…' and 'Would you mind…'. Learners need to know any associated grammatical information: after 'Can you/Could you' use the bare infinitive; after 'Would you mind' use 'V+ing'. Importantly, students need to learn responses that allow them to say yes or no!
General English	English with no specialised focus.
genre	A text type, for example 'essay', 'newspaper article', 'advertisement'. Students need to recognise genres when they read, and follow conventions of genres when they write.
gesture	A body or facial movement that conveys meaning. Gestures differ between cultures, so your students need to learn conventions in the English-speaking world.
gist	See **main idea**.

grading	Selecting language carefully so it will be understood by learners of a particular level. Generally it refers to the spoken language the teacher uses. This is not the same as **foreigner talk**. Graded language should be natural; however, the teacher should plan what she's going to say, choose words she knows her students will understand, and not speak too rapidly. Grading can also refer to choice/adaptation of written texts and activities. Some teachers suggest all students should be exposed to undoctored authentic texts, but the task, rather than the text, should be graded according to the level.
grammar	The rules of a language regarding how words can be put together. Traditionally language teaching has focused heavily on grammar at the expense of other things a student needs to know in order to communicate. See **form, meaning and pronunciation**.
grammar, descriptive	Rules that aim to describe what English speakers actually say, not what they should say. Descriptive grammar is what's relevant to learners. It's concerned with the fact that 'enjoy' is followed by 'V+ing', not with whether we should say 'who' or 'whom'.
grammar, prescriptive	Rules that describe what people should say, not what they do say. It's relevant to native speakers who would like to observe social niceties. A typical prescriptive rule is that we should say 'It is I' rather than 'It's me'.
grammar translation	A traditional method of language teaching involving presentation of grammar rules and vocabulary lists, which students then use to translate written texts. Typically there is no practise of spoken language at all. This method is still used in many classrooms around the world (including foreign language classrooms in English-speaking countries).
grammatical/structural syllabus	A syllabus based on grammar points (e.g. Unit three: present perfect).
group work	Students working in a group of three or more. This is often more advisable than pair work for freer discussion activities, as it will generate a greater variety of ideas, and can still succeed even if one student does not participate.
handout	A worksheet or other activity that a teacher gives out to students in class. (It's a good idea to instruct before you give out handouts to ensure students pay attention to you).
highlighting	Pointing to your fingers as you say individual words in a sentence (so students can distinguish the words without your needing to write the sentence on the board).
icebreaker	See **warmer**.

Idiom — A group of words that has a different meaning from the individual words: 'cool as a cucumber', 'under the weather'.

IELTS — An English proficiency test run by the University of Cambridge, commonly used in the UK, Ireland, Australia and New Zealand for university entrance and migration purposes. Students receive a score on a scale of 1-9 for each skill; most university courses require an average of around 6.5. Compare **TOEFL**.

ILC — Independent Learning Centre. An area in many language schools where students can access books and magazines, audio, video and online resources.

inductive teaching — A learner-centred teaching technique where students work out rules themselves. This contrasts with **deductive teaching**. For example, students could be given a text containing future predictions; they need to underline all the verbs, and in groups work out what structures are used to refer to the future.

informal — See **register**.

information gap ('info gap') — A classic TEFL activity where two students have different information they must share to achieve something. Often students are given different cards which they can't show anybody; they need to share the information orally.

intelligibility — The ability to be understood. Clearly, pronunciation is central to intelligibility, and plays a much bigger role in language teaching now than in the past. Since most learners want to use English as a tool for international communication, and not to live in an English-speaking country, intelligibility is generally more important than acquiring a native-like accent.

interactive whiteboard — A whiteboard connected to a computer and a data projector.

interference — See **transfer**.

intermediate (pre-, upper-) — See **proficiency level**.

intonation — Change in pitch when speaking. The most meaningful change in intonation occurs on the last stress in a sentence. The most common English intonation patterns are rising (e.g. for many yes/no questions) and falling (e.g. for most statements), although fall-rise and rise-fall are also possible.

jagged profile — Having a distinctly different level in different **macroskills**; for example, students from traditional classrooms often have much higher reading than speaking ability.

jigsaw reading — Having students read different parts of a text, or texts, which they then share. This introduces a lot of meaningful speaking practice into a reading lesson, as students attempt to explain what they've read.

kinesthetic learners — **See learning style**.

L1 — First language. The language a person speaks from early childhood. Terms like L1 and **native speaker** are becoming problematic as multilingual societies develop, and English is not considered the property of people who happened to be born in certain countries.

L2 — Second language. A language learnt after a first language.

language analysis — Examining the **form**, **meaning** and **pronunciation** of language, on teacher training courses or in the classroom.

lead-in — An activity before students read or listen to a text in order to engage them with the topic. A successful lead-in makes students want to read or listen. It may involve visuals and realia to generate discussion on the topic, or a headline or pictures from the text which students use to predict the content.

learner — A language student (of any age).

learner-centered — Techniques, activities and materials focused on the needs and interests of students. In a learner-centred classroom, the students have extensive opportunities to participate, and ideally have some control over the content of the course. The teacher, unlike in a traditional classroom, may act as a **facilitator**.

learner's dictionary — A dictionary specially designed for learners. A good learner's dictionary gives authentic examples of words in context, and additional information such as grammatical information and **collocations**.

learning style — The way an individual student likes to learn. We frequently distinguish visual learners (who like to see things written), aural learners (who like to hear language), and kinesthetic learners (who enjoy physical activity). However, it can be difficult to describe a person as one or the other; all learners display some mixture of preferences.

less-controlled practice — A stage in **PPP** between **controlled** and **free practice** where the teacher removes some element of control. For example, in the controlled stage, students may have to repeat a dialogue using substitution prompts on cards; in the less-controlled stage, they put the cards aside and can change information in the dialogue to make it true about them.

lesson plan — An outline of a lesson. A typical lesson plan will state the aims of the lesson, an analysis of any language to be taught, and a stage-by-stage description of the lesson along with expected timing. One risk of a detailed lesson plan is that teachers become too focused on 'getting through' the plan rather than responding to what students need. However, even very experience teachers will jot down a lesson aim and use a **running sheet** to help give the lesson structure.

level — See **proficiency level**.

lexical approach — An approach to teaching that argues that being able to use **chunks** of vocabulary is very important in producing fluent and natural speech.

lexis — Vocabulary.

macroskills — The four primary language skills: speaking and writing (the 'productive skills') and listening and speaking (the 'receptive skills').

main idea (read/listen for) — What a reading or listening text is 'about'. Language learners often find it difficult to work this out, and focus immediately on detail. Therefore when we teach reading and listening it's possible to develop this skill by starting with a very general question (e.g. 'Is it about X or Y?').

meaning — See **form**, **meaning** and **pronunciation**.

metalanguage — Language used to talk about language. Typically in a TEFL class this is the language a teacher uses to describe grammar. The challenge for a teacher is to ensure the meta-language is simpler than the **target language** itself. If we're teaching past simple to elementary students, we can't say 'We use past simple to describe a completed past action'. As an alternative we might ask, 'Is -ed talking about now or the past?'.

method — A way of teaching language, like **approach**. However, a method has a connotation of inflexibility. It may present itself as the 'one true way'. It will probably talk about 'scientific research' that has proven how effective it is. If you come across a teaching program with unusually strict guidelines – e.g. the teacher must play soft music, must repeat each phrase three times, and must use a particular hand gesture – it's fair to call this a 'method'.

Course materials and schools tend to shy away from the term 'method', and even 'approach'. Students need a range of different types of activities. Many in the industry call themselves **eclectic** or 'post-method'.

methodology — A general word to describe classroom practices, without specifying a particular belief. It's often used as a term on teacher training courses to mean 'how to teach' (classroom management, using the whiteboard etc), as opposed to **language analysis**.

microteaching — A practice activity on a teacher training course where trainees teach one small part of a lesson.

midnight run — When a teacher quits a school (and usually leaves the country) without completing a contract.

mingling — An activity where students move about the classroom and talk to different people.

minimal pair — Two words that differ only in the pronunciation of one sound (for example 'thick' and 'sick'). Minimal pair drilling and practice is useful for students having trouble pronouncing the two sounds differently, which will affect their **intelligibility**.

mnemonic — A technique to remember something. For example, we might remember the name of a student Bu because she has a blue jumper. Mnemonic techniques are helpful for students to learn vocabulary. Crazy mental pictures work well. For example, to remember the word 'carpet', a student might picture a pet dog on top of a rolled-up carpet on top of a car.

modelling — When the teacher says a word or sentence for students to hear how it's pronounced (often followed by **drilling**). It's also helpful to **beat the stress** at the same time. Models should always be at natural speed, with natural stress, rhythm, and **reduced vowels**, or we end up teaching students to say something that sounds weird.

Model can also be used to mean **demonstrate** (an activity).

monitor — Move around the classroom as students work in pairs and groups.

monophthong — A single vowel sound such as /ɪ/ in big.

motivation — Wanting to learn. Helping students to want to learn is probably the most important aim of a teacher.

motivation, extrinsic — Motivation through receiving rewards such as chocolate and gold stars. Rewards may work in the short term (and can be fun) but can be counterproductive; they can undermine **intrinsic motivation** by making students think English learning is boring in itself, made bearable only by winning prizes.

motivation, instrumental — Wanting to learn English for a concrete goal such as a promotion, or success in an **IELTS** test.

motivation, intrinsic — Motivation to achieve something for its own sake. This is the most self-sustaining form of motivation. Although it is hard to see intrinsic motivation increase in the short term, an enthusiastic and supportive teacher can have the long-term effect of helping a student to love learning English.

motivation, Integrative — Wanting to learn in order to become a part of an English-speaking community.

native speaker — People who speak English from early childhood. This term is becoming irrelevant as English is used by so many people for so many purposes.

needs analysis — A process to find out what students need in order to design or supplement a course. This generally involves determining the purposes students need English for (e.g. what aspects of their work or study), and an evaluation of their language strengths and weaknesses. Needs analyses are very common (and useful) when starting a program with a group of professionals.

networking — Making contacts in the industry. Networking is crucial in TEFL, as most jobs are never advertised.

neutral — See **register**.

non-native speaker — See **native speaker**.

objectives — A statement of what students can do at the end of a course. Sometimes the term is used interchangeably with **outcomes** and **aims**; however, usually **aim** refers to one class rather than a course.

observation — When someone watches someone else teach. The most useful observations are developmental – when a peer, a supervisor, or a teacher trainer observes a class in order to give constructive feedback, and discuss teaching ideas with the person being observed. However, many schools have compulsory observation to decide whether a teacher should continue to be employed. (Don't be anxious in this situation; remember it happens to everyone, and you probably teach better than the administrator observing!)

OHP/OHT — Overhead Projector/Overhead Transparency. Once central to TEFL, now rapidly becoming old technology.

oral — Related to speaking, such as 'oral test' or 'oral presentation'.

pairwork — Students working in twos. Very useful for structured language practice; not so effective for discussion, as one student might have nothing to say. Remember to change pairing arrangements frequently, to add both interest and language challenge.

passive vocabulary — Words a learner can understand when they listen or read, but not necessarily produce when they speak or write. See **active vocabulary**.

peer correction — An activity where students give feedback to each other. It's an important stage after all correction coming from the teacher in order to develop a learner's language awareness and ability to self-correct.

phoneme — A meaningful sound. For example /b/ and /p/ are phonemes; they help create different words 'ball' and 'Paul'.

phonemic symbols

The symbols that represent **phonemes**. They're helpful for students so they can record how words are pronounced (especially since English spelling is often misleading). It's useful for a teacher to know symbols for at least the sounds students have trouble with, as an efficient way of drawing attention to them on the whiteboard.

phonetic symbols

Much more detailed symbols than phonemic symbols, that represent every detail of pronunciation, whether it's important for expressing meaning or not. Phonetic symbols are generally too detailed for a language classroom.

phonology

The description of meaningful sounds; often used to mean pronunciation.

PPP

Presentation, practice, production. A model in lesson plan design closely associated with communicative language teaching. In the presentation stage, the teacher 'teaches' some new language (grammar, function or vocabulary). In the practice stage (often called **controlled practice**), the students take part in teacher-controlled activities such as drilling and substitution, in order to be able to produce the language correctly and confidently. In the production stage (often called **free practice**), the students interact in a more meaningful and lifelike way, preparing them to use the language in the real world.

While there are many critics of PPP, especially for being too rigid (see **ARC** and **ESA**), many developing teachers like the framework it offers for lesson planning; they can break away from it as their confidence develops. Also, students can find the routine and explicit lesson staging of PPP reassuring.

private tutoring

Teaching one-to-one or to small groups.

productive skills

Speaking and listening. See **macroskills**.

proficiency level

How well a student can use English (which has nothing to do with age or intelligence!). Learners are typically described as:

- **beginner** (has learnt no, or very little, English)
- **elementary** (still acquiring simple tenses; has a vocabulary of 1000 words)
- **pre-intermediate** (can use basic grammar structures; has a vocabulary of 1500 words)
- **intermediate** (can communicate on everyday topics; many errors interfere with meaning; has a vocabulary of 3000 words)
- **upper-intermediate** (can express themselves on a range of topics; frequent errors, but which don't usually interfere with meaning; has a vocabulary of 5000 words)
- **advanced** (can communicate accurately on a wide range of topics; has a vocabulary of 7000 words)

Note a learner can be quite different levels in different **macroskills** (see **jagged profile**).

Introduction How to Teach Activities Lesson Plans Grammar Appendix

prompt

A visual or word to elicit language from students in a presentation or practice activity. It's more useful to elicit language from a prompt (e.g. 'how/do?') than writing up a whole sentence and getting students to read it out ('How do you do?'); both because the students have to do the work themselves, and also because when students read out a sentence, all natural **prosodic features** disappear.

pronunciation

How a learner produces meaningful sounds. Pronunciation can be described as both 'sound-level' (students need to be able to differentiate /p/ and /f/) and 'sentence-level' (students need to use intelligible rhythm, stress and intonation).

prosodic features

Sentence-level pronunciation (stress, rhythm and intonation).

pyramid discussion

An activity where students agree on a number of items (e.g. 'what to take to a desert island') first in pairs, then groups, then as a whole class.

ranking activity

An activity where students work together to agree on a list. An easy and effective way to give some structure to a discussion activity.

rapport

The positive relationship between teacher and students. Any teacher can use their individual personality to build rapport. An outgoing teacher might make a class fun; a quiet teacher might make the class very learner-centred, giving personalised attention to students.

realia

Real objects. Realia, like visuals, makes English meaningful and engaging by bridging the gap between the classroom and the outside world.

receptive skills

Reading and listening. See **macroskills**.

recruiter

See **agent**.

recycling

Returning to language items across days, weeks and months. We tend to focus on making sure each individual lesson is a success, forgetting that students will soon forget what they've learnt without constant recycling.

reduced vowel

An unstressed vowel (generally) pronounced as /ə/. English has strong word **stress** and most unstressed vowels are reduced; for example 'station' is pronounced /ˈsteɪ·ʃən/ not /steɪ·ʃɒn/. Not reducing syllables makes word stress less distinct, in turn reducing intelligibility.

reference grammar

A book where teachers and students can look up grammar rules.

register — Level of formality in spoken and written English, generally categorised as informal, neutral or formal. A good **learner's dictionary** tells us which words are which. Formal language is not always 'better'; certainly you might ask 'Would you care for a drink, madam?' at a formal function, but it would sound very strange at the pub. However, if you use this language in the pub, it could be very socially distancing (basically, people would think you were weird). Learners need to know language for different levels of formality.

resources — Print or electronic materials.

restricted practice — Same as **controlled practice**.

resume — Often used interchangeably with CV, a document showing a person's qualifications and employment history.

rhythm — The timing of **stresses** in a sentence. Languages can be classified as having stress-timed or syllable-timed rhythm. English is stress-timed; there is a regular amount of time between each stress, and if we insert more words between the stresses, it takes virtually no longer to say (e.g. SHE's my NEIGHbour / SHE's my new NEIGHbour / I think SHE's maybe my new NEIGHbour all take about the same amount of time to say, and the unstressed words are squeezed together). This is different from a language like Italian, which is syllable-timed; twice as many words will take around twice as long to say.

role play — Students pretend to do something in real life (e.g. as shop assistant and customer). Successful role plays involve a life-like goal (e.g. customers have to complete their shopping list while spending as little as possible) rather than just going through the motions to 'practise a dialogue'.

running sheet — A brief **lesson plan**, easy to refer to (as opposed to a detailed plan that may be unworkable in class).

scan — Read quickly to find specific information (e.g. names, dates or times). Compare **skim**.

schwa /ə/ — The short vowel sound in the second syllable of 'station' or 'teacher'. The most common sound in English, produced by relaxing the mouth completely, so it's useful to know the name of the sound, and the symbol. Most unstressed syllables in English are pronounced /ə/.

selection criteria — The skills and attributes an employer looks for to choose someone for a job. When applying for work, it's first important to know exactly what the employer needs. Your application letter, resume and interview all need to communicate an identical message: that you can give the employer what they need.

self-access — Materials in a school students can use to practise on their own, often found in an **ILC**.

situation	Teaching from a situation means using a **context** to elicit language. For example, the teacher might draw a shop on the whiteboard, and a shop assistant and a customer. She then asks, 'What does the shop assistant say first'?
skills	Generally a synonym for **macroskills**. However, confusingly, it is often used to refer to receptive skills (reading and listening) only, for example in the term 'a skills lesson'.
skim	Read quickly to understand the **main idea**. Compare **scan**.
SLA	Second Language Acquisition, a field of **applied linguistics**. Essentially SLA means language learning. Sometimes, the term acquisition is contrasted with learning. Acquisition refers to developing language ability through doing real things with language, like children do with their first language; learning refers to the study of explicit rules. Some linguists (such as Krashen) have asserted learning grammar rules is a waste of time, and learners should just be exposed to authentic language. However, this is very controversial; learners seem to need some **focus on form** to develop **accuracy**.
split shift	A work schedule with a long break in the middle (e.g. two hours in the morning and two hours in the evening)
stress	In English, a stressed syllable is a longer, higher pitch and louder. English words of more than one syllable have one main stress (e.g. 'SUpermarket'), and in a sentence **content words** are stressed (for example 'We WENT to the SHOP to BUY some STUFF'). Also, most unstressed vowels are **reduced**. Stress is quite different in different languages; many languages have quite even stress. When English speakers hear learners speak without strong stress, they find it hard to distinguish the important information and follow the message.
structure	Generally synonymous with **grammar**.
STT	Student Talking Time. Most teachers would agree that students need to practise, not just learn rules about English; therefore the more time students can practise in class the better. See **TTT**.
student-centred	See **learner-centred**.
student-student, student-to-student	Happening between or among students e.g. in a student-to-student warmer, students talk to other students (rather than the teacher).
substitution prompts	In a controlled practice activity, words or pictures to elicit variations on a sentence. For example, after students have learnt 'Feel like having a drink?', the teacher writes up a list of other activities (have/dinner, see/movie etc) so the students replace 'having a drink'.

summative assessment	See **assessment**.
supplementary materials	Materials used alongside a coursebook.
syllable	A unit of spoken language often longer than a sound but shorter than a word e.g. dangerous → dan•ge•rous
syllabus	Often used synonymously with curriculum; the plan for a course, that may include a daily or weekly breakdown of topics, grammar, skills work etc.
tapescript	A written text which accompanies listening material; may be used to make cloze passages or for student review.
target language	New language (grammar, function or vocabulary) that students will learn and practise in a lesson.
task-based syllabus	An approach where the aim of a lesson is for students to perform a life-like activity using English, rather than learning and practising a language structure or function. A task might be to make a video, write a letter to a local MP, or create a brochure for visitors to the school. After the task is completed, the teacher may then address any language issues that arose.
teacher forum	A page on a website where teachers post comments about schools and experiences.
teacher talk	Can be used to describe the type of language the teacher uses in class (see **foreigner talk**) or the amount (see **TTT**).
teaching practice (TP)	A lesson a trainee teaches on a teacher training course
TEFL	Teaching English as a Foreign Language (used mainly in the UK); generally English for work or study.
TELL	Technology-Enhanced Language Learning; any use of technology (not only computers) in the classroom, including mobiles and **interactive whiteboards**.
TESOL	Teaching English as a Second or Other Language (used mainly in the US, Australia and New Zealand); any English teaching to non-native speakers.
test, achievement	A test of content students have covered in class, to see how much they've learned. This may not reflect a student's English level, but simply how much they've memorised in the past week.
test, diagnostic	A test to identify students' strengths and weaknesses. A teacher may conduct this in an early lesson to know what she should focus on in class, and how she can help individual students.
test, placement	A test at the start of a course to put students in the right **level**. These tend to be quick and easy (e.g. a short interview) as it's generally possible to move a student if they've been wrongly placed.

test, proficiency — A test (such as **IELTS** or **TOEFL**) that assesses a student's ability in all four macroskills. Proficiency tests are often used to judge readiness for entry into university.

test preparation — A course for students who are going to take an important test such as IELTS or TOEFL.

test teach test — A lesson plan that begins with an activity to evaluate how much students know. The teacher then decides whether she needs to teach new language, or whether students can move straight to more practice.

text — Any sample of written or spoken English students read or listen to for understanding.

timeline — A line drawn on the WB that depicts past, present and future, used to clarify where an event or situation described by a verb structure is located in time. For example:
E.g. 'I bought Essential TEFL yesterday'

past x now future
←————————————————————→

TOEFL/TOEIC — US-based English **proficiency** tests, popular in China, Japan and Korea, used primarily for university entrance. Compare **IELTS**.

topic-based syllabus — Syllabus based on themes or topics rather than grammatical structures.

transfer — Influence of a learner's **L1** on their **L2**. This is especially noticeable in pronunciation, where speakers of a certain L1 have a distinctive accent. Interestingly, grammatical influence of an L1 is much less significant than once thought; studies suggest speakers of any L1, whether German, Korean or Swahili, pick up English grammar in a similar order.

Trinity CertTESOL — The Certificate in TESOL, run by Trinity College, London.

TTT/STT — Teacher Talking Time. TTT is regarded as a sin by many in the field; the more time the teacher talks, the less time the students have to practise (see **STT**). While TTT does provide listening practice for students, it can be argued it is not very rich (or interesting) input; students need to hear English spoken by a wide range of speakers in different contexts.

variety — A type of English used in a particular country or area, with its own pronunciation, grammar and vocabulary; sometimes used interchangeably with 'dialect'. However, 'dialect' has a political connotation, used to describe a language variety without its own country (for example, the 'languages' Norwegian and Swedish are much closer than the 'dialects' Mandarin and Cantonese).

visual learners — See **learning style**.

visualiser — A projector that projects an image of anything including text and three-dimensional objects.

visuals
Pictures used for teaching. Visuals engage students by bringing the real world into the classroom. They are very effective in motivating students in grey, depressing school environments.

vocabulary
Students need an active vocabulary of the most frequent 1500-2000 English words in order to express themselves effectively on even familiar topics. The **lexical approach** has refocused TEFL teaching on the importance of vocabulary and **collocation**.

voice range
The difference between the lowest and highest pitch a speaker produces. In English, a wide voice range shows interest; a narrow range suggests the speaker is bored. This is not the same in all languages, so students need to learn and practise this to interact successfully with English speakers.

vowel
This can refer to either a written letter (i.e. 'a', 'e', 'i', 'o', 'u'), or a sound. A vowel sound is created when the air from the lungs is not blocked coming through the mouth (e.g. /e/ or /i:/). You can see the difference in the word 'myth', which doesn't have a vowel letter, but does have the vowel sound /ɪ/. Compare consonant.

warmer
A student-to-student speaking activity at the start of a class. This aims to relax students, to get them used to moving their mouths to produce English, and to establish a communicative atmosphere for the whole lesson.

word linking/liaison
In English, a word ending in a consonant sound is linked to a word beginning with a vowel sound: so 'an egg' sounds like /ə'neg/.

workbook
A book containing additional practice activities for learners to work on in their own time.

worksheets
Paper-based activities, developed by a teacher or copied from a master.

young learners
Teaching English to children.

.2 Grammar A-Z

abstract noun
A noun you can't touch (*problem, communism*). See **noun**.

action (verb)
See **verb**.

active sentence
A sentence where the subject is the doer of the action: *That kid* (subject) *robbed my brother!* The opposite of **passive**, where the subject receives the action: *My brother* (subject) *was robbed!*

adjective
A word that describes a **noun** or **pronoun** (e.g. *big, expensive*).

adjective, comparative

An adjective formed with '-er' or 'more…' that shows a comparison between two nouns or pronouns: English food is better than French food.

adjective, non-gradable

An adjective that's already an extreme (e.g. 'fantastic', 'terrible') or an absolute (e.g. 'dead') so can't be used after 'very' (e.g. '~~very fantastic~~')..

adjective, superlative

An adjective formed with 'the -est' or 'the most…' that compares three or more nouns or pronouns, and singles one out as special: Mumbai is the most exciting city in the world.

adverb

A word that (usually) describes a **verb**, formed by adding -ly to an adjective: *She speaks Mandarin fluently.*

Adverbs can also describe frequency (e.g. *sometimes, never*) and certainty (e.g. *maybe, definitely*).

Words that go before adjectives and other adverbs (e.g. *very, quite*) are also called adverbs.

adverb, comparative

An adverb formed with 'more … *-ly*' that shows a comparison between two verbs: *Helmut plays piano more beautifully than me*. Unlike a comparative adjective, this is comparing the playing, not how Helmut and I look.

adverb, superlative

An adverb formed with the 'most … *-ly*' that compares three or more verbs: *She speaks the most clearly of any of us.*

agreement

See **subject verb agreement**.

antonym

A word with the opposite meaning (e.g. *big* vs *small*).

article

English has two articles: **indefinite** (a/an) and **definite** (the). They come before nouns, and are a type of **determiner**.

The main difference between them is '*a/an*' is used for new information (the first time you mention something), and the for something the listener knows (the second time you mention something): *I saw a man and a woman there. The woman was wearing black.*

auxiliary verb

A 'helping' verb used before another verb. The verbs '*be*', '*have*' and '*do*' can be used as auxiliaries. English uses *be* and *have* to form continuous and perfect structures ('*I'm thinking*'/'*I've finished*') and do to form questions and negatives ('*Do you like music*'/'*I don't care!*'). See **verb structures** for more detail.

Modals are also called auxiliaries as they 'help' another verb e.g. *I can swim.*

The verb after an auxiliary is called the **main verb**. An auxiliary is **finite** (it takes endings to show the person and tense) and the main verb after an auxiliary is **non-finite** (it doesn't change).

bare infinitive	An infinitive without 'to' e.g. *I should leave*. See **infinitive**.
base form	The form of a verb you find in the dictionary (e.g. *go* vs *not going/goes etc*).
be	A very common but very **irregular** verb; it's the **base form** of present tense *'am'*, *'is'*, *'are'* and past tense *'was'*, *'were'*. The verb 'be', like other **state verbs**, isn't usually used in the continuous: ~~I was being at the station~~ → *I was at the station*.
cardinal number	*One, two, three* etc. See **numeral**.
clause	A group of words containing a **subject** and a **finite verb**. A clause can be a complete sentence (*I went home*) or half a sentence (*because I felt ill*). The first clause, that can stand on its own, is called an **independent clause**. The second, that needs to hang from an independent clause, is called a **subordinate clause**.
common noun	See **noun**.
complement	A noun or adjective after a **linking verb** that tells you more about the subject: '*She's a doctor/The milk went off*'.
complex sentence	A sentence containing a **main clause** and a **subordinate clause**: *I studied French - because - I wanted to work in Paris* (the second clause can't stand on its own; compare **compound sentence**).
compound noun	A noun combined with another noun to create a new word (e.g. *post-office*). People often disagree when compound nouns should be written as one word, two words, or two words with a hyphen; if in doubt it's safest to write two words.
compound sentence	A sentence containing two **independent** clauses: *I studied French - and - she did Japanese* (both clauses can stand on their own; compare **complex sentence**).
concrete noun	See noun.
conditional	Sentences containing *'If…'* There are four types: • zero conditional (for scientific facts): *If you heat plastic it melts* • first (for making future choices): *If the tickets are cheap I'll go* • second (for imaginary situations): *If I had a million dollars, I'd buy a yacht* • third (for regrets): *If I'd studied harder I would have passed* A mixed **conditional** is a combination of two of the above: *If I'd studied harder* (3rd) *I'd be a lawyer now* (2nd).

conjunction

A word like '*and*' or '*but*' that connect words, phrases and clauses.

- Coordinating conjunctions – you can remember them with the acronym FANBOYS: *for/and/nor/but/or/yet/so* – which join two **independent clauses:**
  He's going but she's staying.
- Subordinating conjunctions – all the others, e.g. *because/although//if/when/since* – which join a **main clause** and a **subordinate clause:**
  He's going because it's late.

continuous

See **verb structures**.

coordinating conjunction

See **conjunction**.

copula

Same as **linking verb**.

countable noun

See **noun**.

definite article

'the'. See **articles**

demonstrative

The words *'this', 'that', 'these' and 'those'* used to point things out, and to express how far away something is.

A demonstrative can be a **determiner** or a **pronoun**. As a determiner, it comes before a noun: *I want that car!* As a pronoun, it replaces a noun: *I want that!* (that car).

dependent clause

Same as **subordinate clause**.

determiner

Determiners are special words that come before nouns and limit them in some way. There are five main groups:

- articles: *a(n)/the*
- demonstratives: *this/that/these/those*
- possessives: *my/your etc*
- quantifiers: *some/many etc*
- numerals: *one/two/three etc*

You can see they have something in common with adjectives, in that they in some way 'describe' nouns. However, they have special features: they have to come before nouns (~~That car is my~~) and they don't have a comparative form (~~I need a my-er suit~~).

One rule many learners have trouble with is that a singular countable noun must have a determiner before it (~~She is teacher~~ → She is a/the/our teacher).

direct object

See **indirect object**.

direct speech

The exact words someone says: *He said, "I'm leaving tomorrow"*. This is common in writing, but in speaking we use **reported speech**.

discourse marker

A word or expression like '*anyway*', '*by the way*', '*on the other* hand' that helps a listener follow a conversation.

disjunct
A word or expression like *'unfortunately'*, *'surprisingly'*, *'if you ask me'* that refers to the whole sentence.

embedded question
A question inside another question or statement: *Do you know where the station is?/I'm not sure when the train leaves*. The **word order** in an embedded question is like a statement, not a normal question. See **question**.

ending
Something added to the end of a word for a grammatical purpose: e.g. -s after *he/she/it* or past tense *-ed*. Compare **suffix** which creates a new word: *happy → happiness*.

exception
Not following a pattern: E.g. while most nouns add '-s' in the plural, *child (children)* is an exception. See **regular**.

finite verb
A verb that comes immediately after a **subject** and can add endings to show **person, number** and **tense**.

If a verb structure has two or more words, only the first verb (the finite one) changes. The other verbs do not change, and are called **non-finite**. In the examples *'she was running'* and *'we were running'*, 'was' and 'were' are finite; 'running' is non-finite.

form vs meaning
The name of a verb structure may not have much or anything in common with its meaning. For example, 'present continuous' can be used to describe future arrangements – nothing to do with 'now'. This can be a revelation for teachers and students who are anxious to find some connection. When we teach a structure, we need to differentiate what it's called, and what it means.

fragment
An incomplete sentence: *Because it was cold. / Are going to sleep*. Fragments written by learners are often **subordinate clauses** standing on their own.

future forms
Traditionally, grammar books asserted English had 'one future tense' formed with *'will'* (Some materials still call *will* 'simple future'.) In fact, English has a range of ways to talk about the future, each with a distinct meaning. The most common are:

- *will* for a decision made now: *I think I'll go now*
- *will* for a prediction: *You'll meet someone, I'm sure!*
- *going to* for a plan: *I'm going to start a business*
- *going to* for a prediction from evidence: *Look, it's going to rain*
- present continuous for an arrangement between two people: *I'm seeing a friend after class*
- present simple for a timetabled event: *The party starts at five*

gerund
The '-ing' form of a verb, used as a noun: *Drinking ruins your liver*. You can check if something's a gerund by replacing it with a noun (*Alcohol ruins your liver*). The *-ing* form used in a verb structure (*What are you drinking?*) is called a **present participle**, not a gerund.

gradable adjective
See **adjective, non-gradable**.

grammatical
'Correct', according to what native speakers feel is 'English': *I bought a magazine* is grammatical, whereas ~~*I a magazine bought*~~ and ~~*I buyed a magazine*~~ are ungrammatical. There is a fairly widespread acceptance of what is grammatical, not least because it's backed up by reference grammars, coursebooks, English teachers etc.

imperative
The **base form** of a verb used as a command: *Stand up! Come here!*.

indefinite article
'a(n)'. See **articles**.

independent clause
A clause that can stand on its own. See **clause**.

indirect object
See **object, indirect**.

indirect speech
See **reported speech**.

infinitive
Confusingly used to mean two things: the base form of the verb (e.g. *I must go*), also called the **bare infinitive**, and the base form of the verb with 'to' (e.g. *I need to go*).

An infinitive is a **non-finite** verb. It never comes straight after a subject; it's usually used after an auxiliary like a **modal** in a verb structure: *I can speak Swahili*.

Caution – when you see a base form being used, it may not be an infinitive. The verb in: '*I speak* Arabic', looks the same but is actually present tense, not an infinitive.

interjection
An exclamation: *Ouch! Oh!*

intransitive
A verb that can't have an **object** (e.g. *live, sit, run*). See **transitive**.

invert
Turn around. English questions often invert the subject and auxiliary: *You are studying* → *Are you studying*?
See **question**.

irregular
Not following a pattern: 'went' is an irregular past tense because it doesn't add -ed'. See **regular.**

lexical verb
Same as **main verb**.

linker
Words or expressions like 'in addition', '*however*', '*by contrast*' that link part of a text to something earlier. Similar to **discourse marker** but usually used to describe written language.

linking verb

An **intransitive** verb followed by a noun or adjective (called the **complement**) that tells you more about the subject: *He is a student, She got angry.*

Linking verbs that can be followed by a noun or adjective include '*appear*', '*be*', '*become*', '*feel*', '*remain*', '*seem*', '*stay*'.

Linking verbs usually followed by an adjective are '*get*', '*go*', '*grow*', '*look*', '*smell*', '*sound*', '*taste*', '*turn*'.

A linking verb is like an equals sign; what comes before and after the verb refers to the same thing: '*He is a student*' *means He = student.*

main clause

An **independent clause**, when used in a **complex sentence** along with a **subordinate clause**.

main verb

A verb that has 'meaning'. In the sentence: She has arrived, '*arrived*' is the main verb (not *has*). Confusingly, main verb is not the same as **finite verb**, the one that comes straight after the subject and takes endings.

modal verb

A small group of special **auxiliary** verbs in English e.g. '*will*', '*would*', '*can*', '*could*', '*shall*', '*should*', '*must*', '*may*', '*might*'.

Modals often describe obligation: '*You should quit smoking*,' or a deduction: '*That must be your tenth cigarette today!*'

Modals also behave differently from other verbs:

- they have no third person '-*s*': ~~*He shoulds*~~ → *He should*
- they don't need '*do*' in questions and negatives: ~~*Does he should*~~? → *Should he*?; ~~*He doesn't should*~~ → *He shouldn't*
- they're followed by a bare infinitive: ~~He should going~~ → *He should go*

Their past tenses differ, depending on whether they're expressing obligation or deduction: *She had to go* (obligation)/ *She must have gone* (deduction).

Since there are other non-modal verbs with very similar meanings, sometimes they're taught alongside modals (e.g. 'have to' & 'ought to' for obligation).

modifier

In traditional grammar, an adverb like 'very' or 'so' that adds extra information to adjectives, verbs and other adverbs. See **qualifier**.

negative

A sentence with **not**.
To form a negative:

- if a statement contains be or an auxiliary, you add 'not' after 'be' or the auxiliary: She's a teacher → She isn't a teacher
- Other verbs need to add auxiliary '*do* + *not*': *She lives here* → *She doesn't live here*

non-finite verb

A verb in a verb structure that doesn't come immediately after the subject, and doesn't add endings to show person, number and tense (e.g. *She's swimming* or *She's gone to the beach*). Non-finite verbs can be **participles** or **infinitives**. See **finite verb**.

noun

The name of a thing (e.g. '*table*'), person (e.g. '*woman/ Susan*'), animal (e.g. '*dog/Fido*'), place (e.g. '*Tokyo*') or concept (e.g. '*happiness*').

Proper nouns (as opposed to **common nouns**) are names, written with a capital letter (e.g. *Sue* & *Tokyo*). **Abstract nouns** (as opposed to **concrete nouns**) are nouns you can't touch (e.g. *happiness*).

Countable nouns are nouns you can count: *one banana, two bananas*... **Uncountable nouns** are nouns you can't count: '~~one butter~~', '~~two butters~~'.

number

Singular or plural. See **subject verb agreement**.

numeral

Numerals are classified as **cardinal numbers** (e.g. '*one*', '*two*', '*three*' ...) or **ordinal numbers** (e.g. '*first*', '*second*', '*third*' ...). Cardinal numbers can be used as **determiners**.

object

A noun or pronoun after the verb that 'receives' the action e.g. *We bought a hammer, I hit John*.

Only **active** sentences, and only **transitive** verbs, have an object.

indirect object

Some verbs, like '*give*', have two objects – the thing you give (the direct object), and a person who receives it (indirect object): *They gave me* (indirect object) *a present* (direct object).

ordinal number

'*First*', '*second*', '*third*' etc. See **numeral**.

part of speech

One of around nine categories of words in English: 'noun', 'pronoun', 'adjective', 'verb', 'adverb', 'preposition', 'conjunction', 'determiner' and 'interjection'. Different books may use slightly different categories: quite often **determiners** are broken down into articles, possessives etc.

participle	English has two types of participle: • present participle (often called **V+ing**): e.g. *'living/going'* Present participles just add *'-ing'* to a verb. Present participles are used to form **continuous** structures (e.g. *'I'm thinking'*) and as **adjectives** (e.g. *'running water'*). Note: V+ing can also be used as a **gerund**. • past participle: e.g. *'lived/gone'* Past participles of regular verbs look like past tense: *'live'* → *'lived'* (*past simple*) → *'lived'* (*past participle*). However, past participles of irregular verbs may be different: *'go'* → *'went'* (past simple) → *'gone'* (past participle). Past participles are used to form perfect **structures** e.g. *'I've finished'*, **passives** e.g. *'The house was destroyed'*, and as **adjectives** e.g. *'broken glass'*.
particle	Often used as a name for a small word we're not sure how to categorise! The most common use is to label an adverb (as opposed to a preposition) in a **phrasal verb** – so 'up' is a particle in *'I looked it up'*, but a preposition in *'I walked up the street'*.
passive sentence	A sentence where the subject is not the 'doer', but rather receives an action: e.g. *'Ball point pens* (subject) *were invented in 1938'*. Passives are often used when we don't know, or aren't interested in, who does something. The structure of a passive sentence is 'subject + auxiliary *be* + past participle': e.g. *'Most computers* (subject) *are* (auxiliary be) *made* (past participle) *in China'*.
past	See **tense**.
past participle	See **participle**.
perfect	See **verb structure**.
person	English has three persons: first (I, we), second (you) and third (he/she/it/they). The main issue for learners is that in present simple, third person singular (he/she/it) needs an *'-s'*: *'She works in a bank'*. See subject **verb agreement**.
phrase	A group of words smaller than a **clause**; it has no subject + verb so can't stand on its own (*next to the park, the man over there*).

phrasal verb

A verb made up of two (or occasionally three) parts (e.g. '*pick up*'). Their meaning is often hard to guess (e.g. '*The idea really took off*').

Sometimes an object can come in the middle of a phrasal verb (e.g. '*I sent my application in*); here 'in' is normally called a **particle**. Sometimes the object has to come at the end (~~I went the room in~~ → *I went in the room)*; here 'in' is called a **preposition**.

possessive

Words like '*my*' and '*mine*' that show who owns something.

Possessives can be **determiners** or **pronouns**, and have different forms.

Possessive determiners (*my/your/his/her/its/our/their*) come before a noun: '*Where's my key?*').

Possessive pronouns (*mine/yours/his/hers/its/ours/theirs*) replace a noun: '*Can I borrow yours?*' (= *your key*).

predicate

The part of a clause after the subject (e.g. '*studies hard*' in '*He studies hard*').

prefix

Something added to the start of a word, usually to create a new word with a different meaning (e.g. '*un*' in '*unhappy*').

preposition

A word or phrase like '*on*', '*in*', '*next to*' that usually tells us where or when something is. Prepositions always come before a noun or a pronoun ('*on the table*'; '*next to him*').

The use of prepositions varies greatly between languages, and it's hard to formulate useful rules in English. Students may want to record examples as they come across them in context, and try to work out patterns that are memorable for them.

present

See **tense**.

present participle

See **participle** and **V+ing**.

progressive

Same as **continuous**.

pronoun

A word that replaces a noun, like '*he*' (= my brother) or '*it*' (= the table). English has both **subject** and **object** pronouns:

Subject: *I, you, he/she/it, we, they*
Object: *me, you, him/her/it, us, them*

Object pronouns are also used after prepositions (e.g. '*I sat next to her*') and after '*be*' ('*That was him*').

proper noun

See **noun**.

qualifier

In traditional grammar, an adjective or a phrase that gives extra information to a noun (e.g. 'an *expensive hi-fi*' *and* '*the hi-fi in the shop*'). See **modifier**.

quantifier

A **determiner** like '*some*' or '*many*' that tells us how much or many of a noun there are. Some books classify **numerals** as quantifiers rather than as a separate category.

question

English has five main types of questions:

- yes/no questions: '*Are you Korean?*'
- wh- questions: '*Where's our classroom?*'
- tag questions: '*You're a student, aren't you?*'
- alternative questions: '*Do you want tea or coffee?*'
- embedded questions: '*Do you know when class starts?*'

To form a question:

- if a statement contains 'be' or an auxiliary, you need to **invert** the subject and verb e.g. '*You are a teacher*' → '*Are you a teacher?*'
- with other verbs you need to add auxiliary 'do': e.g. '*She works here*' → '*Does she work here?*'

question tag

A small **yes/no question** added to the end of a statement: e.g. '*You work here, don't you?*'

Question tags contain an auxiliary and subject that mirror the statement. Generally, if the statement is positive, the question tag will be negative; if the statement is negative, the question tag will be positive.

A question tag with falling intonation is used to make small talk, when you're sure the listener will agree with you.

Students find question tags difficult to produce automatically. It may be more useful to focus on other types of questions and ways of making conversation.

regular

This describes something that follows a rule. For example, the verb 'start' has a regular past tense '-ed' and the noun 'job' has a regular plural '-s'.

As a general principle, students should be confident with a rule before they learn **exceptions**. Unfortunately, many of the most common words in English are **irregular** (e.g. 'be' and 'have').

reported speech

Reporting what a person said. Often the **person**, the **tense** and the **time reference** change: e.g. 'He said, *"I'm leaving tomorrow"*' → '*He said he was leaving the next day.*'

sentence

A group of words that convey a complete thought. In writing we show a sentence by beginning with a capital letter and finishing with a full stop (or question mark/ exclamation mark).

In simple terms, a sentence must contain a **subject** and a **finite verb**.

A **simple sentence** comprises one independent clause. A **compound sentence** comprises two independent clauses. A **complex sentence** comprises a main clause and a subordinate clause.

short answer

An answer to a yes/no question such as '*Yes, I have' or 'No, she didn't*'. Short answers contain a subject and auxiliary that mirror the question.

Short answers are more polite than simply saying '*yes*' or '*no*'.

simple

See **verb structure**.

simple sentence

A sentence consisting of one independent clause: '*We're students' and 'She likes music'*.

state (verb)

See **verb**.

subject

A noun or pronoun, before a verb, that 'does' the action e.g. '*The teacher explained the grammar' and 'We didn't understand'*.

A sentence must always contain a subject (and a **finite verb**).

subject verb agreement

A subject, and a **finite** verb that comes immediately after it, must 'agree'. The main difficulty for English learners is that third person singular (after *he/she/it* or a singular noun) needs an '-s' in the present simple: '~~Warsaw have many beautiful streets~~ '→ '*Warsaw has many beautiful streets*'.

subjunctive mood

A term from traditional grammar you may safely shelve. Subjunctive refers to hypothetical events. After *'If ...'*and '*I wish ...*' some people argue we should use a special 'subjunctive' form '*I/he/she/it were'* (not '*was*'). You can teach this as a formal variant without needing to use the term 'subjunctive'.

subordinate clause

A clause that can't stand on its own. See **clause**.

subordinating conjunction

See **conjunction**.

suffix

Something added to the end of a word, usually to change the **part of speech** (e.g. '*-ness*' in '*happiness*', which changes an adjective to a noun). Compare **ending**.

synonym

A word with similar meaning (e.g. '*big*' and '*large*'). True synonyms are rare; usually so-called synonyms have different connotations (e.g. '*skinny*' and '*slim*') or level of formality (e.g. '*buy*' and '*purchase*').

tense

English verbs do two main things:

- they locate an event in time e.g. '*I went there*' just means 'in the past')
- they tell us something else about an action, such as its length, frequency, or effect (e.g. '*I was going down the street*' means 'it went on for some time'; '*He's gone*' means 'his absence is important now')

It's a crucial distinction, because students may wrongly think all verb structures simply locate events in time.

English has only two verb forms that exclusively locate an event in time: present and past. Some teachers and linguists use the term 'tense' to describe these. They therefore suggest it's useful to say English has only two tenses. All the other forms – past continuous, present perfect etc – can be called **verb structures** rather than tenses.

transitive verb

A verb that has an **object** (e.g. '*buy*', '*watch*', '*shoot*'). Some verbs *(e.g. 'eat')* can be used as either a transitive or intransitive verb (e.g. '*I've eaten dinner' and 'I've eaten*'). See **transitive**.

uncountable noun

See **noun**.

verb

As we learnt at school, a 'doing word'. This is slightly simplistic, however, as a verb can describe an **action** (e.g. 'take', 'cook', 'destroy') or a **state** (e.g. 'love', 'have', 'be').

The difference is important because state verbs cannot usually be used in the continuous: e.g. 'I am loving her '→ '*I love her*'. See **verb structures**.

Note: a sentence must include a (finite) verb.

verb structures

English has three main verb structures:

- **simple** (= one word): 'She speaks German'.
- **perfect** (= auxiliary *have* + past participle): 'She has studied Hindi'.
- **continuous** (= auxiliary *be* + V+ing): 'She is learning Spanish.'

Each structure also has a **present** and **past**. This depends on the tense of the **auxiliary**, and has nothing to do with the past participle or V+ing. '*She has studied Hindi*' is called present perfect, because 'has' is present. '*She had studied Hindi*' is called past perfect, because '*had*' is past.

To form **questions** and **negatives**, simple structures need auxiliary do (*Does she speak German/She doesn't speak German.*) Perfect and continuous structures invert the subject and verb in questions (*Has she studied Hindi?/Is she learning Spanish?*) and just add not in negatives (*She hasn't studied Hindi/She isn't learning Spanish*).

verb structures

Traditionally, a sentence like '*She will learn Portuguese*' was called 'simple future'. Now most books refer to '*will* future' instead (this is helpful for our definition of 'simple' as meaning 'one word'). See **future forms**.

What these structures are called, and what they mean, may be completely different (see **form vs meaning**)

V+ing

The '-ing' form of a verb, that can be used as a (present) **participle** (e.g. '*I'm leaving now*') or a **gerund** (e.g. '*Smoking is expensive*').

Some teachers like the term **V+ing** because **present participle** has nothing to do with the tense of a continuous verb structure. '*I was living in Jakarta*' is called past continuous because of 'was'. It's therefore confusing to call 'living' a 'present participle'.

wh- question

A question that asks for information, *not 'yes'* or '*no*' (e.g. '*Where do you live?*' but not 'D*o you live near here?*').

It's normal to answer a 'wh-question' with just the relevant information, and not to repeat the whole question ('~~I live in Hanoi~~' → '*Hanoi*'). See **question**.

word class

See **part of speech**.

word formation

Using prefixes and suffixes to form new words e.g. '*write*' (verb) → '*rewrite*' (different verb), '*writer*' (noun).

word order

The sequence words need to go in. For example, in English an adjective has to go before a noun: e.g. '~~It's a city beautiful~~' → '*It's a beautiful city.*'

English (unlike many others) is called an SVO language (subject + verb + object): you can say 'I (subject) *bought* (verb) *a ticket* (object)', but not '~~I a ticket bought~~' or '~~A ticket I bought~~'.

yes/no question

A question whose answer is 'yes' or 'no' ('*Are you married?*' but not '*Who's your husband?*'). It's polite to answer a yes/no question with a **short answer** (e.g. '*Yes, I am*').

.3 Common Irregular Verbs

base form	past simple (Yesterday I …)	past participle (I have …)
arise	arose	arisen
be (I **am**, he/she/it **is**, you/we/they **are**)	I/he/she/it **was**, you/we/they **were**	been
beat	beat	beaten
become	became	become
begin	began	begun
bend	bent	bent
bet	bet	bet
bind	bound	bound
bite	bit	bitten
bleed	bled	bled
blow	blew	blown
break	broke	broken
breed	bred	bred
bring	brought	brought
broadcast	broadcast/broadcasted	broadcast/broadcasted
build	built	built
burn	burned, burnt	burned, burnt
burst	burst	burst
buy	bought	bought
can (he/she/it *can*)	could	been able
cast	cast	cast
catch	caught	caught
choose	chose	chosen
cling	clung	clung
come	came	come
cost	cost	cost
creep	crept	crept
cut	cut	cut

base form	past simple (Yesterday I ...)	past participle (I have ...)
deal	dealt (/delt/)	dealt (/delt/)
dig	dug	dug
do (He/she/it **does** /dʌz/)	did	done (/dʌn/)
draw	drew	drawn
dream	dreamt (/dremt/), dreamed	dreamt (/dremt/), dreamed
drink	drank	drunk
drive	drove	driven
eat	ate	eaten
fall	fell	fallen
feed	fed	fed
feel	felt	felt
fight	fought	fought
find	found	found
flee	fled	fled
fling	flung	flung
fly	flew	flown
forbid	forbade	forbidden
forecast	forecast/forecasted	forecast/forecasted
forget	forgot	forgotten
freeze	froze	frozen
get	got	got, gotten (US)
give	gave	given
go	went	gone
grind	ground	ground
grip	gripped	gripped
grow	grew	grown
hang	hung	hung
have (He/she/it *has*)	had (he/she/it **has**)	had
hear	heard	heard

base form	past simple (Yesterday I ...)	past participle (I have ...)
hide	hid	hidden
hit	hit	hit
hold	held	held
hurt	hurt	hurt
keep	kept	kept
kneel	knelt	knelt
know	knew	known
lay	laid	laid
lead	led	led
lean	leaned, leant (/lent/)	leaned, leant (/lent/)
leap	leaped, leapt (/lept/)	leaped, leapt (/lept/)
learn	learned, learnt	learned, learnt
leave	left	left
lend	lent	lent
let	let	let
lie	lay	lain
light	lit	lit
lose	lost	lost
make	made	made
mean	meant (/ment/)	meant (/ment/)
meet	met	met
must (He/she/it must)	had to	had to
pay	paid	paid
prove	proved	proved, proven
put	put	put
quit	quit	quit
read	read (/red/)	read (/red/)
ride	rode	ridden
ring	rang	rung
rise	rose	risen
run	ran	run

base form	past simple (Yesterday I ...)	past participle (I have ...)
say	said	said
see	saw	seen
seek	sought	sought
sell	sold	sold
send	sent	sent
set	set	set
sew	sewed	sewn
shake	shook	shaken
shear	sheared	shorn
shine	shone (/ʃɒn/)	shone (/ʃɒn/)
shoot	shot	shot
show	showed	showed, shown
shrink	shrank	shrunk
shut	shut	shut
sing	sang	sung
sink	sank	sunk
sit	sat	sat
slay	slew	slain
sleep	slept	slept
slide	slid	slid
slit	slit	slit
sow	sowed	sown
speak	spoke	spoken
speed	sped	sped
spell	spelled, spelt	spelled, spelt
spend	spent	spent
spill	spilled, spilt	spilled, spilt
spin	spun	spun
spit	spat	spat
split	split	split
spoil	spoiled, spoilt	spoiled, spoilt

base form	past simple (Yesterday I ...)	past participle (I have ...)
spread	spread	spread
spring	sprang	sprung
stand	stood	stood
steal	stole	stolen
stick	stuck	stuck
sting	stung	stung
stink	stank	stunk
strike	struck	struck
swear	swore	sworn
sweep	swept	swept
swell	swelled	swollen
swim	swam	swum
swing	swung	swung
take	took	taken
teach	taught	taught
tear	tore	torn
tell	told	told
think	thought	thought
throw	threw	thrown
tread	trod	trodden
understand	understood	understood
wake	woke	woken
wear	wore	worn
win	won	won
wind	wound	wound
write	wrote	written

5.4 Spelling Rules

Adding plural -s to nouns

With most nouns you can simply add **–s** to pluralise them e.g. dog → dog**s**.
However, here are some exceptions:

Ending	Change	Example
consonant **+ y**	replace **y** with **ies**	countr**y** → countr**ies**
-ch, -s, -sh, -x, -z	add **-es:**	bus → bus**es**
-f or **-fe**	replace **-f** or **-fe** with **-ves**	kni**fe** → kni**ves**
-o (some nouns e.g. potato, hero, echo, volcano)	adds **-es**	tomato → tomato**es**
-is	replace **-is** with **-es**	cris**is** → cris**es**
-us	replace **-us** with **-i** (now commonly add **-es**)	cact**us** → cact**i** (or cactus**es**)

Adding comparative/superlative -er/-est to adjectives

To create most comparatives/superlatives, we can simply add **–er**/**-est** e.g. tall → tall**er**.
Here are some exceptions:

Ending	Change	Example
-e	add **-r**	large → large**r**
consonant + vowel + consonant	double the consonant, add **-er**	big → bi**gger**
-y (two-syllables adjectives)	replace **-y** with **-ier**	pretty → prett**ier**

Adding -ly to adjectives to form adverbs

Most adjectives just need to have **–ly** to be added to make it into an adverb e.g. quick → quick**ly**.
However, here are the exceptions:

Ending	Change	Example
-ll	add **-y**	full → full**y**
consonant + -le	replace **-e** with **-y**	horrible → horribl**y**
-y (two-syllables adjectives)	replace **-y** with **-ily**	happy → happ**ily**

Adding -ing & -ed to verbs

To change the form of a verb, you usually just need to add **–ing** (participle form) or **–ed** (past form) e.g. watch → watch**ing**, watch**ed**. However, here are the exceptions:

Ending	Change	Example
-e	replace **-e** with **-ing** or **-ed**	dance → danc**ing**, danc**ed**
-ie	replace **-ie** with **-ying** or **-ied**	lie → **lying**, **lied**
consonant + vowel + consonant	double the consonant, add **-ed**	stop → sto**pping**, sto**pped**
NB. **consonant + vowel + consonant** (two-syllables adjectives)	double the consonant if the final syllable is stressed	be·'**gin** → begi**nning** (But: '**op**·en → ope**ning**, ope**ned**)*

*UK English can double **-s** or **-l** even when the final syllable is unstressed
('fo·cus → focussing, focussed; 'tra·vel → travelling, travelled)

UK and US Spelling

These are the main differences between UK and US spelling.

Endings

UK	US
-gue (catalogue)	**-g** (catalog)
-ise (minimise)	**-ize** (minimize)
-mme (programme)	**-m** (program)
-our (colour)	**-or** (color)
-re (centre)	**-er** (center)

Words

analyse	analyze
cheque (for money)	check
defence	defense
doughnut	donut
mould	mold
moustache	mustache
mum	mom
offence	offense
pyjamas	pajamas
tyre	tire

5.5 Useful Resources

1. Methodology

Brown, H. Douglas. (2000) *Principles of Language Teaching and Learning* (4th edition). Longman.
Harmer, J. (2007). *How to Teach English: An Introduction to the Practice of English Language Teaching* (2nd Edition). Harlow: Longman.
Johnson, K. (2008) *An Introduction to Foreign Language Learning and Teaching* (2nd edition). Harlow: Longman.
Richards, J. & Schmidt, R. (2010) *Longman Dictionary of Language Teaching and Applied Linguistics* (4th edition). Harlow: Longman.
Scrivener, J. Learning *Teaching: A Guidebook for English Language Teachers* (2nd edition) Oxford: Macmillan.
Thornbury, Scott. (2006) *An A-Z of ELT*. Oxford: Macmillan.

2. Language

Leech, G et al. (2001) *An A-Z of English Grammar and Usage* (2nd edition). Harlow: Longman.
Parrott, M. (2010) *Grammar for English Language Teachers* (2nd edition). Cambridge: Cambridge University Press.
Penston, T. (2005) *A Concise Grammar for English Language Teachers.* TP Publications.
Smith, B. & Swan, M. (eds) (2001) *Learner English: A Teacher's Guide to Interference and Other Problems* (2nd Edition). Cambridge: Cambridge University Press.
Swan, M. (2005) *Practical English Usage* (3rd edition). Oxford: Oxford University Press.
Underhill, A. (2005) *Sound Foundations: Learning and Teaching Pronunciation* (2nd Edition). Oxford: Macmillan.

3. Activities

Hadfield, J. (2000) *Intermediate Communication Games.* (Other levels are also available.) *Pearson.*
Hancock, M. (1996) *Pronunciation Games.* Cambridge: Cambridge University Press.
O'Dell, F. & Head, K. (2003) *Games for Vocabulary Practice: Interactive Vocabulary Activities for all Levels.* Cambridge: Cambridge University Press.
Seymour, D. & Popova, M. *700 Classroom Activities: Instant Lessons for Busy Teachers.* Oxford: Macmillan.
Ur, P. (2009) *Grammar Practice Activities: A Practical Guide for Teachers* (2nd edition). Cambridge: Cambridge University Press. (One of many titles in the Cambridge Handbooks for Language Teachers series)
Zaorob, M.L. (2001) Games for Grammar Practice: A Resource Book of Grammar Games and Interactive Activities. Cambridge: Cambridge University Press.

5.6 Phonemic Symbols

These are the symbols used in the UK, with examples from standard British English; essentially these symbols are universally recognised, but vary slightly from country to country.

- **Vowels**

Monophthongs / **Diphthongs**

Monophthongs				Diphthongs		
/iː/ **ea**t	/ɪ/ **i**t	/ʊ/ t**oo**k	/uː/ t**oo**	/ɪə/ **ea**r	/eɪ/ s**ay**	
/e/ t**e**n	/ə/ stati**o**n	/ɜː/ h**er**	/ɔː/ **or**	/ʊə/ t**ou**r	/ɔɪ/ t**oy**	/əʊ/ n**o**
/æ/ b**a**t	/ʌ/ b**u**t	/aː/ p**ar**t	/ɒ/ p**o**t	/eə/ th**ere**	/aɪ/ m**y**	/aʊ/ n**ow**

- **Consonants**

/t/ **t**o	/d/ **d**o	/p/ **p**en	/b/ **b**in	/k/ **c**ar	/g/ **g**o	/tʃ/ **ch**oose	/dʒ/ **j**ust
/s/ **s**o	/z/ **z**oo	/f/ **f**ar	/v/ **v**an	/θ/ **th**in	/ð/ **th**ere	/ʃ/ **sh**e	/ʒ/ plea**s**ure
/m/ **m**e	/n/ **n**o	/ŋ/ ha**ng**	/h/ **h**e	/l/ **l**ow	/r/ **r**ed	/w/ **w**in	/j/ **y**et

- **Other symbols**

// = phonemic transcription house /haʊs/
• = separates syllables a – part – ment /ə·'paːt·mənt/
' = main stress station /'steɪ·ʃən/
ˌ = secondary stress (long words only) international /ˌɪn·tə·'n·ʃ·nəl/
↗ = rising intonation
↘ = falling intonation

Index

T

U

V

W

Y

Z